VOLUME ONE HUNDRED AND TWELVE

SEMICONDUCTORS AND SEMIMETALS

2D Excitonic Materials and Devices

SERIES EDITORS

VOLUME ONE HUNDRED AND TWELVE

SEMICONDUCTORS AND SEMIMETALS

2D Excitonic Materials and Devices

Edited by

PARAG B. DEOTARE

Electrical Engineering and Computer Science,
University of Michigan,
Ann Arbor, MI, United States

ZETIAN MI

Department of Electrical Engineering and Computer Science,
University of Michigan,
Ann Arbor, MI, United States

ELSEVIER

ACADEMIC PRESS

An imprint of Elsevier

Academic Press is an imprint of Elsevier
50 Hampshire Street, 5th Floor, Cambridge, MA 02139, United States
525 B Street, Suite 1650, San Diego, CA 92101, United States
The Boulevard, Langford Lane, Kidlington, Oxford OX5 1GB, United Kingdom
125 London Wall, London, EC2Y 5AS, United Kingdom

First edition 2023

Notices

Knowledge and best practice in this field are constantly changing. As new research and experience broaden our understanding, changes in research methods, professional practices, or medical treatment may become necessary.

Practitioners and researchers must always rely on their own experience and knowledge in evaluating and using any information, methods, compounds, or experiments described herein. In using such information or methods they should be mindful of their own safety and the safety of others, including parties for whom they have a professional responsibility.

To the fullest extent of the law, neither the Publisher nor the authors, contributors, or editors, assume any liability for any injury and/or damage to persons or property as a matter of products liability, negligence or otherwise, or from any use or operation of any methods, products, instructions, or ideas contained in the material herein.

ISBN: 978-0-443-19392-7
ISSN: 0080-8784

For information on all Academic Press publications
visit our website at https://www.elsevier.com/books-and-journals

Publisher: Zoe Kruze
Acquisitions Editor: Jason Mitchell
Editorial Project Manager: Sneha Apar
Production Project Manager: Abdulla Sait
Cover Designer: Victoria Pearson

Typeset by MPS Limited, India

Contents

Contributors

L.V. Butov
University of California at San Diego, La Jolla, CA, United States

Po-Liang Chen
Institute of Photonics Technologies, National Tsing Hua University, Hsinchu, Taiwan

Alexey Chernikov
Institute of Applied Physics and Würzburg-Dresden Cluster of Excellence ct.qmat, TUD Dresden University of Technology, Dresden, Germany

Kanak Datta
Department of Electrical Engineering and Computer Science; Department of Materials Science and Engineering, University of Michigan, Ann Arbor, MI, United States

Parag B. Deotare
Department of Electrical Engineering and Computer Science; Department of Materials Science and Engineering, University of Michigan, Ann Arbor, MI, United States

Mikhail M. Glazov
Ioffe Institute, Saint Petersburg, Russian Federation

Emmanouil Kioupakis
Department of Materials Science and Engineering, University of Michigan, Ann Arbor, MI, United States

Woncheol Lee
Department of Electrical Engineering and Computer Science, University of Michigan, Ann Arbor, MI, United States

Zhen Lian
Department of Chemical and Biological Engineering, Rensselaer Polytechnic Institute, Troy, NY; Department of Physics, Carnegie Mellon University, Pittsburgh, PA, United States

Chang-Hua Liu
Institute of Photonics Technologies, National Tsing Hua University, Hsinchu, Taiwan

Zetian Mi
Department of Electrical Engineering and Computer Science, University of Michigan, Ann Arbor, MI, United States

Jiajie Pei
College of Materials Science and Engineering, Fuzhou University, Fuzhou, P.R. China

Su-Fei Shi
Department of Chemical and Biological Engineering, Rensselaer Polytechnic Institute, Troy, NY; Department of Physics, Carnegie Mellon University, Pittsburgh, PA, United States

Yuanpeng Wu
Department of Electrical Engineering and Computer Science, University of Michigan, Ann Arbor, MI, United States

Renjing Xu
Microelectronics Thrust, The Hong Kong University of Science and Technology (Guangzhou), Nansha, Guangdong, P.R. China

Shyam Narayan Singh Yadav
Institute of Photonics Technologies, National Tsing Hua University, Hsinchu, Taiwan

Preface

Two-dimensional (2D) quantum materials have emerged as an attractive platform to investigate a plethora of quantum excitations such as exciton, magnons, and phonons. Excitons, hydrogen-like quasiparticles that emerge from the Coulomb attraction between electrons and holes, are excellent carriers of energy at nanoscale. The strong binding energy of excitons in 2D materials together with the spin-optic response and unprecedented light-matter interaction strength provide a unique platform for realizing next-generation devices serving various applications in data communication and processing to sensing. This book focuses on important technological breakthroughs that create potential pathways towards viable opto-excitonic devices based on 2D excitons. This book consists of seven chapters.

Chapter 1 provides an overview of excitons in transition metal dichalcogenides (TMDCs), including interlayer excitons, intervalley dark and spin-forbidden dark excitons, exciton complexes such as charged excitons, biexcitons, and trion-excitons, as well as Moiré excitons. Electric field tunability and valley selection rules are also presented.

Chapter 2 focuses on valley excitonic states in TMDCs. These include bright and dark valley excitons, intravalley and intervalley trions, valley exciton–trion complexes, valley phonon–exciton complexes, and interlayer valley–exciton complexes. Fundamental studies, understanding, and control of these many-body complexes are significant for the emerging spin-valley-polarized devices, valleytronic logic devices, and entanglement-based quantum logic.

Chapter 3 provides an overview of experimental and theoretical studies of exciton propagation in 2D materials. The background knowledge and theory of exciton transport, the linear and nonlinear regimes of propagation, and the role of environment, disorder, carrier density, and temperature on exciton transport are discussed.

Chapter 4 presents studies on the interaction between excitons in TMDCs and the traveling strain and electric field associated with propagation surface acoustic wave. The underlying exciton photophysics and the experimental demonstration of directional exciton transport and exciton modulation by surface acoustic waves are discussed.

Moving the discussions beyond TMDCs, excitonic devices in 2D heterostructures are further presented in Chapter 5. The physics, properties, and transport of spatially indirect excitons, together with a broad range of excitonic devices, including excitonic traps, excitonic lattices, excitonic conveyers, excitonic ramps, excitonic split gates, and excitonic transistors, are discussed.

Chapter 6 discusses the emerging III-nitride optoexcitonics, including the epitaxy, physics, and their device applications, such as ultraviolet LEDs, visible micro-LEDs and ultralow threshold laser diodes.

Chapter 7 presents the recent advances, emerging breakthroughs, and potential challenges of black phosphorus and its alloys, with a special focus on their applications in mid-infrared light emitters. Their integration with various photonic structures for on-chip sensing, communications, and spectroscopy applications is also discussed.

The topics and contents presented in this book are well suited for researchers and students in the fields of low-dimensional semiconductor materials and physics, quantum materials, nanotechnology, and light-matter interaction in nanoscale systems. Engineers in the fields of nanotechnology and photonic, electronic, and excitonic devices will also find this book highly valuable. By presenting a comprehensive overview of excitons in 2D materials, in-depth discussions of some major advances, as well as practical device demonstrations, this book is a must read in the emerging fields of 2D semiconductors and excitonic devices.

Parag B. Deotare
Zetian Mi
University of Michigan, Ann Arbor,
MI, United States

CHAPTER ONE

Excitons in transition metal dichalcogenides (TMDCs)

Zhen Lian[a,b] and Su-Fei Shi[a,b,*]
[a]Department of Chemical and Biological Engineering, Rensselaer Polytechnic Institute, Troy, NY, United States
[b]Department of Physics, Carnegie Mellon University, Pittsburgh, PA, United States
*Corresponding author. e-mail address: shis2@rpi.edu

Contents

Abstract

The strong Coulomb interaction resulting from the reduced screening in two-dimension (2D) leads to tightly bound excitons and even high-order Rydberg excitons. The strong Coulomb interaction also gives rise to many-particle excitonic complexes bound or interacting with other quasiparticles. Excitonic complexes usher in new venues of controlling critical exciton properties such as lifetime and valley degree of freedoms. In this chapter, we present a comprehensive review of the recent studies on the optical properties of various excitonic complexes in TMDCs. We start by introducing the experimental progress on the excitonic Rydberg series, which helps to reveal the effective Coulomb interaction in 2D. We then focus on experimental evidence for exciton complexes in monolayer TMDCs, including charged excitons, biexcitons, dark excitons, and exciton-phonon replicas, with emphasis on their valley and magneto-optical properties. In the third part, we review the recent findings on interlayer excitons in type-II aligned van der Waals hetero-bilayers of TMDCs, which

Semiconductors and Semimetals, Volume 112
ISSN 0080-8784, https://doi.org/10.1016/bs.semsem.2023.09.012

exhibit distinctly different properties than intralayer excitons in exciton dynamics, optical selection rules, and the electric field tunability. Finally, we briefly discuss the moiré excitons. We hope this chapter will guide readers interested in the unique excitonic physics in TMDCs and inspire future development of quantum optoelectronics based on excitons in TMDCs.

1. Overview

Transition metal dichalcogenides (TMDCs) are a group of compounds with the chemical composition MX_2, where M is a transition metal such as tungsten or molybdenum, and X is a chalcogen such as sulfur, selenium, or tellurium. TMDCs can exist in 2H, 1T, 1T′, and 3R phases (Schutte et al., 1987; Keum et al., 2015; Wickramaratne et al., 2014; Manzeli et al., 2017). Among those phases, the 2H phase is one thermal dynamically stable form for most bulk TMDCs crystals and is a semiconducting phase that is promising for electronic and optoelectronic applications. Therefore, in this chapter, we will limit our discussion to the 2H phase. A TMDCs monolayer is composed of two layers of chalcogen atoms sandwiching one triangular lattice of transition metal atoms, with the transition metal atoms located at the centers of the triangles formed by the chalcogen atoms (Schutte et al., 1987; Keum et al., 2015; Wickramaratne et al., 2014; Manzeli et al., 2017), as shown in Fig. 1a.

In monolayer or few-layer TMDCs, the dielectric environment is significantly different from that in a bulk semiconductor due to the reduced dielectric screening effect from the adjacent layers. As a result, 2D excitons in TMDCs exhibit enhanced electron–hole interaction and are thus more tightly bound, giving rise to larger binding energies (typically larger than 100 meV) (Chernikov et al., 2014; He et al., 2014; Zhu et al., 2015) compared to their 3D counterparts (typically 10s meV) (Nam et al., 1976). Excitons in 2D semiconductors can remain robust even at room temperature since their binding energies are often significantly larger than the thermal energy (He et al., 2014). The strong excitonic effects in TMDCs significantly enhance the light–matter interaction in these semiconductors, resulting in enhanced absorption and emission, which make them promising for optoelectronic applications, including light-emitting diodes or lasing devices (Mak and Shan, 2016). Rydberg excitons, i.e., the excitonic analogy of the Rydberg atoms, also emerge due to the giant exciton binding energy, which could possibly leads to enhanced exciton–exciton

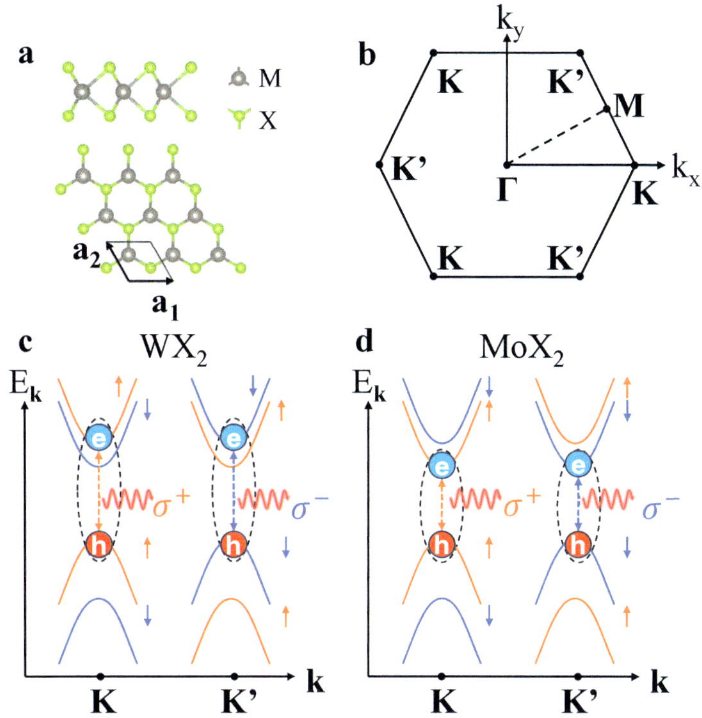

Fig. 1 Intralayer excitons in transition metal dichalcogenides (TMDCs). (a) shows the crystal structures of monolayer TMDCs. (b) is a schematic of the hexagonal Brillouin zone of monolayer TMDCs. (c) and (d) are the spin-valley configurations and the optical selection rules of the optically bright excitons in tungsten-based TMDCs (WX_2) and molybdenum-based TMDCs (MoX_2), respectively.

interactions (Taylor et al., 2022; Urban et al., 2009; Gaëtan et al., 2009; Poddubny and Glazov, 2019).

TMDCs possess hexagonal Brillouin zones, as shown in Fig. 1b. In the monolayer limit, their bandgaps are direct and are located at the K and K′ points of the Brillouin zone. Compared with graphene, the crystal structure of monolayer TMDCs lacks inversion symmetry. As a result, the degeneracy of the electronic state at K and K′ is broken, resulting in the valley degree of freedom (Xu et al., 2014; Schaibley et al., 2016; Xiao et al., 2012). The strong spin–orbital coupling (SOC) in TMDCs further causes their conduction bands and valence bands to split into spin-polarized bands at specific valley, leading to the spin-valley locking effect (Xu et al., 2014). As shown in Fig. 1c and d, the bands at K and K′ carry opposite SOC splitting. The typical magnitude of the SOC splitting is 3–60 meV for the conduction bands and 150–400 meV for the valence bands (Liu et al., 2013). The valley degree of freedom gives rise to the valley-dependent optical selection rules in monolayer TMDCs. The K valley exclusively allows absorption or emission of right circularly polarized light (σ^+), while the K′ valley exclusively couples to left circularly polarized light (σ^-). Interband transitions for photons with in-plane polarizations can only occur between bands with the same spins (Xu et al., 2014; Schaibley et al., 2016; Xiao et al., 2012). The excitons at the K or K′ valley in monolayer TMDCs inherit the valley degree of freedom of the specific valley (Xu et al., 2014; Schaibley et al., 2016), enabling the manipulation of spin or valley degree of freedom using the helicities of photons, a field known as valleytronics.

One important experimental technique to characterize the valley polarization of an excitonic state in TMDCs is helicity-resolved photoluminescence (PL) measurement, which is performed by selectively populating a valley using a certain circularly polarized light (say σ^+) and collecting the PL emissions in σ^+ and σ^- polarized configurations separately. Taking σ^+ excitation for example, the degree of polarization is defined as (Mak et al., 2012; Cao et al., 2012; Zeng et al., 2012; Jones et al., 2013):

$$\rho = \frac{I^+ - I^-}{I^+ + I^-} \tag{1}$$

Where I^+ and I^- denote the PL intensities in σ^+ and σ^- polarizations, respectively. The degree of polarization for σ^- excitation is defined by reversing the sign of Eq. (1).

Another important quantity related to valley polarization is the valley lifetime, which characterizes the decay of ρ (Mak et al., 2012; Cao et al., 2012; Zeng et al., 2012; Jones et al., 2013). For valleytronics applications, it is desirable to realize an excitonic state with a large degree of polarization, a long exciton lifetime and a long valley lifetime. The degree of polarization for the charge neutral bright exciton in monolayer TMDCs, however, is often far from unity due to the intervalley scattering mechanism. In addition, the relatively short lifetime (<10 ps) of the charge neutral bright exciton also poses a great challenge for efficient utilization of their valley degree of freedom (Zhang et al., 2015). This motivated the study of exciton complexes, i.e., multibody quasiparticles resulting from the interaction between an exciton and charges, excitons or phonons, to seek excitonic states that are ideal for valleytronics. In addition, van der Waals heterostructures, the heterostructures created by stacking different monolayer TMDCs on top each other, enables countless possibilities to modify the properties of excitons by engineering the band-alignment, lattice mismatch or twist angle. TMDCs hetero-bilayers with type-II alignment are discovered to host long-lived inter-layer excitons, localized moiré excitons also emerge from TMDCs moiré superlattices. In this chapter, we briefly review the experimental progresses on the emerging excitonic states in monolayer TMDCs and their heterostructures.

2. Intralayer excitons in monolayer TMDCs

An intralayer exciton is an exciton with its electron and hole residing in the same layer of a 2D material. The spatial distribution of the electron and hole in an intralayer exciton gives rise to a finite in-plane dipole moment, making it possible for them to radiate or absorb photons with in-plane polarization (He et al., 2014; Amani et al., 2015). On the other hand, since the electron and hole are confined in the same atomic layer, an intralayer exciton possesses negligible out-of-plane dipole moment, which results in limited DC Stark shift under out-of-plane electric fields (Roch et al., 2018; Verzhbitskiy et al., 2019).

Analogous to a Rydberg atom, an intralayer exciton is a two-body system bound by Coulomb interaction. Solving the Schrödinger equation of this system results in bound states characterized by the principal quantum number n and the angular quantum number l. This leads to a series of discrete energy levels, referred to as the excitonic Rydberg series (Chernikov et al., 2014; Wu et al., 2015a). High-order Rydberg excitons possess large exciton radii, which could enhance the interaction between these states. The Rydberg blockade effect, that is, the prevention of the double-occupancy of the same Rydberg state due to the dipolar interaction (Urban et al., 2009, Gaëtan et al., 2009), could be possibly realized in monolayer TMDCs and potentially enables the quantum information processing based on this interaction (Taylor et al., 2022; Aßmann and Bayer, 2020). Measurements of the fundamental properties of Rydberg excitons in monolayer TMDCs, including their binding energies and radii, are crucial for the accurate modeling of these excitonic states.

2.1 Non-hydrogenic model of excitonic Rydberg series

Earlier studies have used the 2D hydrogen model to describe excitons in 2D semiconductors. This model assumes that the electron and hole are strictly confined in a 2D plane and to describe the attraction between the electron and the hole as (Chernikov et al., 2014; Klingshirn, 2012):

$$V_{eh}(\boldsymbol{r}) = -\frac{e^2}{\varepsilon_{2D}\,|\boldsymbol{r}|} \tag{2}$$

Where $\boldsymbol{r}$ is the relative position from the electron to the hole, ε_{2D} is the dielectric constant describing the Coulomb screening effect.

The Hamiltonian of the system can therefore be written as:

$$H = -\frac{\hbar^2}{2m_r}\nabla_{\boldsymbol{r}}^2 + V_{eh}(\boldsymbol{r}) \tag{3}$$

Where $m_r = (m_e^{-1} + m_h^{-1})^{-1}$ is the reduced mass of the electron and the hole.

Analogous to the energy states of a hydrogen atom, solving the eigen values using the Hamiltonian in Eq. (3) results in the hydrogenic Rydberg series, which is related to the principal quantum n by (Chernikov et al., 2014; Klingshirn, 2012):

$$E_b^n = \frac{\mu e^4}{2\hbar^2\varepsilon^2(n - 1/2)^2} \tag{4}$$

The state corresponding to $n = 1$ is the ground state of the exciton, while the states with $n > 1$ are excited states. The 2D hydrogen model is generally in good agreement with the exciton resonance positions in monolayer TMDCs measured by reflectance contrast spectroscopy for high order states with $n > 3$. Yet it differs from experiment results for the ground state and low-order excited states. Chernikov, A. et al. attribute this observation to the deviation of the Coulomb interaction from Eq. (2) for the low-order states: Since Rydberg excitons with small n possess smaller radii, the thickness of the monolayer is no longer negligible compared to the exciton radius and the Coulomb interaction between electrons and holes experiences more screening effect from the surrounding dielectric environment (Chernikov et al., 2014), as shown in Fig. 2c. The effective electron–hole interaction for the Rydberg states with low n number is better approximated by the Keldysh potential (Berkelbach et al., 2013; Keldysh, 1979), which is given by:

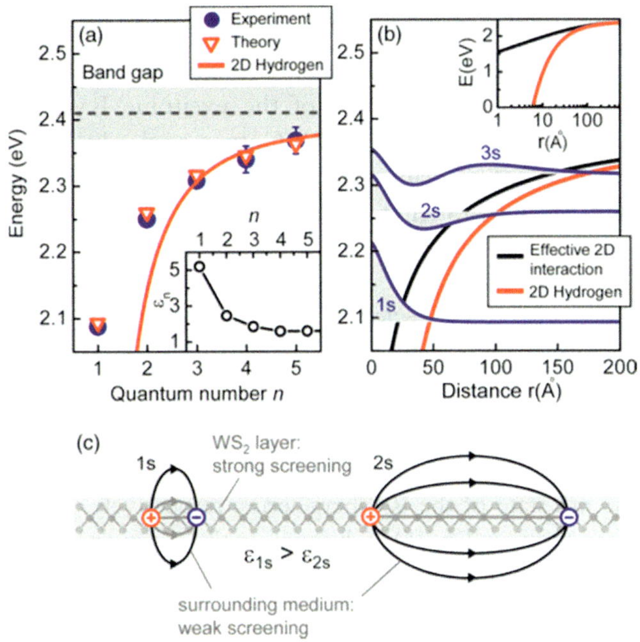

Fig. 2 Non-hydrogenic Rydberg series in monolayer WS_2. (a) shows the comparison between the predicted exciton Rydberg series from the 2D hydrogen model (red line) and the non-hydrogenic model using Keldysh potential as electron–hole interaction (red triangles). The blue spheres indicate the energies of the Rydberg states measured by reflectance contrast. (b) shows the comparison of the magnitudes of the effective 2D interaction (black line) and the 2D hydrogenic interaction (red line) as functions of electron–hole distance. The blue lines are the wavefunctions of Rydberg excitons with $n=1$, 2, and 3. (c) is a schematic of the effective electron–hole interaction for Rydberg excitons with $n=1$ and 2. The 1s state is more tightly bound and therefore experiences stronger screening from the surrounding medium. *Reproduced from Chernikov, A., et al., 2014. Exciton binding energy and nonhydrogenic Rydberg series in monolayer WS2. Phys. Rev. Lett. 113, 076802. Copyright 2014 American Physical Society.*

$$V_{eh}(r) = -\frac{\pi e^2}{2r_0}\left[H_0\left(\frac{r}{r_0}\right) - Y_0\left(\frac{r}{r_0}\right)\right] \tag{5}$$

where H_0 and Y_0 are Struve and Bessel functions, respectively.

As shown in Fig. 2b, for Rydberg excitons with small radii, the effective 2D attractive interaction modeled by Eq. (5) is considerably smaller than that given by the 2D hydrogen model. The non-hydrogenic Rydberg series derived from Eq. (5) match the peak positions of the Rydberg excitons in monolayer WS_2 measured by reflectance contrast measurement up to $n=5$ (Chernikov et al., 2014).

2.2 Experimental determination of exciton radii

Recently, magneto-spectroscopy measurements have been used to reveal the binding energies and the radii of the excitonic Rydberg series in monolayer TMDCs (Stier et al., 2018; Goryca et al., 2019; Wang et al., 2020a). Under an out-of-plane magnetic field, an excitonic state exhibits an energy shift given by (Stier et al., 2018; Goryca et al., 2019; Wang et al., 2020a):

$$E_{ns} = \frac{1}{2}\sigma g \mu_B B + \frac{e^2}{8m_r}\langle r^2 \rangle_{ns} B^2 \tag{6}$$

Where σ, g, μ_B and B are the valley index, the g-factor, the Bohr magneton and the magnetic field, respectively. e, m_r and $\langle r^2 \rangle_{ns}$ are the elementary charge, the reduced mass of the exciton and the expected squared exciton radius of the nth state of the Rydberg series, respectively.

The linear term in Eq. (6) is known as the Zeeman shift, while the quadratic term is referred to as the diamagnetic shift, which is proportional to the square of the exciton radius.

Wang et al. performed helicity-resolved magneto-photocurrent spectroscopy on monolayer WSe_2. Through the valley polarization features of the photocurrent spectra, the magnetic field dependence of the Rydberg state peak positions up to $n = 11$ was revealed (Fig. 3), which allows the extraction of the exciton radii and the binding energies, as shown in Table 1 (Wang et al., 2020a).

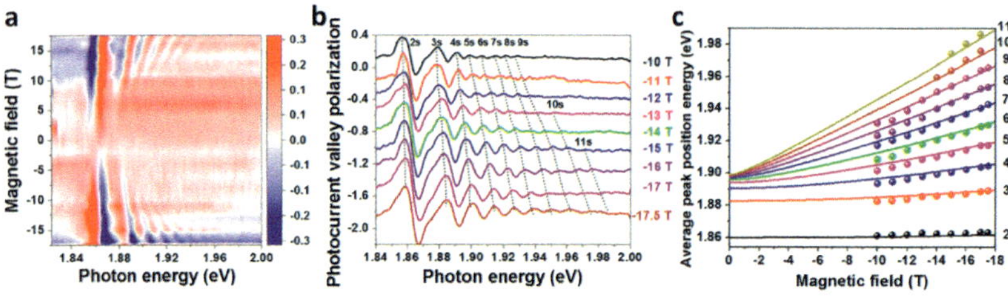

Fig. 3 Excitonic Rydberg series in monolayer WSe_2 revealed by magneto-photocurrent spectroscopy. (a) shows the magnetic field dependence of the degree of valley polarization of the photocurrent spectra measured from a monolayer WSe_2. (b) shows the valley polarization spectra at several magnetic field strengths from (a). (c) shows the extracted Rydberg exciton peak positions at different magnetic fields (spheres). The lines indicate the fitting results to Eq. (6). *Reproduced from Wang, T., et al., 2020a. Giant valley-polarized Rydberg excitons in monolayer WSe2 revealed by magneto-photocurrent spectroscopy. Nano Lett. 20, 7635–7641. Copyright 2020 American Chemical Society.*

Table 1 Exciton binding energies and exciton radii of the ground state and the excited states of the non-hydrogenic Rydberg series of monolayer WSe_2 were measured by magneto-photocurrent spectroscopy.

Rydberg state	Binding energy (meV)	Radius (nm)
1s	168.6	1.75
2s	40.0	6.80
3s	17.4	15.45
4s	9.7	27.68
5s	6.1	43.51

Source: Adapted from Wang, T., et al., 2020a. Giant valley-polarized Rydberg excitons in monolayer WSe2 revealed by magneto-photocurrent spectroscopy. Nano Lett. 20, 7635–7641. Copyright 2020 American Chemical Society.

3. Exciton complexes in monolayer TMDCs

The enhanced Coulomb interaction in monolayer TMDCs not only gives rise to robust excitons, but also enables strong exciton–charge carrier interaction or exciton–exciton interaction, which lead to a plethora of exciton complexes in monolayer TMDCs, including trions (Berkelbach et al., 2013; Mak et al., 2013; Ross et al., 2013), exciton–polarons (Efimkin and MacDonald, 2017, 2018; Sidler et al., 2017), biexcitons (Szyniszewski et al., 2017; Kidd et al., 2016; Mayers et al., 2015; Mostaani et al., 2017; Kylänpää and Komsa, 2015; You et al., 2015; Li et al., 2018; Ye et al., 2018; Chen et al., 2018; Barbone et al., 2018), dark exciton (Zhang et al., 2015; Wang et al., 2017), and phonon replicas (Li et al., 2019a; Liu et al., 2019a; Li et al., 2019b). These features have been revealed by low-temperature high-resolution optical spectroscopy of high-quality TMDCs monolayer samples.

Fig. 4b shows the measured gate-dependent PL spectra from a WSe_2 monolayer encapsulated on both sides by hexagonal boron nitride flakes (schematically shown in Fig. 4a), which shows rich fine-features corresponding different exciton complexes. By changing the voltage (V_{tg}) applied to the top graphite gate, electrons and holes can be injected into the sample, changing its electrostatic doping level. In the charge neutral region of the sample (V_{tg} between -1 and 0 V), the feature with the highest photon energy (X_0) is the charge neutral exciton of WSe_2. The biexciton (XX) is a four-particle complex resulting from the interaction between excitons and thus possesses a

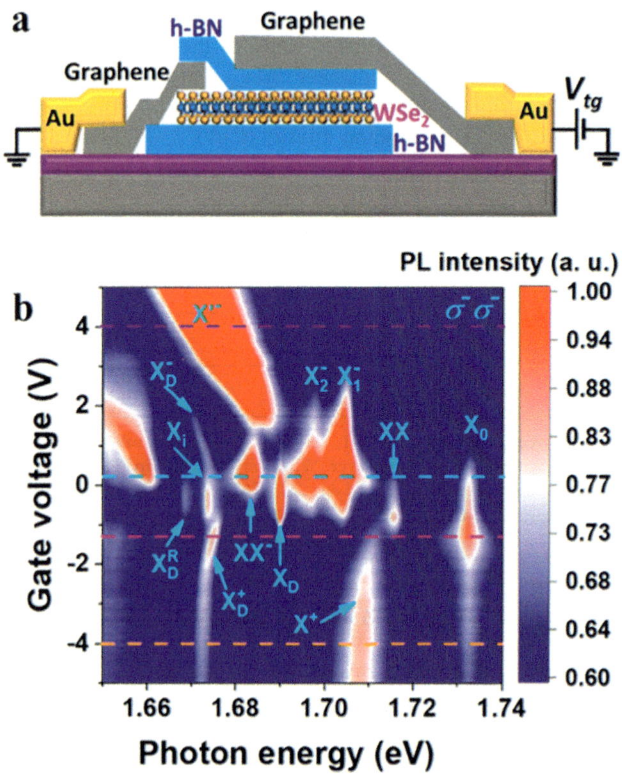

Fig. 4 Fine features in the PL spectra of monolayer WSe_2. (a) shows the schematic of a typical single-gated monolayer WSe_2 for gate-dependent PL study. (b) is the gate dependence of the PL spectra from the device shown in (a). The PL measurement was performed using a 1.879 eV continuous-wave excitation at a temperature of 4.2 K. *Adapted from Li, Z., et al., 2019c. Direct observation of gate-tunable dark trions in monolayer WSe2. Nano Lett. 19, 6886–6893. Copyright 2019 American Chemical Society.*

lower energy (Szyniszewski et al., 2017; Kidd et al., 2016; Mayers et al., 2015; Mostaani et al., 2017; Kylänpää and Komsa, 2015; You et al., 2015; Li et al., 2018; Ye et al., 2018; Chen et al., 2018; Barbone et al., 2018). The peaks X_D and X_D^R correspond to the spin-forbidden dark exciton and its phonon replica (Zhang et al., 2015; Wang et al., 2017; Li et al., 2019a; Liu et al., 2019a). In the electron-doped ($V_{tg} > 0$ V) and the hole-doped ($V_{tg} < -1$ V), the above features disappear and are replaced by newly emerged, redshifted peaks, which indicate the presence of charged exciton complexes. These features include the negatively charged excitons (X_1^- and X_2^-) (Berkelbach et al., 2013; Mak et al., 2013; Ross et al., 2013; Efimkin and MacDonald, 2017, 2018;

Sidler et al., 2017; Back et al., 2017), the exciton–trion complex (XX^-) (Barbone, 2018; Chen et al., 2018; Li, 2018; Ye, 2018), the negatively charged spin-forbidden dark trion (X_D^-) (Liu et al., 2019b; Li et al., 2019c; Tang et al., 2019), the positively charged exciton (X^+) (Berkelbach et al., 2013; Mak et al., 2013; Ross et al., 2013; Efimkin and MacDonald, 2017, 2018; Sidler et al., 2017; Back et al., 2017) and the positively charged spin-forbidden dark trion (X_D^+) (Liu et al., 2019b; Li et al., 2019c; Tang et al., 2019). Note that both the dark trions (X_D^- and X_D^+) are very close in energy to the phonon replica of the intervalley exciton (X_i), but they are distinctly different in terms of radiation patter, due to the out-of-plane radiation nature of the dark trions (inherited from the dark exciton) (Li et al., 2019b; Liu et al., 2019b; Li et al., 2019c; Tang et al., 2019). In this section, we briefly introduce the physics behind the fine features of TMDCs monolayers and review the recent progress to study them.

3.1 Charged excitons

The interaction of an exciton with excessive charges in monolayer TMDCs results in charged excitons, which often result in PL or absorption peaks that are shifted from the charge–neutral excitons. So far, two mainstream theories have been applied to describe exciton–charge interactions: the trion picture and the exciton–polaron picture. The former can be viewed as the low doping limit of the latter.

The former theory views a charged exciton as a three-body particle consists of an exciton and another electron or hole, known as a trion. A trion formed by an exciton and a hole is defined as a p-trion, while an exciton and an electron form a n-trion (schematically shown in Fig. 5c). The attractive interaction between the exciton and the charge reduces the energy of this three-body system from the charge–neutral exciton by an amount E_b^T, defined as the trion binding energy (Berkelbach et al., 2013). Early experimental evidences of trions in monolayer TMDCs were demonstrated by Mak et al. in monolayer MoS_2 and Ross et al. in monolayer $MoSe_2$ using gate-dependent photoluminescence (PL) or reflectance spectra (Fig. 5a and b) in 2013 (Mak et al., 2013; Ross et al., 2013). The binding energy is determined to be 18 meV for the n-trion in MoS_2 and around 20 meV for both the p-trion and the n-trion in $MoSe_2$. These values match the theoretical trion binding energies of 20–30 meV obtained by Berkelbach et al. using ab initio calculations (Berkelbach et al., 2013).

Later spectroscopic studies on high-quality TMDCs samples revealed more fine features of the trions. PL studies performed by Jones et al. and Plechinger et al., together with a reflectance spectroscopy study by

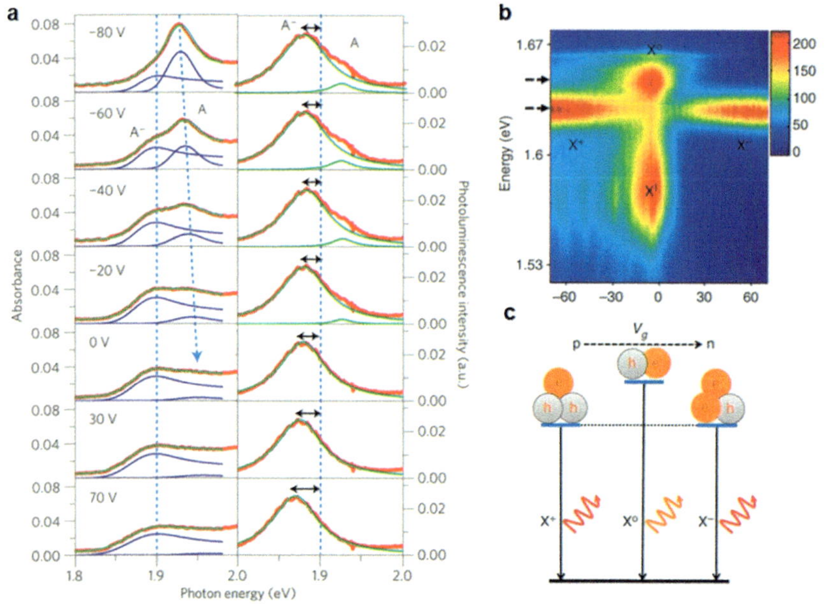

Fig. 5 Evidence of trions in monolayer TMDCs. (a) shows the gate-dependent absorption (left) and PL (right) spectra taken from a monolayer MoS_2. A and A^- denote the charge-neutral exciton peak and the n-trion peak, respectively. (b) shows the gate-dependent PL spectra taken from a monolayer $MoSe_2$. X^0, $X^{+,}$ and X^- denote the charge–neutral exciton peak, the p-trion peak and the n-trion peak, respectively. X^I denotes an impurity-trapped exciton peak. (c) shows a schematic of a p-trion, a charge–neutral exciton, and a n-trion and their relative energies. *Panel (a) adapted from Mak, K.F., et al., 2013. Tightly bound trions in monolayer MoS2. Nat. Mater. 12, 207–211. Copyright 2012 Nature Springer. Panels (b) and (c) adapted from Ross, J.S., et al., 2013. Electrical control of neutral and charged excitons in a monolayer semi-conductor. Nat. Commun, 4, 1474 under a Creative Common license (http://creativecommons.org/licenses/by/4.0/).*

Courtade et al. have revealed a n-trion peak splitting of 6–7 meV and 11 meV for monolayer WSe_2 and WS_2, respectively (Jones et al., 2016; Plechinger et al., 2016; Courtade et al., 2017). The two branches of the n-trions were attribute to the energy splitting of the intravalley trion and the intervalley trion, a result of Coulomb exchange interaction between the charge carriers (Jones et al., 2016).

The theory of exciton–polaron views the interaction between an exciton and charged carriers as a many-body interaction between the exciton and a fermi-sea of charge carriers (Efimkin and MacDonald, 2017, 2018; Sidler et al., 2017). In 2016, Sidler et al. reported the behaviors of the charged excitons in monolayer $MoSe_2$ measured by optical-cavity-coupled transmission and PL spectroscopy. From the gate dependence of the absorption spectra, two spectral features were identified as the electron density in the sample is increased: A low energy branch with its energy close to the energy of trion at low electron density, but blue shifts upon further doping, and a high energy branch which stems from a continuous blueshift of the charge-neutral exciton. These features were attributed to the attractive and the repulsive polarons, respectively (Sidler et al., 2017). Efimkin et al. further examined these observations and concluded that the consideration of many-body interactions is necessary at an intermediate doping level where the Fermi level of the excess charge is comparable to the trion binding energy. Using a many-body model, this work derived the theory of exciton–polaron, which suggest the interaction of an exciton with the carrier Fermi sea in the same/opposite valley results in repulsive/attractive exciton–polarons (Efimkin and MacDonald, 2017) (schematically shown in Fig. 6d). Back et al. performed helicity-resolved magneto-reflectance spectra to study the valley configurations of the exciton polarons in monolayer $MoSe_2$, showing that when the excessive charge electrons exclusively populate the K′ valley, the absorption of the attractive/repulsive polaron is mainly visible with σ^+/σ^- excitation (Fig. 6a–c), a result consistent with the predicted configurations of the exciton polarons (Back et al., 2017).

3.2 Biexcitons and trion–exciton complexes

A Biexcitons is a four-body quasiparticle formed due to the exciton–exciton interaction in a semiconductor, which can be viewed as two excitons bound by Coulomb attractive interaction. The photon emission of a biexciton originates from the recombination of one of its constituent excitons, leaving the other exciton behind. The biexciton binding energy is defined as the difference between the biexciton and the unbound exciton. An early experimental study to search biexcitons in monolayer TMDCs was carried out by You et al. in 2015. In this study, power-dependent PL study was used to reveal a new excitonic peak in monolayer WSe_2, which is located 52 meV below the charge-neutral exciton resonance and exhibits a super-linear power law of 1.39, drastically different from the expected linear or sublinear power dependence of charge–neutral excitons, trions or defect states (You et al., 2015). Similar results were also reported by other research groups in both monolayer WSe_2 and WS_2 (He et al., 2016; Kim et al., 2016; Danovich et al., 2017; Okada et al., 2017; Nagler et al., 2018). However, there is a large discrepancy between the measured binding

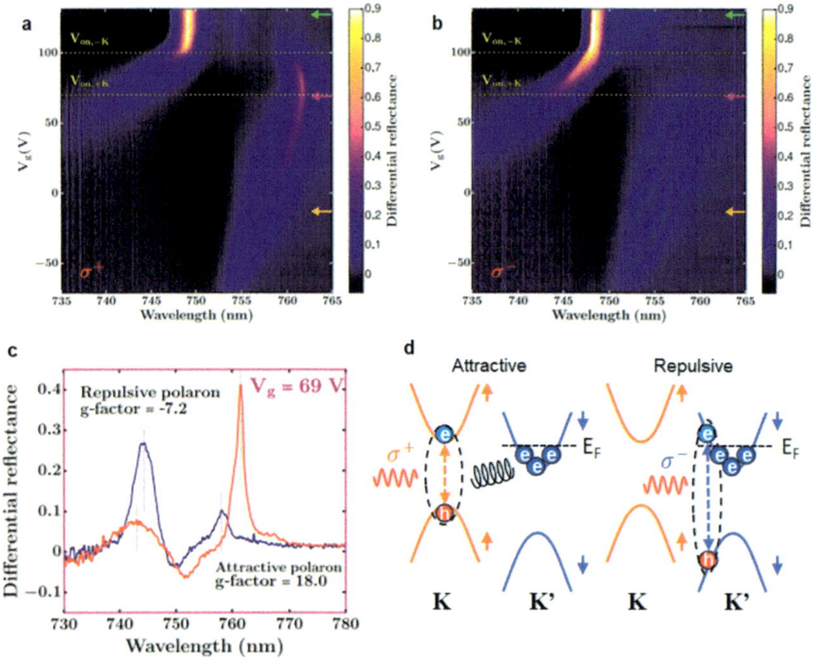

Fig. 6 Attractive and repulsive exciton–polarons in monolayer $MoSe_2$. (a) and (b) are the helicity-resolved, gate-dependent reflectance spectra under an out-of-plane magnetic field of 7 T, using σ^+ and σ^- excitations, respectively. The gate voltage is defined as the voltage difference between the sample and the gate in this work. (c) shows the spectra at a gate voltage of 69 V (moderate electron doping) from (a) and (b). (d) are the configurations of the attractive polaron and the repulsive polaron under a magnetic field. *Panels (a)–(c) adapted from Back, P., et al., 2017. Giant paramagnetism-induced valley polarization of electrons in charge-tunable monolayer MoSe2. Phys. Rev. Lett. 118, 237404. Copyright 2017 American Physical Society.*

energy and the theoretically predicted binding of 20 meV (Szyniszewski et al., 2017; Kidd et al., 2016; Mayers et al., 2015; Mostaani et al., 2017; Kylänpää and Komsa, 2015). Hao et al. and Steinhoff et al. applied polarization-resolved two-dimensional coherent spectroscopy and ultrafast pump-probe spectroscopy to study the high-order states of biexcitons in monolayer TMDCs, yielding a biexciton binding energy of 20 meV in both WSe_2 and $MoSe_2$ (Steinhoff et al., 2018; Hao et al., 2017), in excellent agreement with the theoretical result but different with the previous experimental studies. To explain the inconsistent biexciton binding energies obtained from different experimental techniques, Li et al., Ye et al., Chen et al., and Barbone et al. carried out independent studies by

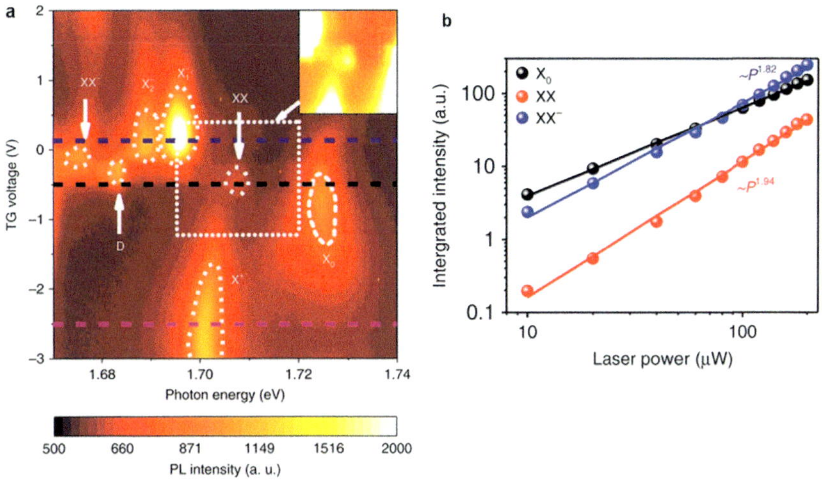

Fig. 7 Biexcitons and trion–exciton complexes in h-BN encapsulated monolayer WSe_2. (a) shows the PL spectra measured at different top gate (TG) voltages in a monolayer WSe_2 sample. XX, XX^- and X_0 denote the biexciton peak, the trion–exciton complex peak, and the charge–neutral exciton peak, respectively. (b) shows the excitation power dependence and the power law fitting of the PL peak intensities of XX, XX^- and X_0. *Adapted from Li, Z., et al. 2018. Revealing the biexciton and trion-exciton complexes in BN encapsulated WSe2. Nat. Commun. 9, 3719 under a Creative Common license (http://creativecommons.org/licenses/by/4.0/).*

performing gate-dependent PL measurements on high-quality, hexagonal boron nitride (h-BN) encapsulated monolayer WSe_2 samples. From the PL spectra (shown in Fig. 7a), they identified an excitonic peak (XX) around 17 meV below the charge–neutral exciton resonance, which only brightens up when the sample is charge–neutral, consistent with the theoretical value. Meanwhile, a peak (XX^-) emerges around 49 meV below the exciton resonance in the slight electron-doped condition. Both XX and XX^- exhibit a super-linear power law (Fig. 7b). Based on the above observations, these works concluded that XX and XX^- are the biexciton and the trion–exciton complex (or charged biexciton), respectively (Li et al., 2018; Ye et al., 2018; Chen et al., 2018; Barbone et al., 2018).

Magneto-PL measurements were also performed to reveal the spin-valley configurations of XX and XX^-. Under an out-of-plane magnetic field, XX exhibits a g-factor of around 4 (Fig. 8a), consistent with the value expected from theoretical calculations. The higher-energy branch of the Zeeman splitting of XX carries more PL intensity than the lower-energy one (Fig. 8b), contrary to the result expected from the quasi-thermal

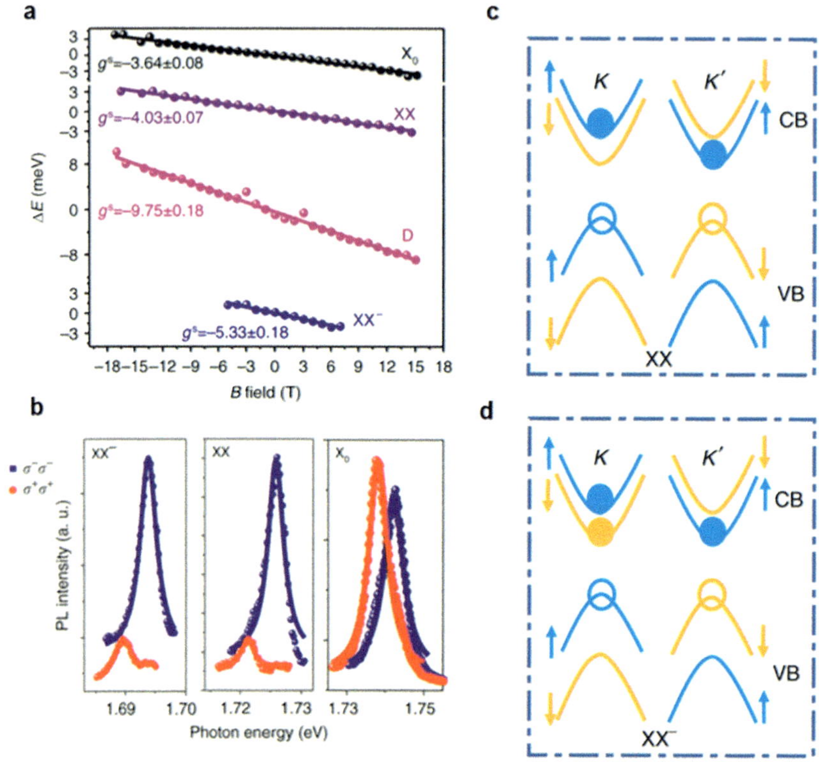

Fig. 8 Spin-valley configurations of the biexciton and the trion–exciton complex. (a) shows the Zeeman shifts and the extracted g-factors of the charge–neutral exciton peak (X_0), the biexciton peak (XX), the dark exciton peak (D), and the trion–exciton complex peak (XX^-). (b) shows the PL spectra of XX^-, XX, and X_0 under different helicity-resolved excitation and detection configurations. (c) and (d) display the spin-valley configurations of XX and XX^-, respectively. *Adapted from Li, Z., et al. 2018. Revealing the biexciton and trion-exciton complexes in BN encapsulated WSe2. Nat. Commun. 9, 3719 under a Creative Common license (http://creativecommons.org/licenses/by/4.0/).*

equilibrium picture, indicating the biexciton consists of a bright exciton in one valley and a spin-dark exciton in another valley (Li et al., 2018; Ye et al., 2018; Chen et al., 2018; Barbone et al., 2018), as shown in Fig. 8c. The inferred configuration of XX^- is also shown in Fig. 8d.

3.3 Spin-forbidden dark excitons and dark exciton complexes

Distinct from the spin configurations of the bright excitons, a spin-forbidden dark exciton in TMDCs originate from the inter-band transition between the bands with antiparallel spins (spin-triplet configuration), as

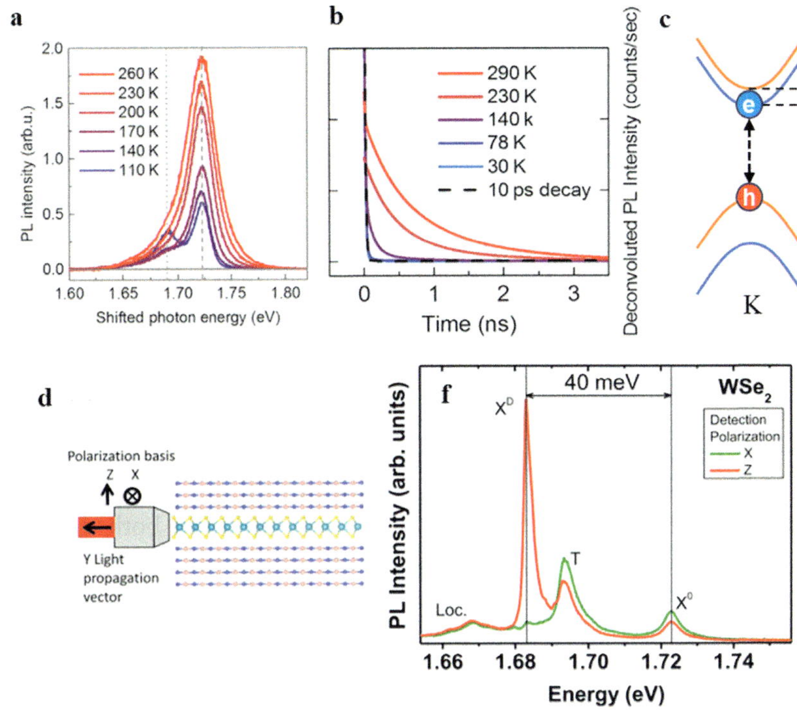

Fig. 9 Spin-forbidden dark excitons in monolayer WSe_2. (a) shows the temperature dependence of the PL spectra measured from a monolayer WSe_2 sample. The peak positions were shifted to align the high-energy peaks. (b) shows the time-resolved PL intensity of the high-energy peak. (c) shows the spin configuration of a spin-forbidden dark exciton in the K valley of WSe_2. *(a) Panel (a) and (b) adapted from Zhang, X.-X., You, Y., Zhao, S.Y.F., Heinz, T.F., 2015. Experimental evidence for dark excitons in monolayer WSe2. Phys. Rev. Lett. 115, 257403. Copyright 2015 American Physical Society. (b) (d) and (f) adapted from Wang, G., et al., 2017. In-plane propagation of light in transition metal dichalcogenide monolayers: optical selection rules. Phys. Rev. Lett. 119, 47401. Copyright 2017 American Physical Society.*

shown in Fig. 9c. A spin-forbidden exciton possesses an out-of-plane optical dipole moment, resulting weak absorption or emission of photons with in-plane polarization. In 2015, Zhang et al. reported the first experimental evidence for the dark excitons in monolayer WSe_2 using PL measurements. As shown in Fig. 9a, the PL intensity of the bright exciton decreases with the lowering temperature, suggesting the existence of a low-lying dark state. Time-resolved photoluminescence (TRPL) intensity of the bright state (Fig. 9b) also shows a weight transfer from a slow decay component to a fast decay component with decreasing temperature, indicating a thermal equilibrium between the bright excitonic state and the dark state (Zhang et al., 2015). In 2017, Wang et al. reported the polarization-resolved PL measurement of the spin-forbidden dark excitons in monolayer TMDCs with a signal collection from the sample edge (Fig. 9d), which confirmed the PL emission is polarized along the out-of-plane direction of the sample (Fig. 9f) (Wang et al., 2017). Other works also demonstrated that the PL emission of the dark state could be brightened for in-plane polarized light by applying an in-plane magnetic field (Zhang et al., 2017), by tip-enhanced PL spectroscopy (Park et al., 2018), or by coupling it to a surface plasmon polaritons (SPPs) (Zhou et al., 2017).

The binding of a spin-forbidden dark exciton with an additional charge can result in a dark trion. In 2019, Liu et al., Li et al., and Tang et al. independently reported the observation of gate-tunable dark trions (Liu et al., 2019b; Li et al., 2019c; Tang et al., 2019). PL spectroscopy performed with large-numerical-aperture objectives shows both the positive and the negative dark trions possess binding energies around 15 meV. Helicity-resolved magneto-PL measurements were also performed to confirm the nature of both the dark trions and the dark exciton, showing the recombination of the dark trions in a specific valley emits light with both σ^+ and σ^- components, a result of the distinct optical selection rule of the dark excitons. G-factors between −8.6 and −9.8 for the dark trions were extracted from the Zeeman shifts of the dark trions, similar to the value of the dark exciton and close to the expected theoretical value of −8 (Liu et al., 2019b; Li et al., 2019c). Using TRPL measurement, Tang et al. further reported a valley lifetime around 4 ns for the dark trions, significantly larger than the reported value of the bright exciton. The ultralong lifetime of dark trions originates from the suppressed electron intervalley scattering process, which involves an additional spin-flip process (Tang et al., 2019).

The long lifetime of the dark excitonic states makes them promising for valleytronics applications. However, their weak coupling to in-plane polarized photons and valley-independent linear polarized emission have prevented the efficient utilization of the valley degree of freedom. Li et al. demonstrated that the coupling of the spin-forbidden dark exciton with a chiral phonon can result in a dark exciton replica state, which couples to circularly polarized photons while preserving the long lifetime of the dark exciton. In this work, an excitonic state (X_D^R) with an energy 21.6 meV below the dark exciton resonance (X_D) was discovered using gate-dependent PL spectroscopy and helicity resolved magneto-PL

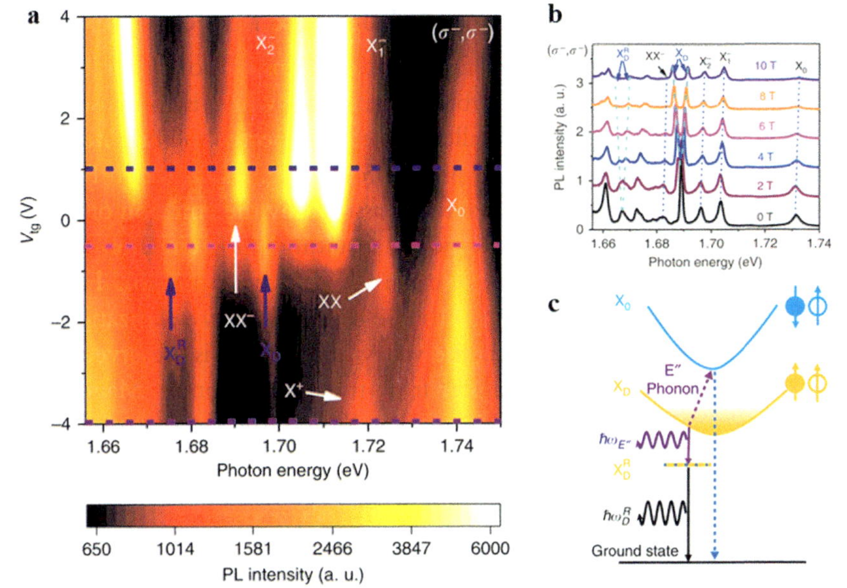

Fig. 10 Dark exciton phonon replica state in monolayer WSe_2. (a) shows the top-gate voltage (V_{tg}) dependence of the PL spectra from a monolayer WSe_2 sample. (b) shows a comparison of the Zeeman shifts of various exciton species in monolayer WSe_2. (c) is a schematic of the recombination process of the dark exciton phonon replica. *Adapted from Li, Z., et al., 2019a. Emerging photoluminescence from the dark-exciton phonon replica in monolayer WSe_2 Nat. Commun. 10, 2469 under a Creative Common license (http://creativecommons.org/licenses/by/4.0/).*

spectroscopy. The energy difference between X_D^R is consistent with the energy of the *E*" phonon mode obtained through first-principles calculation. X_D^R was found to mimic the gate dependence of X_D (Fig. 10a) and possesses a g-factor around −9.4, identical to the g-factor of X_D measured from the same device (−9.3). Remarkably, the relative intensity of the two branches of X_D^R is sensitive to the magnetic field and the detection helicity (Fig. 10b), in sharp contrast to the behaviors of X_D, suggesting a valley polarized selection rule of X_D^R. The above observations can be explained by the recombination process schematically shown in Fig. 10c, which involves the emission of a circularly polarized photon and a chiral *E*" phonon simultaneously. During this process, the dark exciton X_D couples to a bright exciton X_0 in the same valley by emitting a *E*" phonon. The dark exciton replica X_D^R, therefore, inherits the properties of both X_0 and X_D (Li et al., 2019a). Later after this work, Liu et al. also reported the phonon replicas of dark trions (Liu et al., 2019a).

3.4 Intervalley dark excitons and dark exciton complexes

Another possible configuration of the dark excitons in monolayer TMDCs is the intervalley (momentum-forbidden) exciton, which is composed of an electron and a hole in separate valleys. Li et al. reported the brightening of this optically dark state via chiral phonon coupling. As shown in Fig. 11a, the emission of the intervalley exciton is assisted by the emission of a chiral phonon mode LO(E′) at the K point, resulting a PL emission 16 meV below the spin-forbidden dark exciton X_D (X_i in Fig. 11b). Helicity-resolved PL measurement in Fig. 11b also shows a valley polarization ratio as large as 0.36 for X_i, suggesting the valley information is preserved by the pseudo-angular momentum transfer from the chiral phonon. Magneto PL measurement (Fig. 11c) also indicates that X_i possesses a g factor of −12.5,

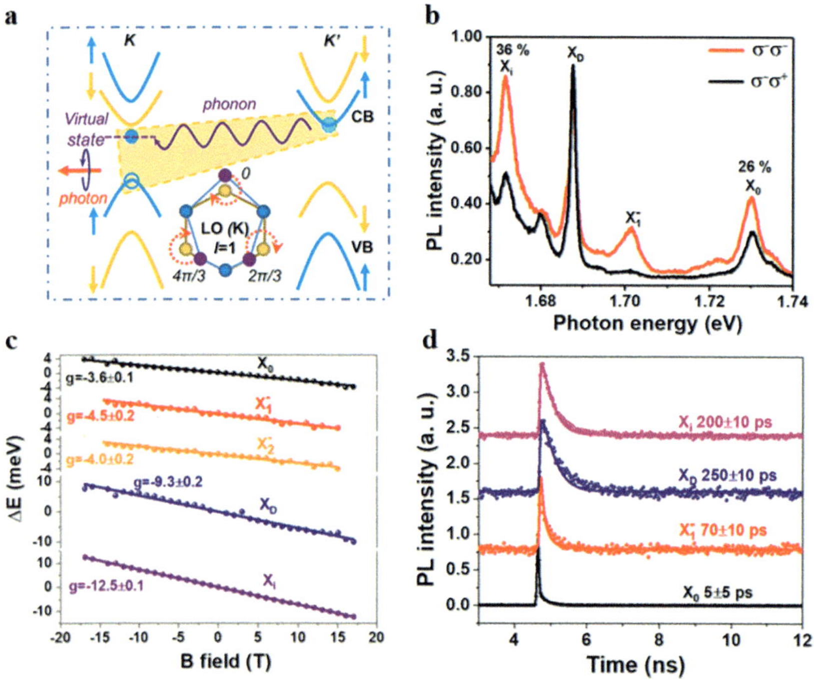

Fig. 11 Intervalley excitons brightened by chiral phonons. (a) shows a schematic of the emission process of the intervalley exciton. (b) shows the helicity-resolved PL spectra from a monolayer WSe_2 device using σ^- polarized excitation and σ^-/σ^+ collection. (c) shows the Zeeman shifts of various excitonic states in monolayer WSe_2 with X_i denoting the intervalley exciton. (d) shows the measured lifetimes of various excitonic states in monolayer WSe_2. *Reproduced from Li, Z., et al., 2019b. Momentum-dark intervalley exciton in monolayer tungsten diselenide brightened via chiral phonon. ACS Nano 13, 14107–14113. Copyright 2019 American Chemical Society.*

consistent with the expected value for the intervalley exciton. Finally, TRPL measurement in Fig. 11d shows X_i possesses a long lifetime around 200 ps, significantly longer than the lifetimes of the bright excitonic states and comparable to that of the spin-forbidden dark excitons (Li et al., 2019b).

4. Interlayer excitons in TMDCs heterostructures

In TMDCs heterobilayers with a type-II band alignment, the conduction band minimum (CBM) and the valence band maximum (VBM) reside in different materials, resulting in efficient charge transfer across the interface, which could occur less than tens of femtoseconds (Hong et al., 2014; Zhu et al., 2017; Ji et al., 2017). The electron–hole separation, together with the strong Coulomb interaction in 2D, result in the formation of interlayer excitons, i.e., excitons with the electron and the hole residing in different layers. The earliest evidence for interlayer excitons was reported by Fang et al. and Chiu et al. in 2014, which emerges as PL emission from MoS_2/WSe_2 heterostructures that is lower than the intralayer excitons of both monolayers (Fang et al., 2014; Chiu et al., 2014). More extensive studies have been carried out later and revealed a plenty of unique properties of interlayer excitons, including their electric-field tunable energies and ultralong lifetimes (Rivera et al., 2015; Jauregui et al., 2019). The long lifetime renders the interlayer excitons promising for realizing exciton condensation (Fogler et al., 2014; Calman et al., 2018; Su and MacDonald, 2017; Wu et al., 2015b). Like intralayer excitons, interlayer excitons also possess the valley degree of freedom. In addition, theoretical and experimental studies have shown that the spin-triplet states of interlayer excitons are brightened for in plane polarized photons at certain local atomic registries and exhibit an emission polarization opposite to that of the spin-singlet states (Yu et al., 2018; Wang et al., 2020b). These properties make interlayer excitons a fascinating platform to explore valleytronics.

4.1 Electric field tunability

Due to the spatial separation of the electron and the hole, an interlayer exciton possesses a permanent out-of-plane dipole moment, which will interact with an out-of-plane electric field, adding a linear Stark shift given by $\Delta E = -e{\cdot}d{\cdot}E_{hs}$ to the energy of the exciton, where e is the elementary charge, d is the vertical distance of the electron–hole pair and E_{hs} is the local out-of-plane electric field in the heterostructure. In 2015, Rivera et al. investigated the PL emission of the interlayer excitons in single-gated $MoSe_2/WSe_2$ heterostructures. A blueshift about 45 meV was observed when the back gate voltage is varied from 100 V to −100 V (Fig. 12c). This energy shift switches sign when the stacking order of the heterobilayer is reversed, indicating it is dependent on the direction of the electric field, consistent with the picture of the Stark effect. The TRPL measurement

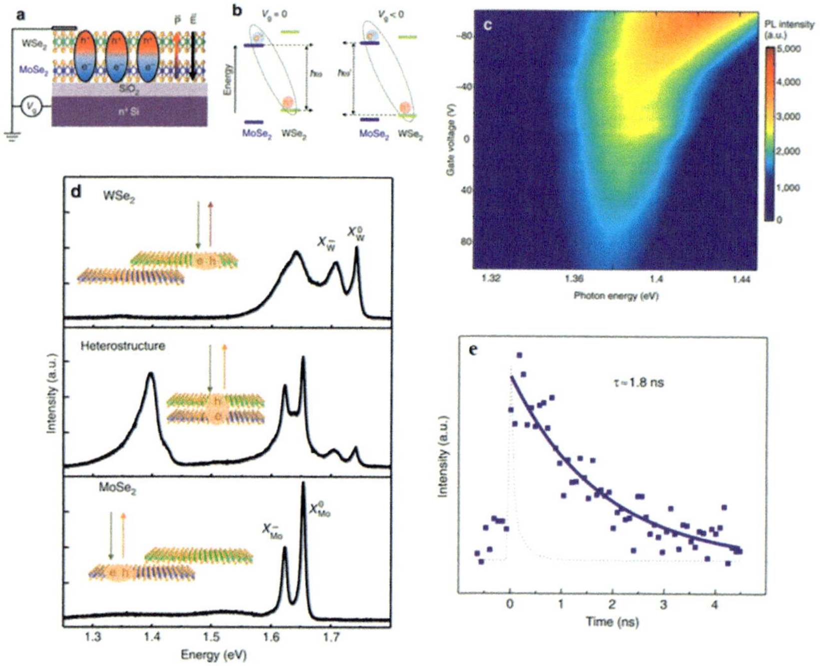

Fig. 12 Long-lived interlayer excitons in $MoSe_2/WSe_2$ heterostructures. (a) shows a schematic of the single-gated $MoSe_2/WSe_2$ device and the dipole moment of the interlayer exciton. (b) is a schematic showing the electric field tuning of the interlayer exciton energy in $MoSe_2/WSe_2$ heterostructures. A negative back gate voltage increases the energy separation between the CBM and the VBM and hence the interlayer exciton energy. (c) shows the interlayer exciton PL spectrum as a function of the back gate voltage. (d) shows the comparison of the PL spectra from the monolayer WSe_2, the $MoSe_2/WSe_2$ heterostructure and the monolayer $MoSe_2$ (e) shows the TRPL measurement of the interlayer exciton lifetime. *Adapted from Rivera, P., et al., 2015. Observation of long-lived interlayer excitons in monolayer $MoSe_2$–WSe_2 heterostructures Nat. Commun. 6, 2015 under a Creative Common license (http://creativecommons.org/licenses/by/4.0/).*

also revealed a ultralong lifetime of 1.8 ns for the interlayer exciton (Fig. 12e), which is a result of the reduced recombination rate originating from the spatial separation of electrons and holes (Rivera et al., 2015).

Jauregui et al. further studied the interlayer exciton PL emission from dual-gated $MoSe_2/WSe_2$ heterostructures, which allow independent control of the out-of-plane electric field and the carrier density (Jauregui et al., 2019). The electric field dependent PL spectra clearly revealed a linear dependence of the PL peak position on the local electric field (Fig. 13a). From the slope of the Stark shift, an interlayer distance d = 0.6 nm was deduced, which matches the interlayer distance of common TMDCs. Remarkably, the recombination dynamic of the interlayer exciton is widely tunable by the electric field: The lifetime decreases when the electric field drives the electron-hole pair towards recombination and increases when the field is in the other direction. This effect results in a tunable lifetime, which could be higher than 500 ns. The PL emission (Fig. 13e) and the lifetime (Fig. 13g) are also sensitive to the carrier density, suggesting that the interlayer excitons also strongly interact with charge carriers, possibly resulting in the formation of charged interlayer excitons (Jauregui et al., 2019).

4.2 Valley selection rules

In TMDCs heterobilayers, a specific valley of a TMDCs monolayer can be aligned with either K or K′ valley of another monolayer based on stacking order (0° or 60°). In addition, the local atomic registry (shown in Fig. 14) further provides a quantum angular momentum to interlayer optical transitions. The combinations of the spin, valley and atomic registry result in diverse optical selection rules. Yu et al. theoretically studied interlayer optical transition rules in TMDCs heterobilayers using symmetry analysis, with results shown in Table 2. Their findings suggest that the spin-triplet interlayer excitons (correspond to $\psi_{v,\uparrow} \leftrightarrow \psi_{c',\downarrow}$ and $\psi_{v,\downarrow} \leftrightarrow \psi_{c',\uparrow}$ transitions) could brighten up at certain lattice sites and emit circularly polarized photons due to the breaking of the out-of-plane symmetry (Yu et al., 2018).

The optical selection rules of the spin-singlet and the spin-triplet interlayer excitons were experimentally studied by helicity-resolved PL measurement and magneto-PL measurement (Wang, 2020b; Zhang, 2019). Zhang et al. reported two interlayer exciton PL peaks with opposite emission helicities in H-stacked $MoSe_2/WSe_2$ heterobilayers (T-state and S-state in Fig. 15b), consistent with the predicted optical transition rules for the spin-triplet and the spin-singlet interlayer excitons with the spin configuration shown in Fig. 15a, localized at the H_h^h lattice site. Remarkably, the spin-triplet states possess a

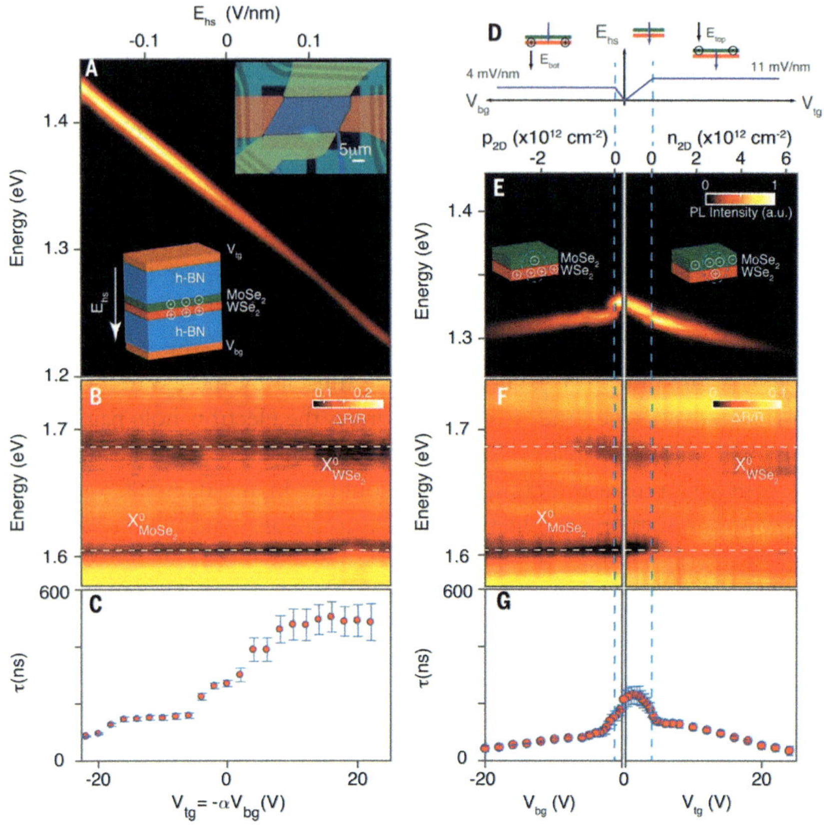

Fig. 13 Electric field tuning of interlayer exciton dynamics in $MoSe_2/WSe_2$ heterostructures. (a) shows the interlayer exciton PL emission as a function of local electric field E_{hs}, with the doping effect canceled out. (b) shows the reflectance contrast spectra measured in the same condition, the charge neutral intralayer excitons of WSe_2 and $MoSe_2$, $X^0_{WSe_2}$ and $X^0_{MoSe_2}$ remain visible, suggesting the sample remains charge-neutral. (c) shows the measured lifetime of the interlayer exciton as a function of E_{hs}. (e) and (f) are the PL spectra and the reflectance contrast spectra measured as functions of carrier density, respectively. (g) shows the interlayer exciton lifetime as a function of carrier density. *Reproduced from Jauregui, L.A., et al., 2019. Electrical control of interlayer exciton dynamics in atomically thin heterostructures. Science 366, 870–875. Copyright 2019 American Association for the Advancement of Science.*

ultralong valley lifetime of 32 ns (Zhang et al., 2019). Wang et al. further studied valley-Zeeman splitting of the interlayer excitons in the same system. Using magneto-PL measurement, the g factors for the triplet state and the singlet state are determined to be around 15.2 and 10.7, respectively, consistent with the expected value of 16 and 12 (Fig. 15d and e) (Wang et al., 2020b).

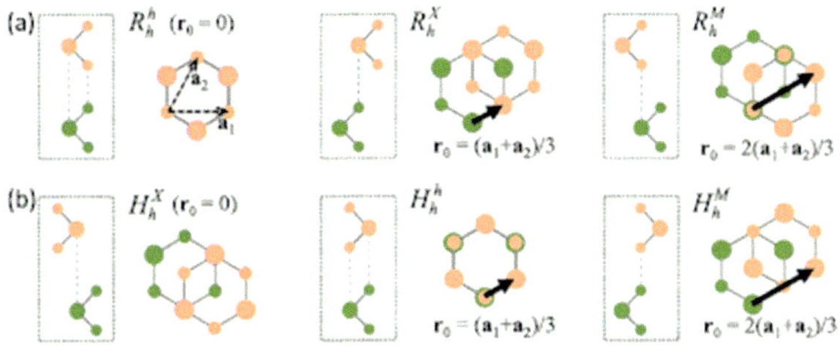

Fig. 14 Local atomic registries in TMDCs heterobilayers. R or H denote 0° and 60° alignment angles, respectively. The subscript h indicates the center of the hexagonal lattice of the hole-layer, while the superscripts denote the atomic registry of the electron layer. h, M or X denote the center of the hexagonal lattice, the transition metal atom and the chalcogen atom, respectively. *Reproduced from Yu, H., Liu, G.-B., Yao, W., 2018. Brightened spin-triplet interlayer excitons and optical selection rules in van der Waals heterobilayers. 2d Mater. 5, 035021. Copyright 2018 IOP Publishing.*

Table 2 Selection rules for interlayer optical transitions in TMDCs heterobilayers. $\psi_{v,\uparrow}$ and $\psi_{v,\downarrow}$ denote the spin-up and the spin-down valence bands in the K valley of the hole layer, respectively. $\psi_{c',\uparrow}$ and $\psi_{c',\downarrow}$ denote the spin-up and the spin down conduction bands of the electron layer, respectively. Copyright 2018 IOP Publishing.

	$\mathbf{R_h^h}$	$\mathbf{R_h^X}$	$\mathbf{R_h^M}$	$\mathbf{H_h^X}$	$\mathbf{H_h^h}$	$\mathbf{H_h^M}$
$\psi_{v,\uparrow} \leftrightarrow \psi_{c',\uparrow}$	σ^+	σ^-	z	σ^+	σ^-	z
$\psi_{v,\uparrow} \leftrightarrow \psi_{c',\downarrow}$	z	σ^+	σ^-	z	σ^+	σ^-
$\psi_{v,\downarrow} \leftrightarrow \psi_{c',\downarrow}$	σ^+	σ^-	z	σ^+	σ^-	z
$\psi_{v,\downarrow} \leftrightarrow \psi_{c',\uparrow}$	σ^-	z	σ^+	σ^-	z	σ^+

Source: Reproduced from Yu, H., Liu, G.-B., Yao, W., 2018. Brightened spin-triplet interlayer excitons and optical selection rules in van der Waals heterobilayers. 2d Mater. 5, 035021.

5. Moiré excitons in TMDCs moiré superlattices

The twist angle of the homojunction and the heterojunction of 2D materials provide another degree of freedom to tune their electronic and optical properties. When two layers of 2D semiconductor with small lattice mismatch or twist angle are stacked onto each other, the interference of the two crystal lattices will result in a reconstructed crystal lattice with much larger lattice constants, forming a so-called moiré superlattice

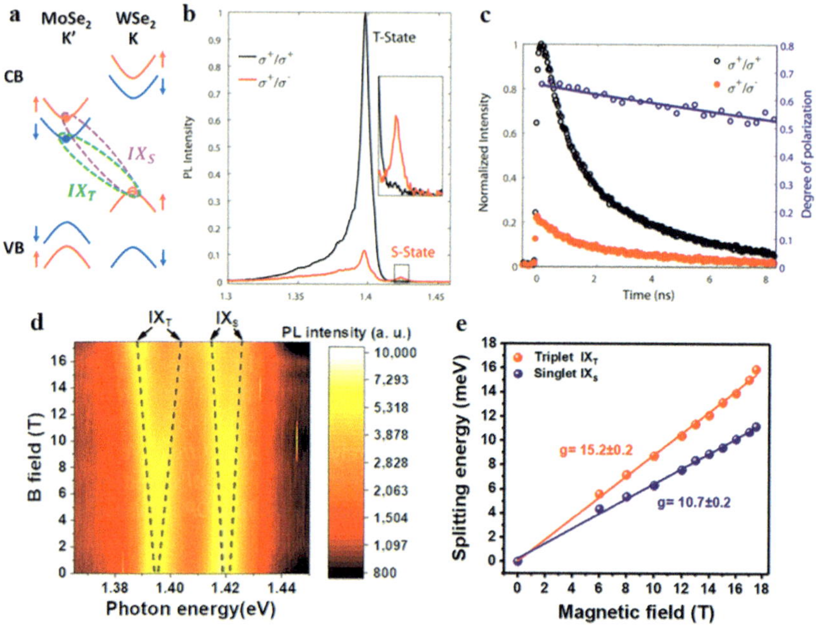

Fig. 15 Spin-triplet and spin-singlet interlayer excitons in $MoSe_2/WSe_2$ heterobilayers. (a) shows the configurations of the spin-triplet (IXT) and the spin-singlet (IXS) interlayer excitons with holes residing in the K valley of WSe_2. (b) shows the helicity-resolved PL spectra of the interlayer excitons. (c) shows the helicity-resolved TRPL measurement of the spin-triplet state. The blue circles are the extracted degree of polarization as a function of time. (d) shows the magneto-PL spectra of the interlayer excitons in a $MoSe_2/WSe_2$ heterobilayer. (e) shows the Zeeman-splittings as functions of magnetic field extracted from (d). *Panels (a), (d), (e) adapted from Wang, T., et al., 2020b. Giant valley-Zeeman splitting from spin-singlet and spin-triplet interlayer excitons in WSe2/MoSe2 heterostructure Nano Lett. 20, 694–700. Copyright 2019 American Chemical Society. (b) Panels (b) and (c) adapted from Zhang, L., et al., 2019. Highly valley-polarized singlet and triplet interlayer excitons in van der Waals heterostructure. Phys. Rev. B 100, 41402. Copyright 2019 American Physical Society.*

(Mak and Shan, 2022; Jin et al., 2019a; Tang et al., 2020; Regan et al., 2020). In TMDCs moiré superlattices with alignment angles close to 0° or 60°, the presence of moiré potential strongly modifies the wavefunctions of both intralayer and interlayer excitons, breaking their translational symmetries, resulting in excitons localized at high symmetry points of the moiré supercells, known as the moiré excitons (Jin et al., 2019a; Yu et al., 2017; Tran et al., 2019). The experimental signatures of these excitons are the strong alignment angle dependence of their resonance energies, which were first reported by Jin et al., Alexeev et al., Seyler et al. and Tran et al. in a variety of moiré systems including WSe_2/WS_2, $MoSe_2/WS_2$ and $MoSe_2/WSe_2$ (Jin et al., 2019a; Tran et al., 2019; Alexeev et al., 2019; Seyler et al., 2019). These moiré excitons carry diverse natures, which could be intralayer (Jin et al., 2019a), interlayer (Tran et al., 2019; Seyler et al., 2019), or interlayer hybridized excitons (Alexeev et al., 2019). Naively speaking, a moiré exciton is composed of electron and hole pair in the presence of moiré minibands, which are often flat. Therefore, it is expected that moiré excitons will have strong interactions with each other and are promising for realizing topological excitons (Wu et al., 2017). Although direct band calculations of moiré excitons are generally difficult to achieve (Naik et al., 2022), quite a few experiments have shown that the moiré exciton interactions are indeed strong (Xiong et al., 2023; Park et al., 2023; Lian et al., 2023a), suggesting possible flat excitonic band. In addition, interlayer excitons at distinct moiré sites also exhibit different optical transition rules, providing a new quantum degree of freedom to the field of valleytronics (Yu et al., 2018, 2017; Jin et al., 2019b). By the time this book is written, the field of moiré exciton is still rapidly developing. In this section, we focus on the intralayer moiré excitons to illustrate the concept of moiré excitons. We refer the readers to recent review articles [such as Huang et al. (2022) and Mak and Shan (2022)] for a better understanding of the frontier research in this field.

Jin et al. studied the alignment-angle-dependent reflectance contrast spectra of WSe_2/WS_2 moiré superlattices, which show a splitting of the WSe_2 intralayer exciton peak into three distinct peaks in near 0° or 60° aligned samples (Fig. 16a). This observation can be explained by a theoretical model considering the localization of the exciton wavefunction induced by the periodic potential in the moiré superlattices (Jin et al., 2019a). Naik et al. studied the wavefunctions of the moiré excitons in WSe_2/WS_2 moiré superlattices by performing large-scale first-principles *GW* and Bethe–Salpeter calculations, which indicate the lowest-energy intralayer moiré exciton is localized at the AA (R_h^h) site of the moiré unit cell, while the highest moiré peak is a charge transfer exciton with its electron at the AA site and the hole at the $B^{Se/W}$ site (Fig. 16b) (Naik et al., 2022). To understand the depth of the moiré potential, Chen et al. performed carrier-density-dependent reflectance contrast measurement in WSe_2/WS_2 moiré superlattices with varying WSe_2 layer numbers (Fig. 16c–e). A strong exciton resonance close to original intralayer exciton resonance of WSe_2 (denoted as X_{IV}^{2L} and X_{IV}^{3L} in Fig. 16d and e) was observed in bilayer WSe_2/monolayer WS_2 and trilayer WSe_2/monolayer

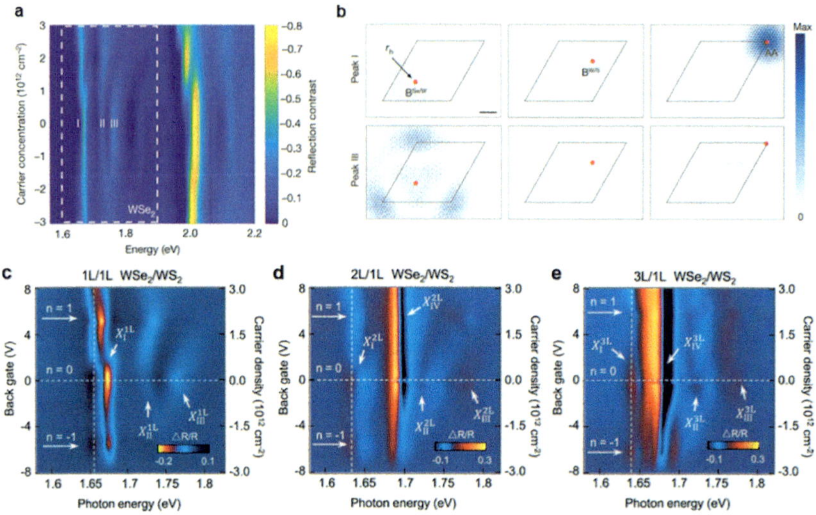

Fig. 16 Intralayer moiré excitons in WSe_2/WS_2 moiré superlattices. (a) shows the reflectance contrast as a function of carrier density in a WSe_2/WS_2 moiré superlattice. (b) shows the electron wave function (blue shade) corresponding to fixed hole positions (red dots) for different moiré exciton peaks obtained through ab initio calculation. (c)–(e) are the carrier density dependence of the reflectance contrast spectra measured from monolayer WSe_2/monolayer WS_2 (1L/1L WSe_2/WS_2), bilayer WSe_2/monolayer WS_2 (2L/1L WSe_2/WS_2) and trilayer WSe_2/monolayer WS_2 moiré superlattices. *Panel (a) reproduced from Jin, C., et al., 2019a. Observation of moiré excitons in WSe_2/WS_2 heterostructure superlattices Nature 567, 76–80. Copyright 2019 Nature Springer. Panel (b) reproduced from Naik, M.H., et al., 2022. Intralayer charge-transfer moiré excitons in van der Waals superlattices. Nature 609, 52–57. Copyright 2022 Nature Springer. Panels (c)–(e) reproduced from Chen, D., et al., 2022. Tuning moiré excitons and correlated electronic states through layer degree of freedom. Nat. Commun. 13, 4810 under a Creative Common license (http://creativecommons.org/licenses/by/4.0/).*

WS_2 moiré superlattices, indicating the excitons away from the interface are barely modified by the moiré potential, revealing the interfacial nature of the moiré coupling (Chen et al., 2022).

6. Conclusion

In summary, TMDCs have provide an exciting platform for the study of the 2D excitons with quantum degrees of freedom. We have seen many progresses, but expect the more exciting results in the future. On the fundamental study side, further control of van der Waals heterostructures will allow us to create more exotic many-particle excitonic complexes such as quadrupolar excitons (Astrakharchik et al., 2021; Slobodkin et al., 2020; Lian et al., 2023b). In addition, the moiré superlattice of TMDCs have been shown to lead to strong electron correlation (Tang et al., 2020; Regan et al., 2020; Wu et al., 2018; Xu et al., 2020; Huang et al., 2021). Considering the strong connection between electrons and excitons in TMDCs, we can envision the investigation and the engineering of exciton correlation for new quantum states (Xiong et al., 2023; Park et al., 2023). Finally, on the application side, we hope to see some of the unique properties of excitons to be utilized for transformative technologies innovations in the near future.

References

Alexeev, E.M., et al., 2019. Resonantly hybridized excitons in moiré superlattices in van der Waals heterostructures. Nature 567, 81–86.

Amani, M., et al., 2015. Near-unity photoluminescence quantum yield in MoS_2. Science 350, 1065–1068.

Aßmann, M., Bayer, M., 2020. Semiconductor Rydberg physics. Adv. Quantum Technol. 3, 1900134.

Astrakharchik, G.E., Kurbakov, I.L., Sychev, D.V., Fedorov, A.K., Lozovik, Yu.E., 2021. Quantum phase transition of a two-dimensional quadrupolar system. Phys. Rev. B 103, L140101.

Back, P., et al., 2017. Giant paramagnetism-induced valley polarization of electrons in charge-tunable monolayer $MoSe_2$. Phys. Rev. Lett. 118, 237404.

Barbone, M., et al., 2018. Charge-tunable biexciton complexes in monolayer WSe_2. Nat. Commun. 9, 3721.

Berkelbach, T.C., Hybertsen, M.S., Reichman, D.R., 2013. Theory of neutral and charged excitons in monolayer transition metal dichalcogenides. Phys. Rev. B 88, 45318.

Calman, E.V., et al., 2018. Indirect excitons in van der Waals heterostructures at room temperature. Nat. Commun. 9, 1895.

Cao, T., et al., 2012. Valley-selective circular dichroism of monolayer molybdenum disulphide. Nat. Commun. 3, 887.

Chen, D., et al., 2022. Tuning moiré excitons and correlated electronic states through layer degree of freedom. Nat. Commun. 13, 4810.

Chen, S.-Y., Goldstein, T., Taniguchi, T., Watanabe, K., Yan, J., 2018. Coulomb-bound four- and five-particle intervalley states in an atomically-thin semiconductor. Nat. Commun. 9, 3717.

Chernikov, A., et al., 2014. Exciton binding energy and nonhydrogenic Rydberg series in monolayer WS_2. Phys. Rev. Lett. 113, 076802.

Chiu, M.-H., et al., 2014. Spectroscopic signatures for interlayer coupling in MoS_2–WSe_2 van der Waals stacking. ACS Nano 8, 9649–9656.

Courtade, E., et al., 2017. Charged excitons in monolayer WSe_2: experiment and theory. Phys. Rev. B 96, 85302.

Danovich, M., Zólyomi, V., Fal'Ko, V.I., 2017. Dark trions and biexcitons in WS_2 and WSe_2 made bright by e-e scattering. Sci. Rep. 7, 45998.

Efimkin, D.K., MacDonald, A.H., 2018. Exciton-polarons in doped semiconductors in a strong magnetic field. Phys. Rev. B 97, 235432.

Efimkin, D.K., MacDonald, A.H., 2017. Many-body theory of trion absorption features in two-dimensional semiconductors. Phys. Rev. B 95, 35417.

Fang, H., et al., 2014. Strong interlayer coupling in van der Waals heterostructures built from single-layer chalcogenides. Proc. Natl Acad. Sci. 111, 6198–6202.
Fogler, M.M., Butov, L.V., Novoselov, K.S., 2014. High-temperature superfluidity with indirect excitons in van der Waals heterostructures. Nat. Commun. 5, 4555.
Gaëtan, A., et al., 2009. Observation of collective excitation of two individual atoms in the Rydberg blockade regime. Nat. Phys. 5, 115–118.
Goryca, M., et al., 2019. Revealing exciton masses and dielectric properties of monolayer semiconductors with high magnetic fields. Nat. Commun. 10, 4172.
Hao, K., et al., 2017. Neutral and charged inter-valley biexcitons in monolayer $MoSe_2$. Nat. Commun. 8, 15552.
He, K., et al., 2014. Tightly bound excitons in monolayer WSe_2. Phys. Rev. Lett. 113, 26803.
He, Y.-M., et al., 2016. Cascaded emission of single photons from the biexciton in monolayered WSe_2. Nat. Commun. 7, 13409.
Hong, X., et al., 2014. Ultrafast charge transfer in atomically thin MoS_2/WS_2 heterostructures. Nat. Nanotechnol. 9, 682–686.
Huang, D., Choi, J., Shih, C.-K., Li, X., 2022. Excitons in semiconductor moiré superlattices. Nat. Nanotechnol. 17, 227–238.
Huang, X., et al., 2021. Correlated insulating states at fractional fillings of the WS_2/WSe_2 moiré lattice. Nat. Phys. 17, 715–719.
Jauregui, L.A., et al., 2019. Electrical control of interlayer exciton dynamics in atomically thin heterostructures. Science 366, 870–875.
Ji, Z., et al., 2017. Robust stacking-independent ultrafast charge transfer in MoS_2/WS_2 bilayers. ACS Nano 11, 12020–12026.
Jin, C., et al., 2019b. Identification of spin, valley and moiré quasi-angular momentum of interlayer excitons. Nat. Phys. 15, 1140–1144.
Jin, C., et al., 2019a. Observation of moiré excitons in WSe_2/WS_2 heterostructure superlattices. Nature 567, 76–80.
Jones, A.M., et al., 2016. Excitonic luminescence upconversion in a two-dimensional semiconductor. Nat. Phys. 12, 323–327.
Jones, A.M., et al., 2013. Optical generation of excitonic valley coherence in monolayer WSe_2. Nat. Nanotechnol. 8, 634–638.
Keldysh, L.V., 1979. Coulomb interaction in thin semiconductor and semimetal films. Sov. J. Exp. Theor. Phys. Lett. 29, 658.
Keum, D.H., et al., 2015. Bandgap opening in few-layered monoclinic $MoTe_2$. Nat. Phys. 11, 482–486.
Kidd, D.W., Zhang, D.K., Varga, K., 2016. Binding energies and structures of two-dimensional excitonic complexes in transition metal dichalcogenides. Phys. Rev. B 93, 125423.
Kim, M.S., et al., 2016. Biexciton emission from edges and grain boundaries of triangular WS_2 monolayers. ACS Nano 10, 2399–2405.
Klingshirn, C.F., 2012. Semiconductor Optics. Springer Berlin, Heidelberg.
Kylänpää, I., Komsa, H.-P., 2015. Binding energies of exciton complexes in transition metal dichalcogenide monolayers and effect of dielectric environment. Phys. Rev. B 92, 205418.
Li, Z., et al., 2019c. Direct observation of gate-tunable dark trions in monolayer WSe_2. Nano Lett. 19, 6886–6893.
Li, Z., et al., 2019a. Emerging photoluminescence from the dark-exciton phonon replica in monolayer WSe_2. Nat. Commun. 10, 2469.
Li, Z., et al., 2019b. Momentum-dark intervalley exciton in monolayer tungsten diselenide brightened via chiral phonon. ACS Nano 13, 14107–14113.

Li, Z., et al., 2018. Revealing the biexciton and trion-exciton complexes in BN encapsulated WSe_2. Nat. Commun. 9, 3719.
Lian, Z., et al., 2023b. Quadrupolar excitons and hybridized interlayer Mott insulator in a trilayer moiré superlattice. Nat. Commun. 14, 4604.
Lian, Z., et al., 2023a. Valley-polarized exitonic Mott insulator WS_2/WSe_2 moiré superlattice Preprint at Arxiv:2308.10799v2.
Liu, E., et al., 2019b. Gate tunable dark trions in monolayer WSe_2. Phys. Rev. Lett. 123, 27401.
Liu, E., et al., 2019a. Valley-selective chiral phonon replicas of dark excitons and trions in monolayer WSe_2. Phys. Rev. Res. 1, 32007.
Liu, G.-B., Shan, W.-Y., Yao, Y., Yao, W., Xiao, D., 2013. Three-band tight-binding model for monolayers of group-VIB transition metal dichalcogenides. Phys. Rev. B 88, 85433.
Mak, K.F., Shan, J., 2016. Photonics and optoelectronics of 2D semiconductor transition metal dichalcogenides. Nat. Photonics 10, 216–226.
Mak, K.F., Shan, J., 2022. Semiconductor moiré materials. Nat. Nanotechnol. 17, 686–695.
Mak, K.F., et al., 2013. Tightly bound trions in monolayer MoS_2. Nat. Mater. 12, 207–211.
Mak, K.F., He, K., Shan, J., Heinz, T.F., 2012. Control of valley polarization in monolayer MoS_2 by optical helicity. Nat. Nanotechnol. 7, 494–498.
Manzeli, S., Ovchinnikov, D., Pasquier, D., Yazyev, O.V., Kis, A., 2017. 2D transition metal dichalcogenides. Nat. Rev. Mater. 2, 17033.
Mayers, M.Z., Berkelbach, T.C., Hybertsen, M.S., Reichman, D.R., 2015. Binding energies and spatial structures of small carrier complexes in monolayer transition-metal dichalcogenides via diffusion Monte Carlo. Phys. Rev. B 92, 161404.
Mostaani, E., et al., 2017. Diffusion quantum Monte Carlo study of excitonic complexes in two-dimensional transition-metal dichalcogenides. Phys. Rev. B 96, 75431.
Nagler, P., et al., 2018. Zeeman splitting and inverted polarization of biexciton emission in monolayer WS_2. Phys. Rev. Lett. 121, 57402.
Naik, M.H., et al., 2022. Intralayer charge-transfer moiré excitons in van der Waals superlattices. Nature 609, 52–57.
Nam, S.B., et al., 1976. Free-exciton energy spectrum in GaAs. Phys. Rev. B 13, 761–767.
Okada, M., et al., 2017. Observation of biexcitonic emission at extremely low power density in tungsten disulfide atomic layers grown on hexagonal boron nitride. Sci. Rep. 7, 322.
Park, H., et al., 2023. Dipole ladders with large Hubbard interaction in a moiré exciton lattice. Nat. Phys. 19, 1286–1292. https://doi.org/10.1038/s41567-023-02077-5.
Park, K.-D., Jiang, T., Clark, G., Xu, X., Raschke, M.B., 2018. Radiative control of dark excitons at room temperature by nano-optical antenna-tip Purcell effect. Nat. Nanotechnol. 13, 59–64.
Plechinger, G., et al., 2016. Trion fine structure and coupled spin–valley dynamics in monolayer tungsten disulfide. Nat. Commun. 7, 12715.
Poddubny, A.N., Glazov, M.M., 2019. Topological spin phases of trapped Rydberg excitons in Cu_2O. Phys. Rev. Lett. 123, 126801.
Regan, E.C., et al., 2020. Mott and generalized Wigner crystal states in WSe_2/WS_2 moiré superlattices. Nature 579, 359–363.
Rivera, P., et al., 2015. Observation of long-lived interlayer excitons in monolayer $MoSe_2$–WSe_2 heterostructures. Nat. Commun. 6, 6242.
Roch, J.G., et al., 2018. Quantum-confined Stark effect in a MoS_2 monolayer van der Waals heterostructure. Nano Lett. 18, 1070–1074.
Ross, J.S., et al., 2013. Electrical control of neutral and charged excitons in a monolayer semiconductor. Nat. Commun. 4, 1474.

Schaibley, J.R., et al., 2016. Valleytronics in 2D materials. Nat. Rev. Mater. 1, 16055.
Schutte, W.J., De Boer, J.L., Jellinek, F., 1987. Crystal structures of tungsten disulfide and diselenide. J. Solid. State Chem. 70, 207–209.
Seyler, K.L., et al., 2019. Signatures of moiré-trapped valley excitons in $MoSe_2/WSe_2$ heterobilayers. Nature 567, 66–70.
Sidler, M., et al., 2017. Fermi polaron-polaritons in charge-tunable atomically thin semiconductors. Nat. Phys. 13, 255–261.
Slobodkin, Y., et al., 2020. Quantum phase transitions of trilayer excitons in atomically thin heterostructures. Phys. Rev. Lett. 125, 255301.
Steinhoff, A., et al., 2018. Biexciton fine structure in monolayer transition metal dichalcogenides. Nat. Phys. 14, 1199–1204.
Stier, A.V., et al., 2018. Magneto-optics of exciton Rydberg states in a monolayer semiconductor. Phys. Rev. Lett. 120, 57405.
Su, J.-J., MacDonald, A.H., 2017. Spatially indirect exciton condensate phases in double bilayer graphene. Phys. Rev. B 95, 45416.
Szyniszewski, M., Mostaani, E., Drummond, N.D., Fal'Ko, V.Isss, 2017. Binding energies of trions and biexcitons in two-dimensional semiconductors from diffusion quantum Monte Carlo calculations. Phys. Rev. B 95, 81301.
Tang, Y., et al., 2020. Simulation of Hubbard model physics in WSe_2/WS_2 moiré superlattices. Nature 579, 353–358.
Tang, Y., Mak, K.F., Shan, J., 2019. Long valley lifetime of dark excitons in single-layer WSe_2. Nat. Commun. 10, 4047.
Taylor, J., et al., 2022. Simulation of many-body dynamics using Rydberg excitons. Quantum Sci. Technol. 7, 035016.
Tran, K., et al., 2019. Evidence for moiré excitons in van der Waals heterostructures. Nature 567, 71–75.
Urban, E., et al., 2009. Observation of Rydberg blockade between two atoms. Nat. Phys. 5, 110–114.
Verzhbitskiy, I., Vella, D., Watanabe, K., Taniguchi, T., Eda, G., 2019. Suppressed out-of-plane polarizability of free excitons in monolayer WSe_2. ACS Nano 13, 3218–3224.
Wang, G., et al., 2017. In-plane propagation of light in transition metal dichalcogenide monolayers: optical selection rules. Phys. Rev. Lett. 119, 47401.
Wang, T., et al., 2020a. Giant valley-polarized Rydberg excitons in monolayer WSe_2 revealed by magneto-photocurrent spectroscopy. Nano Lett. 20, 7635–7641.
Wang, T., et al., 2020b. Giant valley-Zeeman splitting from spin-singlet and spin-triplet interlayer excitons in $WSe_2/MoSe_2$ heterostructure. Nano Lett. 20, 694–700.
Wickramaratne, D., Zahid, F., Lake, R.K., 2014. Electronic and thermoelectric properties of few-layer transition metal dichalcogenides. J. Chem. Phys. 140, 124710.
Wu, F., Lovorn, T., Macdonald, A.H., 2017. Topological exciton bands in moiré heterojunctions. Phys. Rev. Lett. 118, 147401.
Wu, F., Lovorn, T., Tutuc, E., Macdonald, A.H., 2018. Hubbard model physics in transition metal dichalcogenide moiré bands. Phys. Rev. Lett. 121, 26402.
Wu, F., Qu, F., MacDonald, A.H., 2015a. Exciton band structure of monolayer MoS_2. Phys. Rev. B 91, 75310.
Wu, F.-C., Xue, F., MacDonald, A.H., 2015b. Theory of two-dimensional spatially indirect equilibrium exciton condensates. Phys. Rev. B 92, 165121.
Xiao, D., Liu, G.-B., Feng, W., Xu, X., Yao, W., 2012. Coupled spin and valley physics in monolayers of MoS_2 and other group-VI dichalcogenides. Phys. Rev. Lett. 108, 196802.
Xiong, R., et al., 2023. Correlated insulator of excitons in WSe_2/WS_2 moiré superlattices. Science 380, 860–864.

Xu, X., Yao, W., Xiao, D., Heinz, T.F., 2014. Spin and pseudospins in layered transition metal dichalcogenides. Nat. Phys. 10, 343–350.

Xu, Y., et al., 2020. Correlated insulating states at fractional fillings of moiré superlattices. Nature 587, 214–218.

Ye, Z., et al., 2018. Efficient generation of neutral and charged biexcitons in encapsulated WSe_2 monolayers. Nat. Commun. 9, 3718.

You, Y., et al., 2015. Observation of biexcitons in monolayer WSe_2. Nat. Phys. 11, 477–481.

Yu, H., Liu, G.B., Tang, J., Xu, X., Yao, W., 2017. Moiré excitons: from programmable quantum emitter arrays to spin-orbit–coupled artificial lattices. Sci. Adv. 3, e1701696.

Yu, H., Liu, G.-B., Yao, W., 2018. Brightened spin-triplet interlayer excitons and optical selection rules in van der Waals heterobilayers. 2d Mater. 5, 035021.

Zeng, H., Dai, J., Yao, W., Xiao, D., Cui, X., 2012. Valley polarization in MoS_2 monolayers by optical pumping. Nat. Nanotechnol. 7, 490–493.

Zhang, L., et al., 2019. Highly valley-polarized singlet and triplet interlayer excitons in van der Waals heterostructure. Phys. Rev. B 100, 41402.

Zhang, X.-X., et al., 2017. Magnetic brightening and control of dark excitons in monolayer WSe_2. Nat. Nanotechnol. 12, 883–888.

Zhang, X.-X., You, Y., Zhao, S.Y.F., Heinz, T.F., 2015. Experimental evidence for dark excitons in monolayer WSe_2. Phys. Rev. Lett. 115, 257403.

Zhou, Y., et al., 2017. Probing dark excitons in atomically thin semiconductors via near-field coupling to surface plasmon polaritons. Nat. Nanotechnol. 12, 856–860.

Zhu, B., Chen, X., Cui, X., 2015. Exciton binding energy of monolayer WS_2. Sci. Rep. 5, 9218.

Zhu, H., et al., 2017. Interfacial charge transfer circumventing momentum mismatch at two-dimensional van der Waals heterojunctions. Nano Lett. 17, 3591–3598.

CHAPTER TWO

Valley excitons and their many-body complexes

Jiajie Pei[a,*] and Renjing Xu[b]
[a]College of Materials Science and Engineering, Fuzhou University, Fuzhou, P.R. China
[b]Microelectronics Thrust, The Hong Kong University of Science and Technology (Guangzhou), Nansha, Guangdong, P.R. China
*Corresponding author. e-mail address: jjpei@fzu.edu.cn

Contents

Abstract

This chapter presents an overview of the concept of "valleys" in semiconductor physics, which refers to local minima in the energy bands of certain semiconductors. These energy bands are crucial for determining the electronic and optical properties of the material. Specifically, some semiconductor crystals, like transition metal dichalcogenides (TMDCs), exhibit multiple valleys with varying energies, resulting in distinct effective masses and pseudospins for charge carriers. The focus of the chapter is on valley excitonic states that arise from strong interactions between electrons and holes in different valleys. These interactions, along with additional carriers in the electronic band structure, lead to the formation of valley many-body complexes, including intervalley trions, biexcitons, and charged biexcitons. The formation processes, energy valley configurations, fluorescence and absorption peak positions, and changes in polarization characteristics of various valley excitonic states in TMDCs under external fields are explored and summarized in this chapter. These complexes possess diverse optical and electronic properties, making them promising for applications in optoelectronics, quantum information processing, and valleytronics. Additionally, recently discovered moiré exciton states and their relationship with energy valleys in momentum space are discussed. However, research on polarization characteristics in these states is in its early stages, necessitating further exploration.

Semiconductors and Semimetals, Volume 112
ISSN 0080-8784, https://doi.org/10.1016/bs.semsem.2023.09.004

1. Introduction

In semiconductor physics, the "valleys" refer to the local minima in the energy bands of certain semiconductors. These energy bands represent the quantized energy levels available to electrons in the crystal lattice of the semiconductor, and they play a crucial role in determining the electronic and optical properties of the material. In certain semiconductor crystals, such as transition metal dichalcogenides (TMDCs) (Mak and Shan, 2016; Mak et al., 2018), the energy bands exhibit multiple valleys, each with different energies. These different valleys can display different effective masses and pseudospins (Xu et al., 2014; Schaibley et al., 2016), which can affect the behavior of the charge carriers (electrons and holes) in the material.

Valley excitons are excitonic states that arise due to the strong interaction between electrons and holes located in different valleys such as the spin-allowed bright excitons (Chernikov et al., 2014; He et al., 2014), spin-forbidden dark excitons (Park et al., 2018; Zhang et al., 2017; Wang et al., 2017; Zhang et al., 2015a), and momentum-forbidden dark excitons (Bao et al., 2020; Peng et al., 2019; Madéo et al., 2020; Wallauer et al., 2021). The interactions among such excitons and additional carriers in other valleys in the electronic band structure can give rise to the formation of quasiparticles known as valley many-body complexes, such as intervalley trions (Li et al., 2019a; Vaclavkova et al., 2018; Plechinger et al., 2016a; Lyons et al., 2019; Liu et al., 2019a; Pei et al., 2023), biexcitons, charged biexcitons (Barbone et al., 2018; Chen et al., 2018a; Li et al., 2018; Paur et al., 2019; Ye et al., 2018), and even more complex structures, which exhibit rich optical and electronic properties (Fig. 1) (Li et al., 2019a). Valley excitons and their many-body complexes have piqued the interest of researchers due to their unique properties and potential applications in optoelectronics, quantum information processing, and valleytronics (Mak et al., 2018; Schaibley et al., 2016). Moreover, the study of valley excitons and their many-body complexes has opened up new avenues for exploring semiconductor physics and has also led to the development of novel device concepts with potential for enhanced performance.

2. Bright and dark valley excitons

Bright and dark valley excitons are two types of excitonic states that can be formed in transition metal dichalcogenides (TMDCs). Bright valley

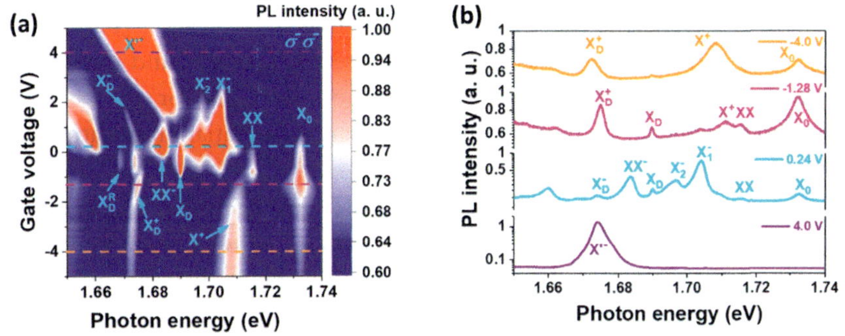

Fig. 1 PL spectroscopy of the BN encapsulated monolayer WSe_2 device showing the complexity and tunability of the many-body states. (a) Gate-dependent PL mapping. The color represents the PL intensity. The labels depict the bright exciton (X_0), dark exciton (X_D), positive bright trion (X^+), negative bright trion (X_1^- and X_2^-), positive dark trion (X_D^+), negative dark trion (X_D^-), biexciton (XX), charged biexciton (XX^-), etc. (b) PL spectra at the gate voltages corresponding to the dashed lines in panel (a). *Reprints with permission from Li, Z., et al., 2019a. Direct observation of gate-tunable dark trions in monolayer WSe2. Nano Lett. 2019, 19 (10), 6886–6893.*

excitons are formed when an electron and a hole have the same momentum and are excited from and recombine into the same valley. Bright valley excitons have strong optical transitions and are optically active, implying they can emit or absorb light efficiently. These excitons are commonly observed in experiments and play a significant role in the optical properties of 2D materials (Chernikov et al., 2014; He et al., 2014; Ugeda et al., 2014). Dark valley excitons, on the other hand, have weaker optical transitions and are optically inactive, meaning they do not emit or absorb light efficiently. Dark valley excitons can be divided into spin-forbidden dark excitons and momentum-forbidden dark excitons (Zhang et al., 2017; Madéo et al., 2020). In the former case, the exciton is formed when an electron and a hole have anti-parallel spins. In the latter case, the electron and hole have different momentum states and are excited from and recombine into different valleys. As a result, they are more challenging to detect and study experimentally compared to bright excitons.

The distinction between bright and dark valley excitons is mainly determined by the selection rules governing optical transitions in the specific material system. Bright excitons satisfy the selection rules and have a higher probability of emitting or absorbing photons, whereas dark excitons violate these rules and have a suppressed emission or absorption. However, it's worth noting that in certain cases, external factors such as defects, strain, or an applied electric field can modify the selection rules and allow dark excitons to become optically active or vice versa (Park et al., 2018; Zhang et al., 2017). The interplay between bright and dark valley excitons in 2D materials has important implications for their optical properties, energy relaxation dynamics, and potential applications in optoelectronics and quantum technologies.

Recent progress in understanding valley excitons has been largely driven by theoretical studies. Computational methods such as density functional theory and many-body perturbation theory have been used to investigate the electronic and optical properties of valley excitons (Peng et al., 2019; Kormányos et al., 2015; Malic et al., 2018; Jiang et al., 2021a). These studies have provided insights into the exciton-valley interaction and the diversity of valley excitons. The Coulomb interaction in monolayer transition metal dichalcogenides (TMDCs) is exceptionally strong, resulting in tightly bound excitons. Malic et al. provide the excitonic landscape of the most researched TMDCs materials, including the spectral position of momentum- and spin-forbidden excitonic states, based on the Wannier equation solution (Malic et al., 2018). TMDCs' electronic band structure has four distinct minima in the energetically lowest conduction band (K, K′, Λ and Λ′) and three maxima in the highest valence band (K, K′, and Γ). Within the K valley, Coulomb-bound electron–hole pairs can develop, resulting in bright K–K excitons (yellow ovals in Fig. 2a, b). They can, however, be produced by combining electrons and holes in distinct valleys, resulting in momentum-forbidden dark excitonic states (red and orange ovals in Fig. 2a, b). In addition to momentum-forbidden dark states, there is a spin-based class of dark excitons known as spin-forbidden (or spin-unlike) excitons, which are optically inaccessible because photons cannot produce a spin-flip process (purple oval in Fig. 2a, b). Since the spin–orbit coupling can be positive or negative depending on the TMDCs material, there are two possible unique spin state orderings: While electrons in the lowest conduction band of molybdenum-based TMDCs (MoS_2 and $MoSe_2$) have the same spin as those in the highest valence band, electrons in the lowest conduction band of tungsten-based TMDCs (WS_2 and WSe_2) have the opposite spin ordering, as seen in Fig. 2a, b.

Fig. 2c–f depict the excitonic dispersion, specifically highlighting the lowest A_{1s} states, in the four extensively studied TMDCs materials, namely MoS_2, $MoSe_2$, WS_2, and WSe_2. For each exciton, the hole resides at the K point, while the electron can be found at either the K, Λ, or K′ point. All dark excitonic states have been included, accounting for both momentum- and spin-forbidden cases. It is revealed that tungsten-based TMDCs exhibit

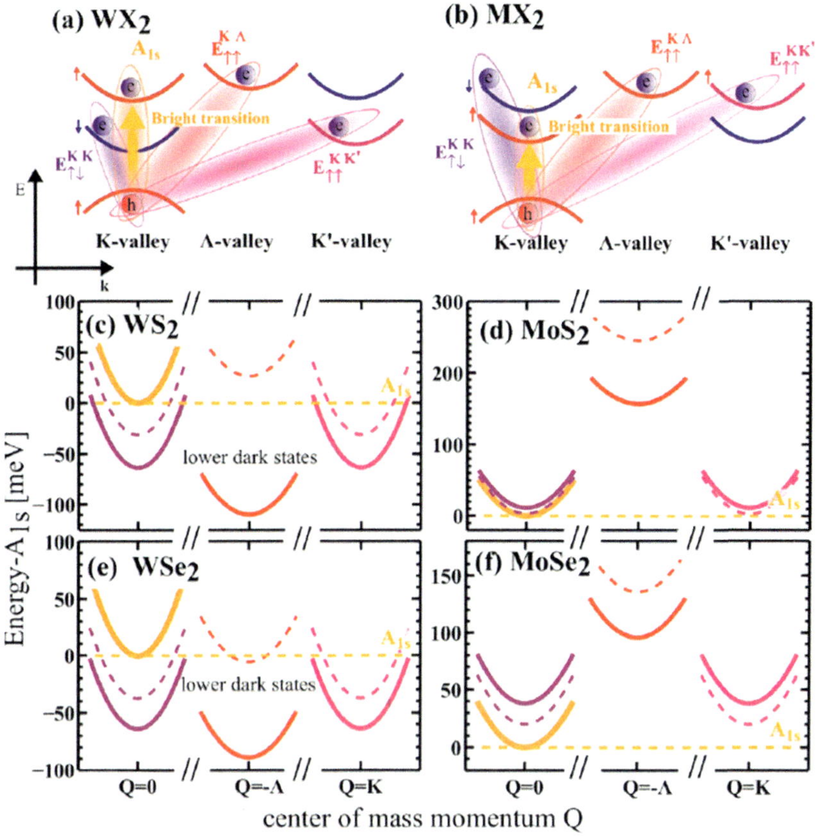

Fig. 2 Direct and indirect excitons with the hole located at the K valley. (a) Schematic electronic dispersions around the K and Λ valley for tungsten-based TMDCs. (b) Schematic electronic dispersions for molybdenum-based TMDCs. (c–f) Exciton dispersions calculated by solving the Wannier equation for WS_2, MoS_2, WSe_2, and $MoSe_2$, respectively. The dashed lines represent the relative band ordering based on a free-particle model, disregarding the influence of excitonic binding energies. *Reprints with permission from Malic, E., et al. 2018. Dark excitons in transition metal dichalcogenides. Phys. Rev. Mater. 2 (1), 014002.*

momentum dark excitons with lower energy levels than the bright K–K transition (indicated by orange dashed lines). Moreover, these materials also exhibit spin-forbidden excitonic states situated well below the bright A_{1s} exciton, as illustrated in Fig. 2c and e. Specifically, the spin-forbidden K–K exciton, formed by Coulomb-bound electrons and holes situated in the K valley with opposite spins, is approximately 50 meV lower in energy compared to the bright state for WS_2 and WSe_2. It is important to note that the momentum-forbidden spin-like K–K′ excitons possess equal energy levels to the spin-forbidden K–K excitons. However, due to the Coulomb electron–hole exchange coupling that solely affects spin-like states, they may be separated by a few meV. The momentum-forbidden K–Λ excitons, characterized by the presence of electrons and holes in different valleys, exhibit a strong dependence on the strength of the Coulomb interaction. These excitons can be externally manipulated by means of substrate-induced screening or doping. The lower energy position of these excitons is attributed to the significantly larger effective mass of the Λ valley, which is three times greater than that of the K valley. This discrepancy in effective mass directly influences the Wannier equation, leading to a larger exciton binding energy for K–Λ excitons. This compensates for the initially higher energy level associated with the Λ valley. On the other hand, in molybdenum-based TMDCs, the energy gap between the K and Λ valleys is too large. Consequently, even when excitonic effects are considered (specifically, if only K–hole excitons are taken into account), these TMDCs still retain their character as direct semiconductors. This observation can be verified by Fig. 2d and f.

The diversity of valley excitons has been demonstrated in experiments involving photoluminescence and absorption spectroscopy. The spin components of the exciton states in monolayer WSe_2 can be mixed by applying an in-plane magnetic field (Zhang et al., 2017). This perturbation leads to the enhancement of the spin-dark exciton states, enabling the direct observation of light emission from these otherwise inaccessible states. By obtaining a spectroscopic signature of the dark excitons, the researchers were able to accurately measure the splitting between the bright and spin-dark states. A different method involves utilizing the in-plane optical excitation and detection geometry (Wang et al., 2017), as illustrated in Fig. 3a. There is advantage due to the predicted anisotropic nature of band-to-band transitions when light propagates parallel to the monolayer's plane. A distinct luminescence line arises when the polarization is perpendicular to the 2D material plane (Fig. 3b), signifying a z-dipole transition. In traditional linear optical spectroscopy, this transition is prohibited under normal incidence due to the transverse nature of the electromagnetic field (Fig. 3c). By conducting these measurements, the bright-dark exciton splitting has been directly determined, for which 40 ± 1 meV and 55 ± 2 meV in WSe_2 and WS_2 monolayer were measured, respectively (Fig. 3b). Another study utilizing the coupling to the out-of-plane transition dipole moment by surface plasmon polaritons has also observed similar characteristics of spin-dark excitons

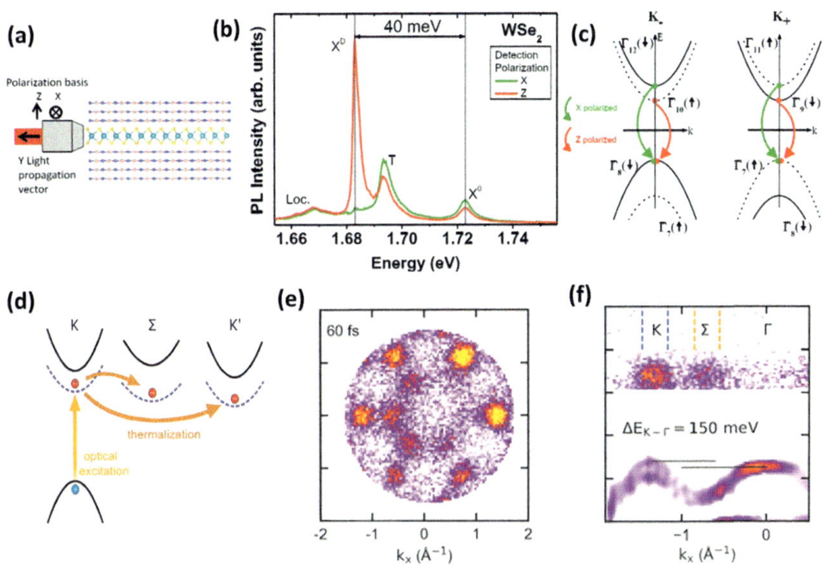

Fig. 3 Experimental observation of spin- and momentum-dark excitons. (a) Schematics of the excitation or detection geometry of the PL for light propagating parallel to the ML plane, detection of the PL from the edge of the sample. (b) The detected PL spectra from the edge of the sample at different polarizations. (c) Sketch of the band structure of monolayer WSe_2 with arrows indicating the spin-allowed and spin-forbidden transitions. (d) Schematic of the optical excitation and subsequent formation processes of momentum-forbidden excitons in WS_2. (e) Momentum maps of the electronic states after optical excitation. (f) Corresponding energy-momentum maps along Γ–K showing the occupied states for the same delays. *Panels (a–c) adapted with permission from Wang, G., et al., 2017. In-plane propagation of light in transition metal dichalcogenide monolayers: optical selection rules. Phys. Rev. Lett. 119 (4), 047401. Panels (d–f) adapted with permission from Wallauer, R., et al., 2021. Momentum-resolved observation of exciton formation dynamics in monolayer WS2. Nano Lett. 21 (13), 5867–5873.*

(Zhou et al., 2017). At both low and higher temperatures, no X_D line's spectral signature can be detected in monolayer $MoSe_2$. This observation aligns with current understanding of the band structure of monolayer $MoSe_2$. According to theoretical and experimental studies (Zhang et al., 2015a; Arora et al., 2015; Wang et al., 2015), it is anticipated that the dark exciton occupies a higher energy level compared to the bright exciton in monolayer $MoSe_2$. This disparity can be attributed to the alteration of spin–orbit splitting sign in the conduction band among these distinct materials (Kośmider and Fernández-Rossier, 2013).

The momentum-dark excitons have been directly visualized in experiment by using the time-resolved XUV micro-angle resolved photoemission spectroscopy for WSe_2 (Madéo et al., 2020) and WS_2 (Wallauer et al., 2021). For monolayer WSe_2, the band structure obtained from measurements exhibited spin-split valence band extrema at the K and K′ valleys, which demonstrated remarkable consistency with GW calculations. Upon resonant photoexcitation of the A exciton, a significant photoemission signal was observed below the bandgap, specifically at the K and K′ valleys (located at $\pm$ 1.26 Å^{-1}), as well as at the Q (also known as Λ or Σ) valleys (located at $\pm$ 0.75 Å^{-1}) at positive time delay. Under above-gap excitation, it is found the exciton relaxation pathways, particularly for the dark Q-valley excitons, were substantially different from those observed for resonant excitation (Madéo et al., 2020). Similar results were observed in a monolayer WS_2; the formation of a momentum-forbidden dark K–Σ exciton (Fig. 3d) was found to be a few tens of femtoseconds after optical excitation. The high temporal resolution of the setup allows researchers to trace the early-stage exciton dynamics on its intrinsic time scale (Fig. 3e–f). The consecutive population of K and Σ valleys serving as a clear signature of the delayed formation of a momentum-forbidden K–Σ exciton after optical excitation at K. Moreover, by tuning the excitation energy, the temporal evolution of the coherent excitonic polarization is manipulated, and its influence on the dark exciton formation could be observed (Wallauer et al., 2021). These studies highlight the significance of momentum conservation in the behavior of dark excitons. By directly visualizing these excitons, the researchers gained insights into the dynamics of dark excitons, on the basis of which they studied the formation, decay, and interactions with other excitons, thus shedding light on the fundamental processes underlying these excitonic behaviors. Accordingly, the binding energies for the K-valley and Q-valley excitons were estimated as ~390 and ~480 meV (with respect to the conduction band minima) (Madéo et al., 2020), respectively. Additionally, it is found these dark excitons dominated the excited-state distribution within hundreds of femtoseconds after above-gap excitation underscores their importance in atomically thin semiconductors.

3. Intravalley and intervalley trions

Trions are quasiparticles that arise when an extra electron (or hole) is added to a system that already contains an electron–hole pair, resulting in a three-particle complex (Ross et al., 2013; Mak et al., 2013). Trions have attracted tremendous interests because they exhibit unique electronic and optical properties. They can interact with each other and with their environment, leading to phenomena such as energy transfer, scattering, and recombination processes. Intravalley trions are formed when an extra electron (or hole) occupies the same valley as the existing electron–hole pair, i.e., zero momentum mismatch. Conversely, intervalley trions are formed when the extra electron (or hole) occupies a different valley from the existing electron–hole pair, meaning that those particles are distributed across different valleys in the momentum space.

Molybdenum compounds (MoS_2, $MoSe_2$) typically exhibit two distinct excitonic species, excitons and trions, in their photoluminescence (PL) spectra (Ross et al., 2013; Mak et al., 2013; Pei et al., 2017, 2015). This is attributed to the aligned spin configuration in the conduction band minimum (CB) and the valence band maximum (VB). In contrast, tungsten compounds (WS_2, WSe_2) have an opposite spin arrangement in the conduction band minimum and the valence band maximum (Zhang et al., 2015a; Malic et al., 2018; Selig et al., 2016), leading to a significant impact on their excitonic emissions. With the coupling of the electron and hole spins and their interaction with the valley pseudospins, the trion fine structure arises in the tungsten compounds. This interaction leads to the splitting of the trion peak into multiple components, each corresponding to a different combination of spin and valley states. Experimental studies have shown that the trion fine structure can be resolved through various spectroscopic techniques, such as photoluminescence spectroscopy or magneto-optical measurements. The observation and characterization of the trion fine structure provide valuable insights into the electronic and spin properties of 2D TMDCs. For instance, it has been observed in low temperature PL measurements for monolayer WS_2 that the peak energies of intravalley and intervalley trions locate at 2.06 and 2.07 eV, respectively, and the splitting energy of them is about 10 meV (Fig. 4a). The Pauli exclusion principle governs intravalley trions, necessitating that the two electrons occupy distinct spin states and reside in the upper and lower spin-split conduction bands, creating a spin singlet (Fig. 4b). Conversely, intervalley trions allow the possibility of the two electrons having the same spin but differing valley indices, resulting in a spin triplet (Fig. 4c). Both types of optically bright trions obey the optical selection rules, causing the electron in the higher-energy spin-split conduction band to recombine with the hole in the corresponding valley, while retaining the electron in the lower spin-split conduction band.

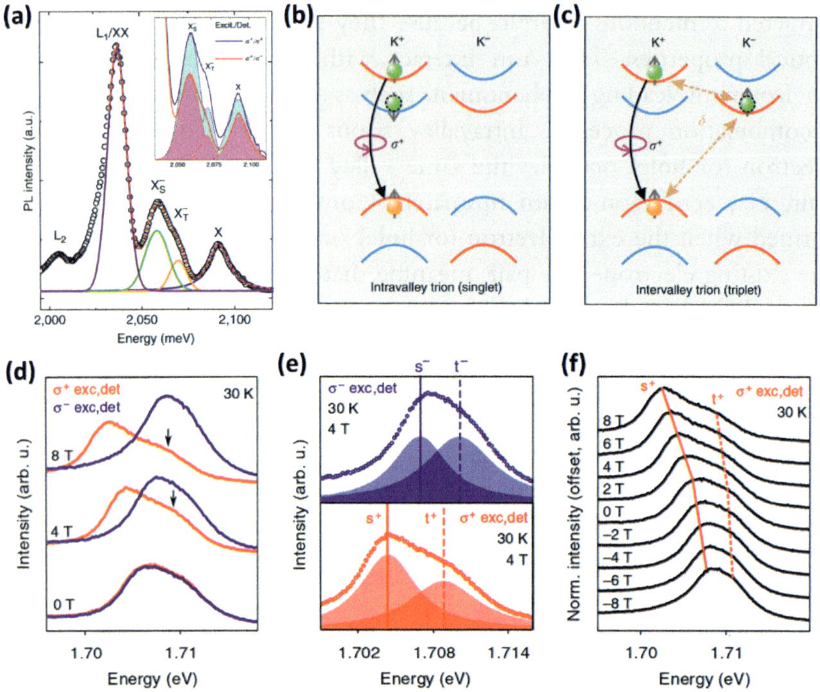

Fig. 4 Scheme and PL spectra of intravalley and intervalley trions in WS_2 and WSe_2. (a) PL spectrum of monolayer WS_2 at a temperature of 4 K. The triplet X_T (orange curve) and singlet trions X_S (green curve) were resolved by multipeak fitting. Inset: helicity-resolved PL spectra of the same sample. (b–c) Schematic illustration of intravalley (singlet) and intervalley (triplet) trion configurations for WS_2. (d) Circular polarization resolved trion PL spectra at B= 0 T, 4 T, and 8 T, at 30 K. (e) The fitted trion fine structure at B = 4 T in each photon helicity. (f) B-field dependent trion PL in σ^+ polarization. *Panels (a–c) adapted with permission from Plechinger, G., et al., 2016a. Trion fine structure and coupled spin-valley dynamics in monolayer tungsten disulfide. Nat. Commun. 7, 12715. Panels (d–f) adapted with permission from Lyons, T. P., et al., 2019. The valley Zeeman effect in inter- and intra-valley trions in monolayer WSe2. Nat. Commun. 10 (1), 2330.*

In order to account for the energy disparity between the two types of trions, the exchange interaction between the excess electron (or hole) and the electron–hole pair in the opposite valley must be considered, particularly in intervalley trions where all three particles possess the same spin orientation. This phenomenon was initially explored by Yu et al. (2014). primarily in relation to monolayer WSe_2. Due to Coulomb repulsion, the exchange interaction between electrons in the intervalley trion is significantly smaller compared to the exchange interaction between the electron and the hole. As a result of this exchange interaction, the energy of the triplet state experiences an increase by an estimated value δ, approximately 6 meV for WSe_2 (Yu et al., 2014). Thus, the overall consequence is the division of the trion dispersion into two branches: an upper branch for triplet states and a lower branch for singlet states, separated by the energy difference δ.

Additional evidence supporting the presence of two distinct trion species can be observed in helicity-resolved photoluminescence (PL) spectra. In Fig. 4a inset, the PL spectra of the same flake are depicted for co-circular (solid blue line) and contra-circular (solid red line) excitation and detection. Both exciton and trion PL emissions exhibit greater intensity under co-circular excitation and detection, indicating a pronounced and optically induced valley polarization. A direct comparison of the spectra reveals that the emission of the higher-energy triplet trion is significantly more diminished for contra-circular excitation and detection compared to the lower-energy singlet trion. To quantitatively assess this observation, the integrated intensity of each peak function has been calculated (represented by the shaded areas under the curves) to determine the PL intensity of all three peaks. From these values, the circular polarization degree (P_{Circ}) of the different trion species is determined using the equation $P_{Circ} = (I_{Co} - I_{Contra})/(I_{Co} + I_{Contra})$. Notably, it is observed that the circular polarization degree varies by a factor of nearly two for the two trion species: the peak associated with the singlet trion exhibits a value of approximately 19%, similar to that of the neutral exciton (21%), while the peak corresponding to the triplet trion shows a significantly larger value of 34%. This observation aligns qualitatively with the recent findings on trion fine structure in WSe_2 (Jones et al., 2015), which was attributed to the distinct dispersions of the two trion species and hence a narrower spread in momentum space for the higher-energy triplet trion.

The fine structure of trion in WS_2 monolayers has also been investigated by conducting helicity-resolved PL experiments by embedding them between two flakes of hexagonal BN (Vaclavkova et al., 2018). It has been found that the encapsulation of the WS_2 monolayer in hBN leads to the observation of two well-resolved emission peaks at low temperatures which have been attributed to two spin configurations of the negatively charged exciton, i.e., intervalley spin-triplet and intravalley spin-singlet. It is demonstrated through polarization-resolved experiments that the helicity of the excitation light is better preserved in the emission spectrum of the triplet trion compared to that of the singlet trion. Moreover, the singlet (intravalley) trions are detectable even under ambient conditions, whereas the emission associated with the triplet (intervalley) trions is only efficient at low temperatures.

Due to the different spin configurations, magneto-optical measurements have uncovered inequivalent valley Zeeman g-factors for the two fine structure components in monolayer WS_2 (Plechinger et al., 2016b) and WSe_2 (Lyons et al., 2019). In more detail, at zero external magnetic field, an asymmetrical trion characteristic in photoluminescence (PL) is observed (0 T trace shown in Fig. 4d), which consists of a singlet trion peak at lower energy and a triplet trion peak at higher energy. As illustrated in Fig. 4e–f, when no external magnetic field is applied, the emissions from K^+ and K^- occur at the same energy level, while there exists a small energy difference δ of a few meV between the triplet and singlet emissions due to the intervalley exchange coupling. When a perpendicular external magnetic field is applied to the sample, ranging from $B = 0$ T to $B = 8$ T, the emission components σ^+ and σ^- exhibit spectral shifts, moving away from each other. Specifically, the σ^- emission shifts to higher energy, which aligns with the behavior expected from the valley Zeeman effect. However, the traces at 4 T and 8 T in Fig. 4d reveal a simultaneous transformation of the trion feature's lineshape under the external field. It becomes apparent that as the emission shifts to lower energy, the singlet and triplet states progressively increase their energy separation. This is supported by the emergence of a distinct shoulder in the σ^+ emission at $B > 0$, as indicated by black arrows. Conversely, as the emission shifts to higher energy, the singlet and triplet peaks display a reduced energy gap, resulting in an overall brighter and narrower emission profile observed in the σ^- emission at $B > 0$. These findings lead to a crucial conclusion: the measured value of the actual valley Zeeman effect will be influenced by the varying contributions of the four fine structure components, each of which experiences a different rate of shift.

The existence of an additional type of intervalley trion, commonly referred to as the dark trion (Li et al., 2019a; Liu et al., 2019a), is typically not observable using out-of-plane excitation and detection methods (Fig. 5a). The energy peak of these dark trions is approximately 16 meV lower than that of the dark exciton (Fig. 5b), indicating that the binding energies of the dark trions are slightly lower compared to the binding energies of the bright trions (~30 meV). The dark trion has a longer lifetime compared to the bright trion, with a duration of approximately 1.3 ns, which is two orders of magnitude greater than the bright trion's lifetime of around 10 ps. By utilizing gating techniques, it is possible to freely tune the

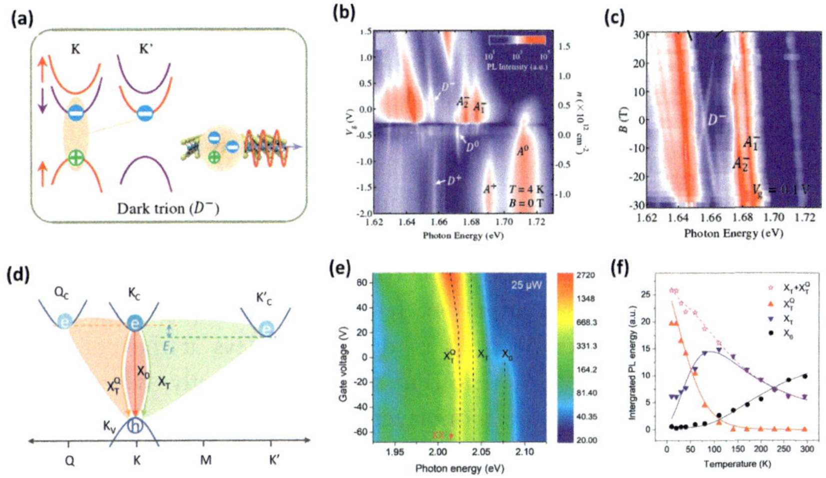

Fig. 5 K-valley dark trion and Q-valley bright trion. (a) Illustration of intervalley dark trion in monolayer WSe_2 emitting vertically polarized light in the in-plane direction. (b) Gate-dependent PL map of BN-encapsulated monolayer WSe_2 at $T = 4$ K and $B = 0$ T. (c) The B-dependent PL map at the electron side ($V_g = 0.1$ V) corresponding to slight n doping. (d) Configurations of the K–K′ intervalley trion (X_T) and K–Q intervalley trion (X_T^Q). (e) Color plot of the measured PL spectra for monolayer WS_2 as a function of back-gate voltage. (f) Integrated PL intensity of X_0, X_T, and X_T^Q emissions in dependence of temperature. *Panels (a–c) adapted with permission from Li, Z., et al., 2019a. Direct observation of gate-tunable dark trions in monolayer WSe2. Nano Lett. 19 (10), 6886–6893. Panels (d–f) adapted with permission from Pei, J., et al., 2023. Switching of K-Q intervalley trions fine structure and their dynamics in n-doped monolayer WS2. Opto-Electron. Adv. 6, 220034.*

dark trion's lifetime within the range of 0.4–1.3 ns. Owing to their spin-triplet configuration, dark trions are expected to have significantly longer lifetimes than bright trions due to the suppression of radiative decay. The different spin configurations between dark and bright trions can be differentiated through the Zeeman effect. Fig. 5b–c depict the gate-dependent photoluminescence (PL) map at B $= 0$ T and the magnetic field-dependent PL maps ranging from $B = -31$ to 31 T. The spectra of bright excitons and trions generally remain unchanged, except for a Zeeman shift. In contrast, the D^- and D^+ peak splits into two peaks with the same energy separation, which is a characteristic feature of the dark states. D^+ is the case where one electron is at the bottom of the conduction band and the other two holes are at the top of the valence bands of K and K′, respectively. Conversely, the bright positive trion A^+ features an electron situated in the uppermost conduction band.

In the interpretation of valley trions, the focus has typically been on the direct K and K′ valleys. However, the indirect Q valley (also known as Λ or Σ) in the first Brillouin zone of monolayer WS_2 has received less attention, despite having the same spin and a very similar energy level to the K valley (Roldán et al., 2014; Lo et al., 2021). Recent findings have shown that this valley plays a significant role in the distribution of excited states when subjected to above-gap excitation (Madéo et al., 2020; Wallauer et al., 2021). Therefore, the role of Q valleys in the formation of intervalley trions deserves attention. It is crucial to explore this state by manipulating the carrier densities in different valleys at the conduction band edges with electrostatic field (Fig. 5d). Consequently, when exposed to high electron doping, the monolayer WS_2 displays a distinct emission feature that competes with the population of the K–K′ trion, manifesting itself as approximately 20 meV lower in energy (Pei et al., 2023). Fig. 5e presents a color contour plot depicting the PL spectra at various doping densities, indicated by back-gate voltages ranging from −60 to 60 V. With an increase in electron doping density, the intensities of the X_T peak and the X_T^Q peak experience a slow increase from −60 to 0 V. As the back gate voltage rises from 0 V to 60 V, the intensity of the X_T peak rapidly declines. In contrast, the intensity of the X_T^Q peak exhibits a remarkable growth, ultimately becoming the dominant feature in the spectrum at 60 V due to the elevated Fermi level. The transition between these two types of intervalley trions follows a Boltzmann distribution (Pei et al., 2023) and can also be observed with changes in temperature. As the temperature decreases from 295 K, the intensity of the X_T peak initially increases, followed by a rapid decay after reaching 110 K. Concurrently, the emergence of the X_T^Q peak occurs around 110 K. However, the overall emission from both the X_T and X_T^Q peaks shows a monotonic increase with temperature (as shown by the violet curve in Fig. 5f), which can be attributed to the energy variation between the K and Q valleys induced by thermalization strain.

4. Valley exciton–trion complexes

Valley exciton–trion complexes refer to composite particles formed by the interaction of excitons and trions in a material with valley degrees of freedom. The combination of these composite particles leads to new electronic and optical properties, influenced by both the valley index and the charge state of the complex. Valley biexcitons and charged biexcitons are commonly observed composite particles that involve two excitons in materials with valley degrees of freedom (Ye et al., 2018). The valleys represent distinct momentum states in the material's reciprocal lattice, and the interaction between the excitons can result in the formation of a valley biexciton, which is a bound state of two excitons in a material with distinct valleys in its band structure. A charged biexciton is a biexciton in which one or both of the constituent excitons carry an excess charge. It can form when an additional charge carrier (such as an electron or a hole) is captured by one of the excitons in the biexciton. The additional charge alters the charge state of the biexciton and introduces electrostatic interactions between the charged constituents. Both valley biexcitons and charged biexcitons are important in the study of excitonic and valley physics. They offer opportunities for exploring novel quantum phenomena and manipulating valley-dependent properties.

By employing low-temperature photoluminescence spectroscopy in BN-encapsulated single-layer WSe_2, both the biexciton and (negative) trion–exciton complexes have been revealed in experiment (Barbone et al., 2018; Chen et al., 2018a; Li et al., 2018). The presence of the biexciton state is exclusive to charge-neutral WSe_2, achieved by effectively controlling electrostatic gating. In the case of lightly electron-doped WSe_2, a single free electron combines with a biexciton, resulting in the formation of a trion–exciton complex. Power-law analysis serves as a commonly employed technique for characterizing the biexciton characteristic. As depicted in Fig. 6a, an increase in excitation power reveals the emergence of additional photoluminescence (PL) peaks at 1.723 and 1.691 eV. In Fig. 6b, the integrated PL intensity of these two peaks, alongside the exciton PL intensity, is plotted against the excitation power. The power law relationship between the integrated PL intensity (denoted as I) and the excitation power (represented as P), expressed as $I \propto P^{\alpha}$, distinctly differs for the two newly emerged peaks in comparison to the exciton. By fitting the experimental data shown in Fig. 6a, $\alpha = 1.94$ for the peak at 1.723 eV and $\alpha = 1.82$ for the peak at 1.691 eV are derived, respectively. These values are considerably higher than that of the exciton ($\alpha = 1.29$) and are close to the anticipated value for the biexciton ($\alpha = 2$). The binding energy of the biexciton, denoted as Δ_{XX}, is determined by subtracting the emitted photon energy of the biexciton from that of the free exciton. Theoretical calculations have previously predicted the biexciton-binding energy to be approximately 20 meV (Szyniszewski et al., 2017; Mayers et al., 2015; Kylänpää and Komsa, 2015; Zhang et al., 2015b). Notably, the newly

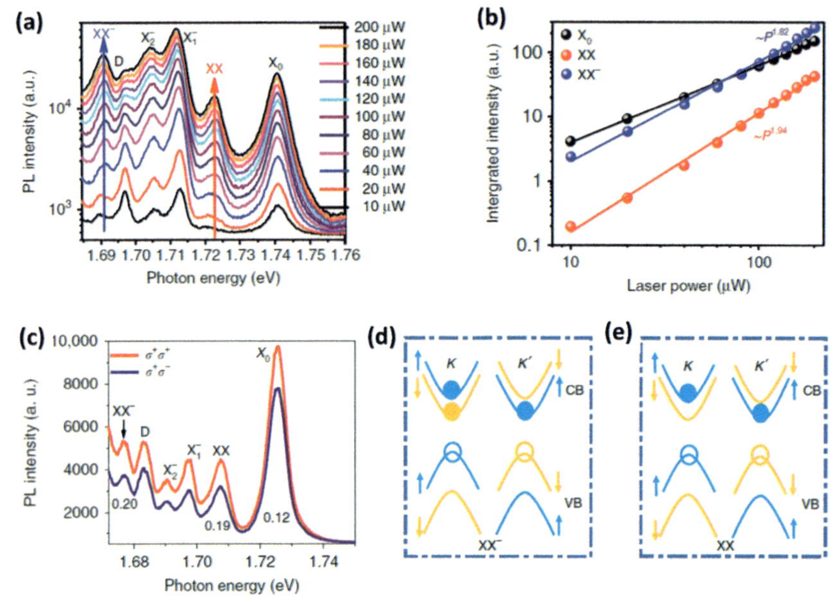

Fig. 6 Power-dependent and valley-polarized PL spectra of neutral and charged biexcitons and their valley configurations. (a) PL spectra of the monolayer WSe_2 at different excitation powers. (b) Integrated PL intensity of the X_0, XX, and XX^- peaks as a function of the excitation power. (c) Helicity-resolved PL spectra of the sample. (d–e) Spin-valley configurations for the neutral and charged biexcitons. *Reprints with permission from Li, Z., et al., 2018. Revealing the biexciton and trion-exciton complexes in BN encapsulated WSe2. Nat. Commun. 9 (1), 3719.*

observed photoluminescence (PL) peak at 1.723 eV is positioned 17 meV lower than the exciton peak, aligning remarkably well with the predicted biexciton binding energy. In contrast, the second newly observed photoluminescence (PL) peak at 1.691 eV, positioned 49 meV lower than the exciton peak, has been interpreted as a negatively charged bound biexciton (XX^-) based on the analysis of PL spectra in relation to the gate voltage (Li et al., 2018). Since the photoluminescence of the biexciton (XX) is predominantly observed in the charge-neutral region, the PL associated with XX^- does not manifest in the charge-neutral region but becomes apparent at higher gate voltages, corresponding to the slightly n-doped region.

By utilizing circularly polarized photoluminescence (PL) spectroscopy, it is possible to explore the intricate valley configuration of XX and XX^-. This technique enables the selective excitation and probing of a specific valley in WSe_2, thereby providing insights into the detailed characteristics of these biexciton states. This can be quantified through the measurement of valley polarization, which is defined as the ratio between the right circular and left circular components of the emitted photoluminescence (PL). Intriguingly, both XX and XX^- exhibit higher valley polarizations of 0.19 and 0.20, respectively, surpassing the valley polarization of the exciton (0.12) as depicted in Fig. 6c. This higher valley polarization suggests that both XX and XX^- involve a bright exciton residing in the K valley. Considering XX^- as a five-particle complex, the lowest energy configuration that maintains the observed valley polarization mostly correspond to the arrangement depicted in Fig. 6d. This configuration can be interpreted as a biexciton bound to a free electron or, equivalently, a trion bound to a dark exciton. Therefore, the negatively charged bound biexciton effectively represents a trion–exciton complex. On the other hand, the biexciton state of XX is composed of four particles, and the possible configuration is illustrated in Fig. 6e. In this configuration, one bright exciton occupies the K valley, while the dark exciton occupies the K′ valley.

The degeneracy of the K and K′ valleys could be lifted with an out-of-plane magnetic field. This allows selective access to the spectra of the K (K′) valley by exciting with σ^+ (σ^-) polarization and detecting with σ^+ (σ^-) configuration. The PL peaks in the spectra exhibit a Zeeman splitting due to opposite Zeeman shifts in the different valleys (Fig. 7a). This Zeeman splitting, denoted as Δ, can be described as $\Delta = \Delta_{\sigma+\sigma+} - \Delta_{\sigma-\sigma-} = g\mu_B B$, where μB represents the Bohr magneton, approximately 58 μeV/T, and $\Delta_{\sigma+\sigma+}$ ($\Delta_{\sigma-\sigma-}$) represents the Zeeman shift in the $\sigma^+\sigma^+$ ($\sigma^-\sigma^-$) configuration. By linear fitting the experimentally obtained Zeeman splitting Δ as a function of the magnetic field, the g-factor for the exciton can be determined as -3.64 ± 0.08 (indicated by black dots in Fig. 7b). Furthermore, the obtained g-factors for XX and XX^- are -4.03 ± 0.07 and -5.33 ± 0.18, respectively, which agree well with the theoretical expectation. Since the PL emission of XX and XX^- involves the recombination of a single bright exciton, their expected g-factor should be the same as that of the exciton. The PL intensity ratio of the two Zeeman-split states ($\sigma^+\sigma^+$ and $\sigma^-\sigma^-$) in the magnetic field provides valuable information about the detailed configuration of the biexciton. The Zeeman splitting lifts the energy degeneracy at the K and K′ valleys, resulting in two states with an energy difference of $\Delta = g\mu_B B$, which leads to an unequal population of the two states, thereby causing distinct PL emission intensities (Fig. 7a).

Under a positive magnetic field, the higher-radiative energy transition in the K′ valley is occupied by the bright exciton component of XX_0,

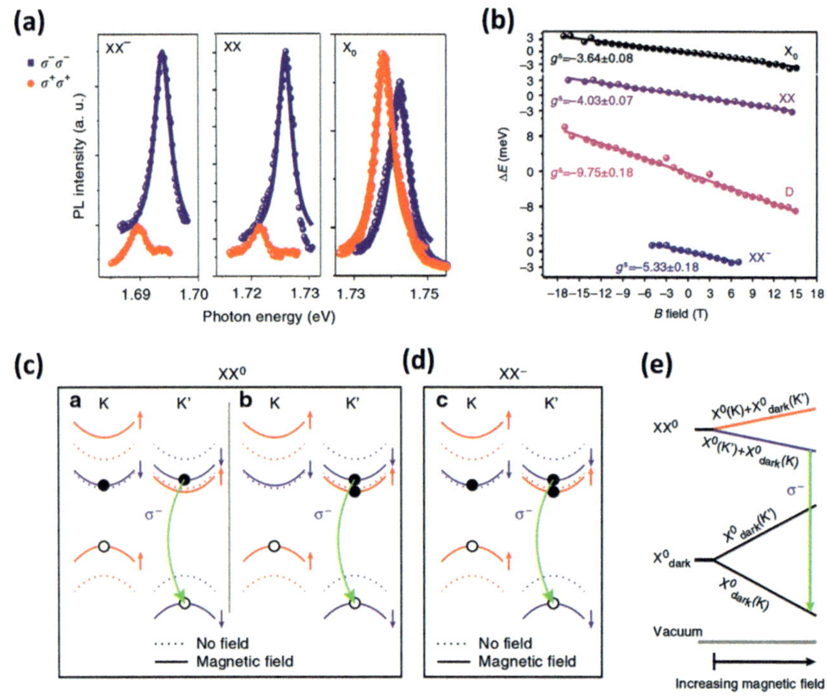

Fig. 7 Magneto-PL spectra of the monolayer WSe_2 sample and. (a) Helicity-resolved PL spectra for exciton (X_0), biexciton (XX), and charged biexciton (XX^-) at B=17 T, respectively. (b) The calculated g-factors for different peaks from the Zeeman splitting between the $\sigma^-\sigma^-$ and $\sigma^+\sigma^+$ states as a function of B field. (c–d) Single-particle picture of the internal structure of XX_0 and XX^- with applied magnetic field. (e) Many-body picture of the magnetic field effect on XX_0, comprising a bright and a dark exciton. *Panels (a–b) adapted with permission from Li, Z., et al., 2018. Revealing the biexciton and trion-exciton complexes in BN encapsulated WSe2. Nat. Commun. 9 (1), 3719. (b) Panels (c–e) adapted with permission from Barbone, M., et al., 2018. Charge-tuneable biexciton complexes in monolayer WSe2. Nat. Commun. 9 (1), 3721.*

resulting from the thermalization of the photogenerated electrons. Simultaneously, the electron in the dark component of XX_0 can be found in either the opposite valley or the same valley (Fig. 7c) as the bright exciton component, which leads to an energy difference between these two configurations and hence is responsible for the observed fine structure of XX_0 in another study. Fig. 7d illustrates the single-particle arrangement of XX^- under a magnetic field, while Fig. 7e presents a qualitative many-body depiction of XX_0 under a magnetic field, where XX_0 consists of a bright and a dark exciton component in opposite valleys. Since the total Zeeman splitting of XX_0 depends on both the bright and dark components, and the dark exciton has a larger g-factor and opposite sign than X_0, XX_0 exhibits reversed energy ordering compared to its bright exciton component and dissociates into a dark exciton and a photon. The distribution of biexciton states resembles the scenario near thermal equilibrium, which explains the different circularly copolarized emission intensity under σ^+ or σ^-.

5. Valley phonon–exciton complexes

Phonon–exciton complexes refer to composite entities formed by the interaction between excitons and phonons in a material. Phonons are quantized units of lattice vibrations or collective atomic motions in a solid. They can be thought of as particles representing the vibrational energy of the crystal lattice. Phonons can influence the dynamics of excitons by providing channels for energy relaxation or scattering processes. In materials, excitons, on the other hand, can interact with phonons through various mechanisms, which can lead to the formation of exciton–phonon complexes. The interaction can occur through processes such as the phonon emission or absorption by excitons, phonon-assisted exciton relaxation, or phonon-mediated exciton scattering. Exciton–phonon complexes are characterized by the coupling between the excitonic and phononic degrees of freedom in the material. The interaction between excitons and phonons plays a significant role in phenomena such as energy transfer, exciton dissociation, and the thermalization of excited states in semiconductors and other materials.

The distinctive properties associated with different valleys not only have an influence on quasiparticles but also have the potential to introduce novel physical phenomena involving zone edge phonons, also known as valley phonons. These collective lattice oscillations occur at the corners of the hexagonal Brillouin zone ($\pm$K points). It has been predicted that these phonons play a significant role in facilitating spin and valley pseudospin relaxation through phonon-assisted intervalley scattering (Dery and Song, 2015; Song and Dery, 2013). Moreover, valley phonons can exhibit chirality, possessing intrinsic pseudo-angular momentum, which has recently garnered considerable interest (Li et al., 2019b; Liu et al., 2019b; Chen et al., 2018b; Zhu et al., 2018; Zhang and Niu, 2015). Xu and colleagues conducted a study where they observed the presence of multiple valley phonons in the monolayer semiconductor WSe_2. These valley phonons possess momentum vectors pointing towards the corners of the hexagonal Brillouin zone, causing the formation of exciton complexes. The researchers discovered that these valley phonons play a crucial role in facilitating efficient intervalley scattering of quasiparticles during both exciton formation and relaxation processes. This phenomenon gives rise to a series of PL peaks, which are identified as valley phonon replicas of dark trions. By leveraging the identified valley phonons, the researchers were able to recognize an intervalley exciton in proximity to charge neutrality. This study not only unveils several previously unknown 2D excitonic species but also establishes monolayer WSe_2 as an ideal candidate for investigating the interactions between spin, pseudospin, and zone-edge phonons.

Fig. 8 shows the valley phonon replicas of positively charged dark trion. The four peaks identified as D^+_{K3}, $D^+_{\Gamma 5}$, D^+_{K1}, D^+_{K2} in Fig. 8a are attributed to the bright replicas of the positive dark trion, and the corresponding valley phonon replicas exhibit energy differences of 26, 22, 18, and 13 meV below D^+, respectively. These peaks originate from the coupling of D^+ with valley phonons K_3, K_1, and K_2, as well as the zone-center phonon Γ_5. Although intervalley recombination is naturally prohibited for delocalized

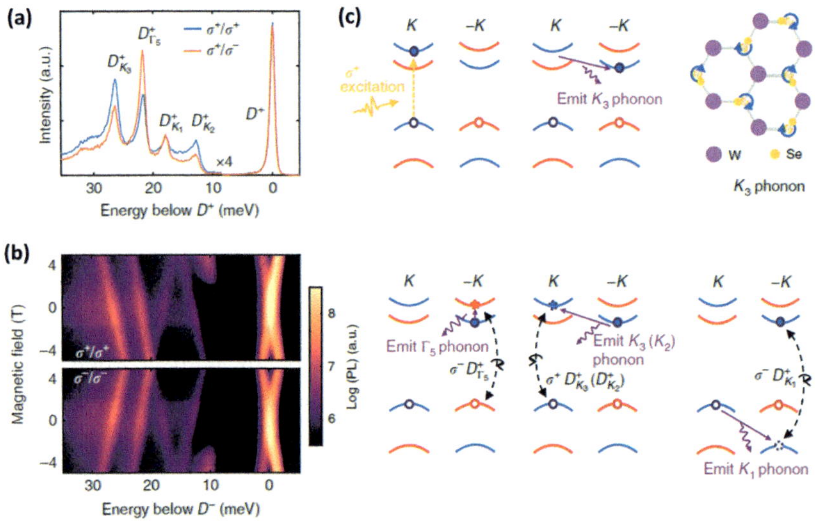

Fig. 8 Valley phonon replicas of positively charged dark trion. (a) Circular polarization resolved PL of the quadruplet PL peaks on the hole doping side. (b) Circular polarization resolved magneto PL of the quadruplets with σ^+/σ^+ and σ^-/σ^- excitation/detection. (c) Schematic of formation processes for D^+ and phonon-assisted emission process of its valley phonon replicas. *Reprints with permission from He, M., et al., 2020. Valley phonons and exciton complexes in a monolayer semiconductor. Nat. Commun. 11 (1), 618.*

exciton complexes due to significant momentum mismatch, emission of a valley phonon can provide the necessary momentum, enabling phonon-assisted luminescence of these otherwise dark states. Magneto-PL measurements were conducted to determine the effective Landé g-factors of different states and to ascertain the spin and valley indices of the constituent electrons and holes in the excitonic states (Lyons et al., 2019; Aivazian et al., 2015; Srivastava et al., 2015; Li et al., 2014; MacNeill et al., 2015). The circular polarization-resolved PL intensity as a function of out-of-plane magnetic field is depicted in Fig. 8b, where the Zeeman shifts observed in the quadruplet PL peaks provide evidence that they originate from phonon replicas of the dark positive trion D^+. When electron–hole recombination occurs within the same valley, the g-factor magnitude is approximately 4 for parallel spins and around 9 for anti-parallel spins. Conversely, if recombination involves electron and hole pairs from opposite valleys with parallel spins, a g-factor value of approximately 13 is expected. The extracted g-factors of −13.4, 12.2, and −13.0 for D^+_{K3}, D^+_{K1}, and D^+_{K2}, respectively, align with intervalley recombination of electrons and holes (Fig. 8c). The positive g-factor observed for D^+_{K1} indicates the involvement of the hole and K_1 valley phonon in their interaction. In accordance with the selection rules derived from group theory, K_1 represents the only phonon mode capable of facilitating spin-conserving intervalley transitions between the valence-band states at K and −K points (Song and Dery, 2013). Fig. 8a shows that $D^+_{\Gamma5}$ exhibits cross-circular polarization. The optical selection rules specific to the valley state indicate that σ^+ excitation results in the creation of an electron and hole in the +K valley, while σ^- polarized emission is only possible through spin-conserved electron–hole recombination in the −K valley. Moreover, the value of g $(D^+_{\Gamma5}) = -9.7$ confirms the nature of intravalley electron–hole recombination for this peak.

The valley phonon replicas associated with the negative dark trion D^- can be analyzed using a similar approach as described above. Fig. 9a presents the PL with helicity resolution under σ^+ polarized excitation. Notably, the spectral features labeled as T1, D^-_{K3}, and $D^-_{\Gamma5}$ at lower energies exhibit significant co-circular polarization. Fig. 9b illustrates the PL intensity as a function of magnetic field and photon energy, with $\sigma^-\sigma^-$ polarized excitation and detection. The extracted g-factor for D^-_{K3} is −12.5, indicating intervalley recombination, with an energy 26 meV below D^-. Therefore, it can be deduced that D^-_{K3} originates from the interaction between an electron in the −K valley and the K_3 phonon, as depicted in

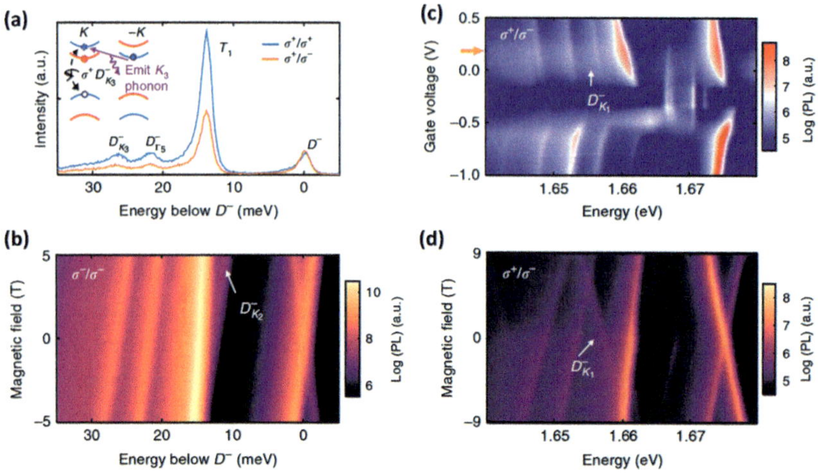

Fig. 9 Valley phonon replicas of negatively charged dark trion. (a) Circular polarization resolved PL of the triplet PL peaks under electron-doped conditions. (b) Magneto PL of the triplets with σ^- excitation and copolarized detection. (c) Gate-dependent PL with σ^+ resonant pumping of the bright exciton and cross-polarized (σ^-) detection. (d) Magneto PL at the gate voltage indicated by the orange arrow in c. *Reprints with permission from He, M., et al., 2020. Valley phonons and exciton complexes in a monolayer semiconductor. Nat. Commun. 11 (1), 618.*

the inset of Fig. 9a. Furthermore, the g-factor of $D^-_{\Gamma5}$ (−9.9) is nearly identical to that of D^- (−9.5), and the peak appears 22 meV below D^-, the same as that for the case of $D^+_{\Gamma5}$.

The PL spectrum under electron doping shows a different triplet peak pattern compared to the quadruplet pattern observed under hole doping. The intensity of T_1 is greater than D^-, and its g-factor is approximately −4.5, suggesting that it is unlikely to be a phonon replica of D^-. The origin of T_1 is unknown, but it overlaps with the spectral range of the K_2 and K_1 valley phonon replicas, obscuring them completely. However, the significant difference in g-factor between T_1 and D^- enables the identification of D^-_{K2} in magneto-PL since the states shift apart. In Fig. 9b, a faint line appears on the high energy side of T_1 under high magnetic field, which is attributed to D^-_{K2} based on its g-factor of around −13.6 and its energy difference of 13 meV from D^-, both of which align well with expectations for the K_2 phonon replica. Similarly, D^-_{K1} becomes noticeable when the excitation laser is resonant with the bright neutral exciton (X_0). Fig. 9c illustrates the PL intensity in relation to gate voltage and photon energy, with σ^+ polarized excitation laser in resonance with X_0 (1.733 eV) and σ^- polarized detection. At $V = 0.2$ V (arrow on the side of Fig. 9c), D^-_{K1}

becomes evident, appearing 18 meV below D^-. Fig. 9d demonstrates magneto-PL with the same cross-polarized polarization (σ^+/σ^-), where a positive g-factor of 12.2 is observed for D^-_{K1}, which is identical to that of D^+_{K1}. Therefore, despite the initial differences, the four phonon replica states observed under hole-doping conditions also manifest under electron-doping conditions. The valley phonon-assisted momentum relaxation mechanism generates phonon replicas for neutral dark excitons. However, not all phonon replicas of excitons are observable. Only the Γ_5 phonon replica of intravalley dark exciton (D_0) and the K_1 and K_3 phonon replicas of intervalley excitons (I_0) have been observed (He et al., 2020).

6. Interlayer valley exciton complexes

Interlayer exciton refers to an excitonic state that arises in a two-dimensional (2D) material system composed of two or more stacked layers, such as a bilayer or trilayer system. In a bilayer system, the two layers can have different electronic properties, such as band structures or bandgaps, which can lead to the formation of interlayer excitons. Specifically, interlayer valley excitons are excitonic states that arise due to the presence of valleys in the band structure of the individual layers. When a bilayer system is formed by stacking two TMDCs layers with different valley properties, interlayer valley excitons can form due to the coupling of electrons and holes in different valleys of the two layers. These interlayer valley excitons possess unique properties, such as long lifetimes and large binding energies, which make them interesting for various applications in optoelectronic devices. Moreover, the properties of interlayer valley excitons can be tuned by applying external electric or magnetic fields, making them potential candidates for exploring novel quantum phenomena and developing new device functionalities.

Van der Waals (vdW) heterostructures, which are composed of transition metal dichalcogenides (TMDCs), typically exhibit a type-II band alignment. This alignment promotes the creation of interlayer excitons between individual monolayers, with electrons and holes accumulating in the layer possessing the lower conduction band minimum (CBM) and the higher valence band maximum (VBM) (Fig. 10a), respectively. The interlayer separation, determined by the thickness of atomic layers, ensures strong Coulomb interactions between adjacent layers, leading to the formation of tightly bound electron–hole pairs with significant

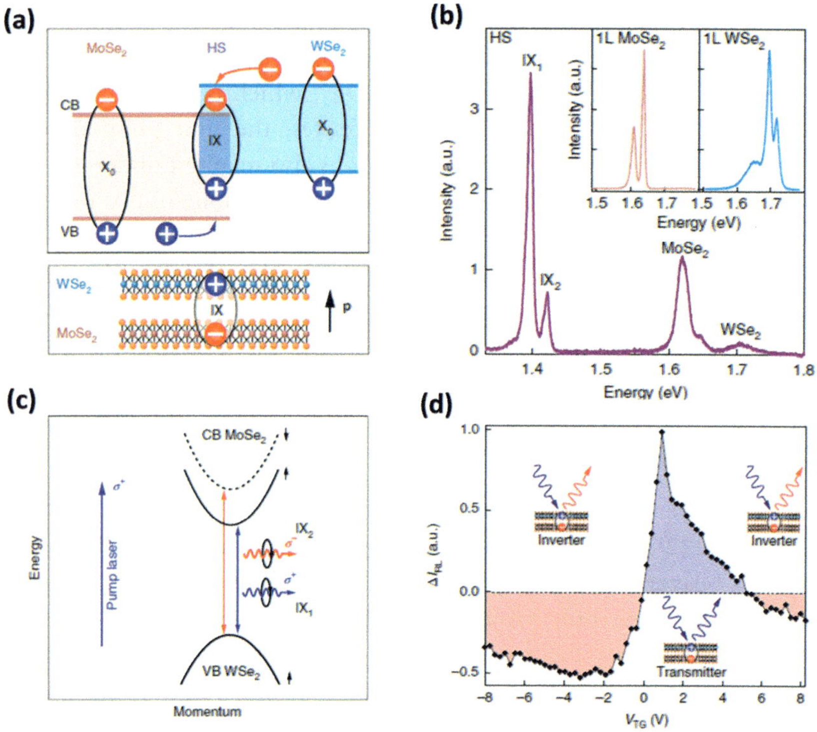

Fig. 10 Polarization switching and electrical control of interlayer excitons. (a) Upper panel: band alignment in the $MoSe_2/WSe_2$ heterobilayer. Lower panel: artistic representation of the interlayer (IX) exciton with its dipole moment *p*. (b) PL spectrum from the heterostructure showing emission from monolayer $MoSe_2$, monolayer WSe_2 and interlayer excitons. (c) Schematic illustration of the spin-conserving (IX_1) and spin-flipping (IX_2) transitions in the K valley with circularly polarized light excitation. (d) Gate voltage tunable polarization switching under circularly polarized light excitation. *Reprints with permission from Ciarrocchi, A., et al., 2019. Polarization switching and electrical control of interlayer excitons in two-dimensional van der Waals heterostructures. Nat. Photonics 13 (2), 131–136.*

binding energies. This results in the emergence of an additional peak at a lower energy in the PL spectrum, indicative of the formation of an interlayer exciton. Simultaneously, the PL intensity of the constituent monolayers decreases, demonstrating the phenomenon of PL quenching. These characteristics have been observed in various representative TMDCs vdW heterostructures (Liu et al., 2022; Jiang et al., 2021b; Rivera et al., 2015, 2018). When the sample was directly exposed to the SiO_2 substrate, a broad peak typically appeared in this region. However, by encapsulating the heterostructure with h-BN, it is able to observe distinct and bright PL peaks from individual monolayers. Additionally, interlayer excitons can be observed with an energy separation of approximately 25 meV between them (Fig. 10b), which corresponds to the conduction band splitting of $MoSe_2$ (Fig. 10c). These interlayer excitons exhibit intrinsic valley-contrasting physics inherited from the constituent monolayers, and their characteristics can be optically manipulated using circularly polarized light (Zeng et al., 2012; Jones et al., 2013; Cao et al., 2012; Mak et al., 2012). In monolayers, the spin-flipping transition is typically not observable, but the presence of a moiré pattern in the heterobilayer brightens this transition due to the selection rules dictated by local atomic registry. Furthermore, the spatial separation between charges leads to the creation of a permanent electrical dipole moment in the out-of-plane direction, enabling the electrical control of the optical properties along the dipole axis. By modulating their relative intensities electrically, a polarization switch with tunable emission intensity and wavelength can be achieved (Ciarrocchi et al., 2019), as shown in Fig. 10d.

What makes it even more fascinating is the emergence of a moiré pattern resulting from the lattice mismatch and rotational angle between adjacent layers. This moiré pattern exhibits locally varying interlayer atomic registries in a periodic manner and generates a periodic potential capable of confining interlayer excitons with specific optical selection rules. As a result, a new type of excitons known as moiré interlayer excitons is formed (Jin et al., 2019; Liu et al., 2021; Tran et al., 2019). Enhanced interlayer hybridization occurs when the band edges of the two layers are aligned in terms of both energy and momentum (Alexeev et al., 2019). This enhancement is achieved through resonant tunneling between the layers. When the resonance condition is met, interlayer hopping facilitates strong hybridization between inter- and intralayer excitons. These hybrid excitons combine the significant oscillator strength of the intralayer excitons with the large electric dipole moment characteristic of interlayer excitons (IXs). The extent of hybridization can be adjusted by the twist angle, as momentum conservation plays a crucial role. Notably, the degree of hybridization is maximized when the twist angles close to 0° or 60°. In the momentum domain, as depicted in Fig. 11a, when there is perfect lattice alignment, the valley mismatch ΔK is minimized. Under these circumstances, the A and B excitons of $MoSe_2$ are brought into close proximity in momentum space with interlayer excitons consisting of $MoSe_2$ holes and WS_2 electrons. This close proximity allows for their hybridization through tunneling in the interlayer conduction

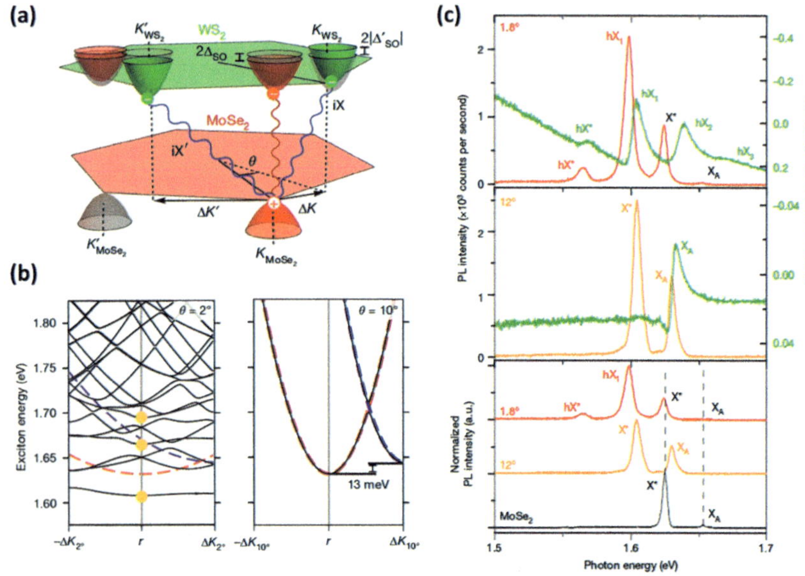

Fig. 11 Resonantly hybridized moiré excitons in $MoSe_2/WS_2$ heterobilayers with aligned energy and momentum. (a) MoSe2 and WS2 electronic band structures and Brillouin zone alignment for twist angle θ. (b) Hybridized exciton bands along the mini-Brillouin zone path for twist angles of 2° and 10°. (c) PL spectra and reflectance contrast spectra of the misaligned ($\theta = 12°$) and aligned ($\theta \approx 1.8°$) $MoSe_2/WS_2$ heterobilayers and $MoSe_2$ monolayers at $T = 10$ K. *Reprints with permission from Alexeev, E. M., et al., 2019, Resonantly hybridized excitons in moire superlattices in van der Waals heterostructures. Nature 567 (7746), 81–86.*

band. Moreover, the presence of the moiré pattern leads to a folding of the dispersion within the mini-Brillouin zone. This folding effect occurs as a result of the moiré superlattice potential, which generates new gaps at the crossing points. In Fig. 11b, the hybridized exciton bands are displayed along the mini-Brillouin zone path for twist angles of 2° and 10°. Notably, for twist angles approaching 0°, three distinct exciton fine structures in close proximity to the energies of X_A and X_B were predicted.

By utilizing absorption spectroscopy, direct observations can be made regarding transitions between distinct excitonic minibands in the case of intralayer moiré excitons (Fig. 11c). Significant enhancements in both reflectance contrast and PL spectral resolution were detected in high-quality $MoSe_2/WS_2$ heterostructures with hBN encapsulation. A comparison between normalized PL spectra obtained from monolayer $MoSe_2$ and $MoSe_2/WS_2$ heterobilayers with twist angles of approximately 12° and 1.8° is shown. The PL spectrum of the monolayer region exhibits a prominent trion peak, denoted as X^*, along with a weaker neutral exciton labeled X_A. In the case of the misaligned structure with a twist angle of 12°, both of these peaks experience a redshift of approximately 20 meV. Conversely, the PL spectrum of the highly aligned heterobilayer ($\theta \approx 1.8°$) displays four distinct peaks. Fig. 11c also showcases the reflectance contrast spectrum (green curves) of the well-encapsulated heterobilayer with a twist angle of 12°, revealing a single strong resonance corresponding to the XA transition. In stark contrast, the closely aligned sample ($\theta \approx 1.8°$) exhibits four clearly discernible features, namely hX^*, hX_1, hX_2, and hX_3. The hX^* and hX_1 peaks directly correspond to the identically labeled peaks in the PL spectrum, whereas the newly observed hX_2 and hX_3 features are unrelated to the X^* and X_A peaks observed in the monolayer $MoSe_2$'s PL spectrum. These additional peaks are attributed to the manifestation of moiré miniband states of hybridized excitons.

The nature of electronic excitations in van der Waals heterostructures is determined by the localization of electrons and holes, leading to either interlayer or intralayer characteristics. Theoretical and experimental investigations have extensively examined both types of excitations in the context of neutral electron–hole pairs known as excitons. However, the exploration of trions has primarily focused on intralayer configurations, with limited attention given to interlayer trions. Empirical confirmation of interlayer trions is still lacking, although first-principles calculations have predicted the presence of bound interlayer trions below neutral interlayer excitons in MoS_2/WS_2 heterostructures (Deilmann and Thygesen, 2018). These calculations indicate binding energies of 18/28 meV for positive/negative interlayer trions with electrons/holes in the same layer. In contrast, trions composed of two equally charged particles on different layers exhibit negligible binding energy. Nevertheless, experimental observations have recently identified indications of moiré trions in a charge-tunable $WSe_2/MoSe_2$ heterobilayers (Liu et al., 2021; Brotons-Gisbert et al., 2021). The moiré trions exhibit a distinctive behavior characterized by the presence of multiple well-defined emission lines, demonstrating a complex charge-density dependence that contrasts sharply with the characteristics of conventional trions. The analysis suggests that the emission of trions originates from radiative decay processes, with the remaining carrier occupying different moiré minibands, thus contributing to the observed spectral features. Due to the sensitivity of moiré minibands to sample inhomogeneity and variability, variation of the trion features was observed in different devices and sample areas (Liu et al., 2021). In another study, the continuous evolution from a few trapped IXs to an ensemble of IXs with both triplet- and singlet-spin configurations was reported (Brotons-Gisbert et al., 2021). Observations of dipolar interactions in the IX ensemble regime, in conjunction with magneto-optical spectroscopy, unveiled that the emission from IXs, manifesting as both narrow quantum-dot-like and broad ensemble characteristics, originates from their entrapment in moiré potentials featuring the same atomic registry. The introduction of electron doping resulted in the emergence of three distinct types of localized negative trions, distinguished by their varying spin-valley configurations. Notably, both intervalley and intravalley IX trions with spin-triplet optical transitions were detected among these trion species. The interlayer moiré biexciton (formed by two IXs) was demonstrated to be observable in a free-standing WS_2/WSe_2 heterobilayer (Sun et al., 2022), where the decreased dielectric screening in a freestanding heterobilayer not only increased the efficiency of exciton formation significantly but also amplified dipole-dipole interactions to a great extent. This enhancement enables the exploration of many-body correlations among pristine excitons in the quantum regime of two dimensions.

7. Conclusion

This chapter introduces the valley excitonic states in TMDCs that result from strong interactions between electrons and holes in different valleys. These include bright and dark valley excitons, intravalley and intervalley trions, valley exciton–trion complexes, valley phonon–exciton complexes, and interlayer valley exciton complexes. The focus is on discussing their formation process, energy valley configuration, fluorescence and absorption peak positions, and changes in polarization characteristics under external fields. The chapter also discusses moiré exciton states discovered recently, but their relationship with energy valleys in momentum space and research on polarization characteristics are still in early stages, demanding further exploration.

While significant progress has been made in controlling the spin-valley polarization of these exciton complexes, research on their application in spin-valley-polarized devices is still relatively limited. Such devices have the potential to enable the development of valleytronic logic gates, facilitating the creation of low-power information processing devices. Furthermore, valley many-body complexes hold promise for achieving the crucial correlation and entanglement properties for new quantum logic devices. However, current research on such complexes primarily focuses on identification and characterization of specific species, with limited reports on energy-valley coupling correlation characteristics within them. In addition, next-generation intelligent optoelectronic devices face the challenge of high-speed and efficient massive data processing. The ultra-fast response and data processing capabilities of these devices, however, heavily rely on the transient conversion dynamics of many-body complexes, which are not yet fully understood, necessitating further research.

References

Aivazian, G., Gong, Z., Jones, A.M., Chu, R.-L., Yan, J., Mandrus, D.G., et al., 2015. Magnetic control of valley pseudospin in monolayer WSe2. Nat. Phys. 11 (2), 148–152.

Alexeev, E.M., Ruiz-Tijerina, D.A., Danovich, M., Hamer, M.J., Terry, D.J., Nayak, P.K., et al., 2019. Resonantly hybridized excitons in moire superlattices in van der Waals heterostructures. Nature 567 (7746), 81–86.

Arora, A., Koperski, M., Nogajewski, K., Marcus, J., Faugeras, C., Potemski, M., 2015. Excitonic resonances in thin films of WSe2: from monolayer to bulk material. Nanoscale 7 (23), 10421–10429.

Bao, D., del Águila, A.G., Thu Ha Do, T., Liu, S., Pei, J., Xiong, Q., 2020. Probing momentum-indirect excitons by near-resonance photoluminescence excitation spectroscopy in WS2 monolayer. 2D Mater. 7 (3), 031002.

Barbone, M., Montblanch, A.R., Kara, D.M., Palacios-Berraquero, C., Cadore, A.R., De Fazio, D., et al., 2018. Charge-tuneable biexciton complexes in monolayer WSe2. Nat. Commun. 9 (1), 3721.

Brotons-Gisbert, M., Baek, H., Campbell, A., Watanabe, K., Taniguchi, T., Gerardot, B.D., 2021. Moiré-trapped interlayer trions in a charge-tunable WSe2/MoSe2 heterobilayer. Phys. Rev. X 11 (3), 031033.

Cao, T., Wang, G., Han, W., Ye, H., Zhu, C., Shi, J., et al., 2012. Valley-selective circular dichroism of monolayer molybdenum disulphide. Nat. Commun. 3, 887.

Chen, S.Y., Goldstein, T., Taniguchi, T., Watanabe, K., Yan, J., 2018a. Coulomb-bound four- and five-particle intervalley states in an atomically-thin semiconductor. Nat. Commun. 9 (1), 3717.

Chen, X., Lu, X., Dubey, S., Yao, Q., Liu, S., Wang, X., et al., 2018b. Entanglement of single-photons and chiral phonons in atomically thin WSe2. Nat. Phys. 15 (3), 221–227.

Chernikov, A., Berkelbach, T.C., Hill, H.M., Rigosi, A., Li, Y., Aslan, O.B., et al., 2014. Exciton binding energy and nonhydrogenic Rydberg series in monolayer WS2. Phys. Rev. Lett. 113 (7), 076802.

Ciarrocchi, A., Unuchek, D., Avsar, A., Watanabe, K., Taniguchi, T., Kis, A., 2019. Polarization switching and electrical control of interlayer excitons in two-dimensional van der Waals heterostructures. Nat. Photonics 13 (2), 131–136.

Deilmann, T., Thygesen, K.S., 2018. Interlayer trions in the MoS2/WS2 van der Waals heterostructure. Nano Lett. 18 (2), 1460–1465.

Dery, H., Song, Y., 2015. Polarization analysis of excitons in monolayer and bilayer transition-metal dichalcogenides. Phys. Rev. B 92 (12), 125431.

He, K., Kumar, N., Zhao, L., Wang, Z., Mak, K.F., Zhao, H., et al., 2014. Tightly bound excitons in monolayer WSe2. Phys. Rev. Lett. 113 (2), 026803.

He, M., Rivera, P., Van Tuan, D., Wilson, N.P., Yang, M., Taniguchi, T., et al., 2020. Valley phonons and exciton complexes in a monolayer semiconductor. Nat. Commun. 11 (1), 618.

Jiang, X., Zheng, Q., Lan, Z., Saidi, W.A., Ren, X., Zhao, J., 2021a. Real-time GW-BSE investigations on spin-valley exciton dynamics in monolayer transition metal dichalcogenide. Sci. Adv. 7 (10), eabf3759.

Jiang, Y., Chen, S., Zheng, W., Zheng, B., Pan, A., 2021b. Interlayer exciton formation, relaxation, and transport in TMDCs van der Waals heterostructures. Light Sci. Appl. 10 (1), 72.

Jin, C., Regan, E.C., Yan, A., Iqbal Bakti Utama, M., Wang, D., Zhao, S., et al., 2019. Observation of moiré excitons in WSe2/WS2 heterostructure superlattices. Nature 567 (7746), 76–80.

Jones, A.M., Yu, H., Ghimire, N.J., Wu, S., Aivazian, G., Ross, J.S., et al., 2013. Optical generation of excitonic valley coherence in monolayer WSe2. Nat. Nanotechnol. 8 (9), 634–638.

Jones, A.M., Yu, H., Schaibley, J.R., Yan, J., Mandrus, D.G., Taniguchi, T., et al., 2015. Excitonic luminescence upconversion in a two-dimensional semiconductor. Nat. Phys. 12 (4), 323–327.

Ko´smider, K., Fern´andez-Rossier, J.W.G.A.J., 2013. Large spin splitting in the conduction band of transition metal dichalcogenide monolayers. Phys. Rev. B 88, 245436.

Kormányos, A., Burkard, G., Gmitra, M., Fabian, J., Zólyomi, V., Drummond, N.D., et al., 2015. k·p theory for two-dimensional transition metal dichalcogenide semiconductors. 2D Mater. 2 (2), 022001.

Kylänpää, I., Komsa, H.-P., 2015. Binding energies of exciton complexes in transition metal dichalcogenide monolayers and effect of dielectric environment. Phys. Rev. B 92 (20), 205418.

Li, Y., Ludwig, J., Low, T., Chernikov, A., Cui, X., Arefe, G., et al., 2014. Valley splitting and polarization by the Zeeman effect in monolayer MoSe2. Phys. Rev. Lett. 113 (26), 266804.

Li, Z., Wang, T., Jin, C., Lu, Z., Lian, Z., Meng, Y., et al., 2019b. Emerging photoluminescence from the dark-exciton phonon replica in monolayer WSe(2). Nat. Commun. 10 (1), 2469.

Li, Z., Wang, T., Lu, Z., Jin, C., Chen, Y., Meng, Y., et al., 2018. Revealing the biexciton and trion-exciton complexes in BN encapsulated WSe2. Nat. Commun. 9 (1), 3719.

Li, Z., Wang, T., Lu, Z., Khatoniar, M., Lian, Z., Meng, Y., et al., 2019a. Direct observation of gate-tunable dark trions in monolayer WSe2. Nano Lett. 19 (10), 6886–6893.

Liu, E., Barre, E., van Baren, J., Wilson, M., Taniguchi, T., Watanabe, K., et al., 2021. Signatures of moire trions in WSe2/MoSe2 heterobilayers. Nature 594 (7861), 46–50.

Liu, E., van Baren, J., Lu, Z., Altaiary, M.M., Taniguchi, T., Watanabe, K., et al., 2019a. Gate tunable dark trions in monolayer WSe2. Phys. Rev. Lett. 123 (2), 027401.

Liu, E., van Baren, J., Taniguchi, T., Watanabe, K., Chang, Y.-C., Lui, C.H., 2019b. Valley-selective chiral phonon replicas of dark excitons and trions in monolayer WSe2. Phys. Rev. Res. 1 (3), 032007R.

Liu, Y., Elbanna, A., Gao, W., Pan, J., Shen, Z., Teng, J., 2022. Interlayer excitons in transition metal dichalcogenide semiconductors for 2D optoelectronics. Adv. Mater. 34 (25), e2107138.

Lo, P.-Y., Peng, G.-H., Li, W.-H., Yang, Y., Cheng, S.-J., 2021. Full-zone valley polarization landscape of finite-momentum exciton in transition metal dichalcogenide monolayers. Phys. Rev. Res. 3 (4), 043198.

Lyons, T.P., Dufferwiel, S., Brooks, M., Withers, F., Taniguchi, T., Watanabe, K., et al., 2019. The valley Zeeman effect in inter- and intra-valley trions in monolayer WSe2. Nat. Commun. 10 (1), 2330.

MacNeill, D., Heikes, C., Mak, K.F., Anderson, Z., Kormanyos, A., Zolyomi, V., et al., 2015. Breaking of valley degeneracy by magnetic field in monolayer MoSe2. Phys. Rev. Lett. 114 (3), 037401.

Madéo, J., Man, M.K.L., Sahoo, C., Campbell, M., Pareek, V., Wong, E.L., et al., 2020. Directly visualizing the momentum-forbidden dark excitons and their dynamics in atomically thin semiconductors. Science 370 (6521), 1199–1204.

Mak, K.F., He, K., Lee, C., Lee, G.H., Hone, J., Heinz, T.F., et al., 2013. Tightly bound trions in monolayer MoS2. Nat. Mater. 12 (3), 207–211.

Mak, K.F., He, K., Shan, J., Heinz, T.F., 2012. Control of valley polarization in monolayer MoS2 by optical helicity. Nat. Nanotechnol. 7 (8), 494–498.

Mak, K.F., Shan, J., 2016. Photonics and optoelectronics of 2D semiconductor transition metal dichalcogenides. Nat. Photonics 10 (4), 216–226.

Mak, K.F., Xiao, D., Shan, J., 2018. Light–valley interactions in 2D semiconductors. Nat. Photonics 12 (8), 451–460.

Malic, E., Selig, M., Feierabend, M., Brem, S., Christiansen, D., Wendler, F., et al., 2018. Dark excitons in transition metal dichalcogenides. Phys. Rev. Mater. 2 (1).

Mayers, M.Z., Berkelbach, T.C., Hybertsen, M.S., Reichman, D.R., 2015. Binding energies and spatial structures of small carrier complexes in monolayer transition-metal dichalcogenides via diffusion Monte Carlo. Phys. Rev. B 92 (16), 161404R.

Park, K.D., Jiang, T., Clark, G., Xu, X., Raschke, M.B., 2018. Radiative control of dark excitons at room temperature by nano-optical antenna-tip Purcell effect. Nat. Nanotechnol. 13 (1), 59–64.

Paur, M., Molina-Mendoza, A.J., Bratschitsch, R., Watanabe, K., Taniguchi, T., Mueller, T., 2019. Electroluminescence from multi-particle exciton complexes in transition metal dichalcogenide semiconductors. Nat. Commun. 10 (1), 1709.

Pei, J., Liu, X., del Águila, A.G., Bao, D., Liu, S., Amara, M.-R., et al., 2023. Switching of K-Q intervalley trions fine structure and their dynamics in n-doped monolayer WS2. Opto-Electron. Adv. 6, 220034.

Pei, J., Yang, J., Wang, X., Wang, F., Mokkapati, S., Lu, T., et al., 2017. Excited state biexcitons in atomically thin MoSe2. ACS Nano 11 (7), 7468–7475.

Pei, J., Yang, J., Xu, R., Zeng, Y.H., Myint, Y.W., Zhang, S., et al., 2015. Exciton and trion dynamics in bilayer MoS2. Small 11 (48), 6384–6390.

Peng, G.H., Lo, P.Y., Li, W.H., Huang, Y.C., Chen, Y.H., Lee, C.H., et al., 2019. Distinctive signatures of the spin- and momentum-forbidden dark exciton states in the photoluminescence of strained WSe2 monolayers under thermalization. Nano Lett. 19 (4), 2299–2312.

Plechinger, G., Nagler, P., Arora, A., Granados Del Aguila, A., Ballottin, M.V., Frank, T., et al., 2016b. Excitonic valley effects in monolayer WS2 under high magnetic fields. Nano Lett. 16 (12), 7899–7904.

Plechinger, G., Nagler, P., Arora, A., Schmidt, R., Chernikov, A., Del Aguila, A.G., et al., 2016a. Trion fine structure and coupled spin-valley dynamics in monolayer tungsten disulfide. Nat. Commun. 7, 12715.

Rivera, P., Schaibley, J.R., Jones, A.M., Ross, J.S., Wu, S., Aivazian, G., et al., 2015. Observation of long-lived interlayer excitons in monolayer MoSe2-WSe2 heterostructures. Nat. Commun. 6, 6242.

Rivera, P., Yu, H., Seyler, K.L., Wilson, N.P., Yao, W., Xu, X., 2018. Interlayer valley excitons in heterobilayers of transition metal dichalcogenides. Nat. Nanotechnol. 13 (11), 1004–1015.

Roldán, R., Silva-Guillén, J.A., López-Sancho, M.P., Guinea, F., Cappelluti, E., Ordejón, P., 2014. Electronic properties of single-layer and multilayer transition metal dichalcogenides MX2 (M=Mo, W and X=S, Se). Ann. Phys. 526 (9–10), 347–357.

Ross, J.S., Wu, S., Yu, H., Ghimire, N.J., Jones, A.M., Aivazian, G., et al., 2013. Electrical control of neutral and charged excitons in a monolayer semiconductor. Nat. Commun. 4, 1474.

Schaibley, J.R., Yu, H., Clark, G., Rivera, P., Ross, J.S., Seyler, K.L., et al., 2016. Valleytronics in 2D materials. Nat. Rev. Mater. 1 (11), 16055.

Selig, M., Berghauser, G., Raja, A., Nagler, P., Schuller, C., Heinz, T.F., et al., 2016. Excitonic linewidth and coherence lifetime in monolayer transition metal dichalcogenides. Nat. Commun. 7, 13279.

Song, Y., Dery, H., 2013. Transport theory of monolayer transition-metal dichalcogenides through symmetry. Phys. Rev. Lett. 111 (2), 026601.

Srivastava, A., Sidler, M., Allain, A.V., Lembke, D.S., Kis, A., Imamoğlu, A., 2015. Valley Zeeman effect in elementary optical excitations of monolayer WSe2. Nat. Phys. 11 (2), 141–147.

Sun, X., Zhu, Y., Qin, H., Liu, B., Tang, Y., Lu, T., et al., 2022. Enhanced interactions of interlayer excitons in free-standing heterobilayers. Nature 610 (7932), 478–484.

Szyniszewski, M., Mostaani, E., Drummond, N.D., Fal'ko, V.I., 2017. Binding energies of trions and biexcitons in two-dimensional semiconductors from diffusion quantum Monte Carlo calculations. Phys. Rev. B 95 (8), 081301R.

Tran, K., Moody, G., Wu, F., Lu, X., Choi, J., Kim, K., et al., 2019. Evidence for moiré excitons in van der Waals heterostructures. Nature 567 (7746), 71–75.

Ugeda, M.M., Bradley, A.J., Shi, S.-F., da Jornada, F.H., Zhang, Y., Qiu, D.Y., et al., 2014. Giant bandgap renormalization and excitonic effects in a monolayer transition metal dichalcogenide semiconductor. Nat. Mater. 13, 1091.

Vaclavkova, D., Wyzula, J., Nogajewski, K., Bartos, M., Slobodeniuk, A.O., Faugeras, C., et al., 2018. Singlet and triplet trions in WS2 monolayer encapsulated in hexagonal boron nitride. Nanotechnology 29 (32), 325705.

Wallauer, R., Perea-Causin, R., Munster, L., Zajusch, S., Brem, S., Gudde, J., et al., 2021. Momentum-resolved observation of exciton formation dynamics in monolayer WS2. Nano Lett. 21 (13), 5867–5873.

Wang, G., Robert, C., Glazov, M.M., Cadiz, F., Courtade, E., Amand, T., et al., 2017. In-plane propagation of light in transition metal dichalcogenide monolayers: optical selection rules. Phys. Rev. Lett. 119 (4), 047401.

Wang, G., Robert, C., Suslu, A., Chen, B., Yang, S., Alamdari, S., et al., 2015. Spin-orbit engineering in transition metal dichalcogenide alloy monolayers. Nat. Commun. 6, 10110.

Xu, X., Yao, W., Xiao, D., Heinz, T.F., 2014. Spin and pseudospins in layered transition metal dichalcogenides. Nat. Phys. 10 (5), 343–350.

Ye, Z., Waldecker, L., Ma, E.Y., Rhodes, D., Antony, A., Kim, B., et al., 2018. Efficient generation of neutral and charged biexcitons in encapsulated WSe2 monolayers. Nat. Commun. 9 (1), 3718.

Yu, H., Liu, G.B., Gong, P., Xu, X., Yao, W., 2014. Dirac cones and Dirac saddle points of bright excitons in monolayer transition metal dichalcogenides. Nat. Commun. 5, 3876.

Zeng, H., Dai, J., Yao, W., Xiao, D., Cui, X., 2012. Valley polarization in MoS2 monolayers by optical pumping. Nat. Nanotechnol. 7 (8), 490–493.

Zhang, D.K., Kidd, D.W., Varga, K., 2015b. Excited biexcitons in transition metal dichalcogenides. Nano Lett. 15 (10), 7002–7005.

Zhang, L., Niu, Q., 2015. Chiral phonons at high-symmetry points in monolayer hexagonal lattices. Phys. Rev. Lett. 115 (11), 115502.
Zhang, X.X., Cao, T., Lu, Z., Lin, Y.C., Zhang, F., Wang, Y., et al., 2017. Magnetic brightening and control of dark excitons in monolayer WSe2. Nat. Nanotechnol. 12 (9), 883–888.
Zhang, X.X., You, Y., Zhao, S.Y., Heinz, T.F., 2015a. Experimental evidence for dark excitons in monolayer WSe2. Phys. Rev. Lett. 115 (25), 257403.
Zhou, Y., Scuri, G., Wild, D.S., High, A.A., Dibos, A., Jauregui, L.A., et al., 2017. Probing dark excitons in atomically thin semiconductors via near-field coupling to surface plasmon polaritons. Nat. Nanotechnol. 12 (9), 856–860.
Zhu, H., Yi, J., Li, M.Y., Xiao, J., Zhang, L., Yang, C.W., et al., 2018. Observation of chiral phonons. Science 359 (6375), 579–582.

CHAPTER THREE

Exciton diffusion in 2D van der Waals semiconductors

Alexey Chernikov[a,*] and Mikhail M. Glazov[b]
[a]Institute of Applied Physics and Würzburg-Dresden Cluster of Excellence ct.qmat, TUD Dresden University of Technology, Dresden, Germany
[b]Ioffe Institute, Saint Petersburg, Russian Federation
*Corresponding author. e-mail address: alexey.chernikov@tu-dresden.de

Contents

Abstract

This chapter provides an overview of experimental and theoretical studies of exciton propagation in 2D materials. Discussed is the diffusive regime of exciton transport, including linear and non-linear phenomena. It further covers the impact of environmental disorder, exciton-carrier mixtures, and finite temperatures. The focus is placed on monolayer transition-metal dichalcogenide semiconductors with a short summary of some of the recent findings for heterobilayer systems.

Semiconductors and Semimetals, Volume 112
ISSN 0080-8784, https://doi.org/10.1016/bs.semsem.2023.09.001

1. Introduction

The absorption of light in a semiconductor leads to an excitation of electrons from the valence to the conduction band and creates hole quasi-particles in the valence band (Ivchenko, 2005; Haug and Koch, 2009). The photoexcited free charge carriers thus store the energy of the photons, are often mobile, and can propagate through the crystal. This process of energy transport in solids is of key importance from the perspectives of both fundamental condensed matter physics and optoelectronic applications. For example, rapid propagation is highly desirable for photodectors and photovoltaics, while localization is usually favorable for efficient light emission and single photon sources. Importantly, in a wide range of materials, the energy transport is facilitated not by the free charge carriers, but by Coulomb-bound electron-hole pairs, known as excitons (Frenkel, 1931; Gross and Karrjew, 1952). This applies in particular for two-dimensional semiconductors, such as mono- and few-layers of transition-metal dichalcogenides (TMDCs) (Wilson and Yoffe, 1969; Novoselov et al., 2005; Mak et al., 2010; Splendiani et al., 2010) that host tightly bound excitons due to strong quantum confinement and weak dielectric screening (Xiao et al., 2017; Wang et al., 2018). Excitons in TMDCs exhibit a number of intriguing properties including spin-valley locking (Xu et al. 2014), electrical (Mueller and Malic 2018), magnetic (Arora, 2021), dielectric (Raja et al. 2017), and strain-tunability (Blundo et al. 2021). In close analogy to the physics of excitons in quantum well systems (Smith et al., 1988; Hillmer et al., 1988; Steininger et al., 1996; Butov et al., 1994; Rapaport et al., 2004), the excitons in TMDCs are in principle free to move across the material plane. Consequently, the topic of exciton transport in two-dimensional (2D) materials attracted considerable attention.

First studies of the exciton transport in TMDCs monolayers illustrated diffusive propagation across many 100's of nm at ambient conditions (Kumar et al., 2014b; Kato and Kaneko, 2016; Yuan et al., 2017). These findings, shown for both $MoSe_2$ and WS_2, were highly encouraging, since much shorter diffusion lengths were initially expected (van der Zande et al. 2013). It was also quickly realized that the exciton propagation depends on the environment of the monolayers (Yuan et al., 2017; Cadiz et al., 2018; Fu et al., 2019; Raja et al., 2019; Zipfel et al., 2020), especially with respect to the dielectric screening. Of particular importance was the role of disorder suppressing exciton transport and its mitigation by encapsulation of TMDCs in high-quality hexagonal boron nitride. The latter provided homogeneous environment leading to reduced inhomogeneous broadening (Cadiz et al., 2017; Ajayi et al., 2017) and faster diffusion (Raja et al. 2019). In parallel, non-linear phenomena associated with exciton-exciton interactions were demonstrated together with consequences of multi-valley bandstructures for the exciton transport (Mouri et al., 2014; Kulig et al., 2018; Wang et al., 2019; Glazov, 2019; Perea-Causín et al., 2019; Goodman et al., 2020b; Rosati et al., 2020b; Qiu et al., 2021). A density-dependent process of exciton-exciton annihilation was recognized to result in an increase of the observed diffusion coefficients as well as in local heating of the excitons; these effects were shown to lead to additional drift currents. Taking advantage of the improved techniques to control external fields, studies focused on electrical tunability and exciton-electron interactions followed (Uddin et al., 2020; Soubelet et al., 2021; Cheng et al., 2021; Beret et al., 2023; Wagner et al., 2023) as well as investigations of temperature-dependent diffusion in hBN-encapsulated samples (Hotta et al., 2020; Wagner et al., 2021). Among especially effective and useful methods to control exciton propagation was the application of strain that allowed for guiding the excitons and realizing exciton currents (Cordovilla Leon et al., 2018; Shahnazaryan et al., 2019; Hao et al., 2020; Rosati et al., 2021b; Dirnberger et al., 2021; Datta et al., 2022). Most recently, intriguing properties of interlayer excitons in heterobilayers came into focus, exploring the interplay of localization, free propagation, and interactions in complex environments, highlighting an active, rapidly evolving field of research (Rivera et al., 2016; Calman et al., 2018; Jauregui et al., 2019; Unuchek et al., 2018; Yuan et al., 2020; Choi et al., 2020; Li et al., 2021; Wang et al., 2021; Sun et al., 2022; Choi et al., 2023).

Theoretical understanding of exciton propagation is based on well-known approaches to transport phenomena in semiconductors (Ziman, 2001; Shklovskii and Efros, 1984; Gantmakher and Levinson, 1987) with considerable extensions taking into account (i) extreme-two-dimensional nature of the TMDCs-based systems, (ii) electric neutrality of excitons, and (iii) specifics of disorder and exciton-phonon interaction in 2D semiconductors. The first step was justification of diffusive regime of exciton propagation under typical experimental conditions. It was based on the comparison of microscopic time scales of exciton dynamics, i.e., scattering or hopping times, that are typically much shorter than the exciton population lifetimes. The second step is the identification of the appropriate regime of exciton propagation depending on the strength of disorder, exciton-phonon interactions, and presence of external potentials, including elastic strain and moiré-type superlattices: whether exciton propagation is related to hopping between localized states or to semiclassical free propagation interrupted by exciton-phonon or static disorder scattering. To date, the most developed are semiclassical models of exciton transport based on Boltzmann kinetic equation (Zipfel et al., 2020; Glazov, 2019; Perea-Causín et al., 2019; Rosati et al., 2021a, 2021b; Choi et al., 2023) in line with the semiclassical propagation regime observed in the state-of-art monolayer structures. Using the Greens function approach and diagram technique, semiclassical approach can be further extended to account key quantum effects in exciton transport related to weak localization (Glazov, 2020a; Glazov et al., 2022). Alternative approaches are based on fully microscopic ab initio picture of exciton wavepacket or density distribution dynamics (Qiu et al., 2021; Cohen et al., 2023) that also enable to study exciton propagation beyond the diffusion paradigm. In this context it is worth mentioning prospects of achieving hydrodynamic and superfluid regimes of exciton propagation based on Navier-Stokes-like equations or hydrodynamic equations for superfluids (Pitaevskii and Stringari, 2003). These approaches await their development for 2D TMDCs.

The aim of this chapter is to present and discuss the physics of exciton propagation in 2D TMDCs, including experimental observations and theoretical descriptions. The focus is placed on linear and non-linear regimes of free propagation. The physics of directional exciton transport and exciton currents will be covered in a separate chapter of this book. We also note that the intriguing phenomena associated with the propagation of exciton-polaritons in 2D TMDCs (e.g., Basov et al. 2016) are beyond the scope of this chapter. Finally, while the presented material is naturally affected by the involvement of the authors in this area of research, it is based on the findings within the broader field with the intended focus on phenomena that are largely accepted in the community. Nonetheless, due to the ongoing activities it also includes more recent results that should be primarily understood as a motivation for future directions. The chapter is organized as follows: Section 2 contains necessary preliminaries needed to study exciton propagation both from the theoretical and experimental standpoints. In particular, we present a brief description of exciton transport in Section 2.1, illustrate different propagation regimes in Section 2.2, and discuss generic experimental approaches to measure exciton propagation in Section 2.3. Main experimental observations are then presented in Section 3. It starts with the discussion of the observations at fixed exciton densities, the role of the environment and disorder in Section 3.1. In Section 3.2 we focus on the non-linear exciton propagation dynamics arising from exciton-exciton interactions. Section 3.3 then addresses the impact of finite free carrier densities and temperature on the exciton diffusion. An overview of recent findings for propagating interlayer excitons in heterobilayers is given in Section 3.4. In Section 4 we discuss the theory of the exciton transport in more detail. Section 4.1 presents the semiclassical approach to calculate exciton diffusion coefficients. Further we elaborate on quantum effects in Section 4.2 and address nonlinear phenomena in Section 4.3. We conclude theoretical analysis with a brief discussion of effects beyond diffusive propagation in, Section 4.4. A brief outlook is given at the end of the chapter in Section 5.

2. Exciton diffusion in 2D transition-metal dichalcogenides

Studies of exciton diffusion in TMDCs follow extensive experimental and theoretical research in the broader area of exciton transport, from bulk inorganic and organic materials, such as CuO_2 (Tikhodeev et al., 1998; Warren et al., 2000) and molecular crystals (Akselrod et al., 2014) to III-V and II-VI quantum wells (Smith et al., 1988; Erland et al., 1993; Butov et al., 2004) as well as inorganic-organic hybrid perovskites (Deng et al., 2020; Delor et al., 2020; Seitz et al., 2020). As a consequence, there is a substantial overlap with respect to the experimental methods and theoretical descriptions, while TMDCs also offer both an interesting platform and new perspectives regarding exciton propagation phenomena. In this section we introduce the general framework for exciton diffusion and provide an overview of the main findings and observations as well as offer phenomenological descriptions prior to more detailed theoretical discussion in Section 4.

2.1 General description

It is convenient to describe the exciton transport starting from a macroscopic model for the dynamics of the exciton density $n(\boldsymbol{r}, t)$ as a function of the in-plane position $\boldsymbol{r} = (x, y)$ and time t. The density n accounts for contributions of all excitonic species, including momentum direct bright and dark ones as well as momentum indirect excitons. Its decomposition into individual contributions can be carried out, if necessary, see Refs. (Wang et al., 2018; Zipfel et al., 2020). The exciton density satisfies the continuity equation

$$\frac{\partial n}{\partial t} + \nabla \cdot \boldsymbol{i} + R\{n\} = G, \tag{1}$$

where $\boldsymbol{i}$ is the exciton flux (current) density, $\nabla = (\partial/\partial x, \partial/\partial y)$, $R\{n\}$ is the operator describing the recombination of excitons, and $G \equiv G(\boldsymbol{r}, t)$ is their generation rate. Eq. (1) states that the exciton density in a given point of space changes due to the generation, G, and recombination, R, processes, as well as due to the balance of the in- and out-flow of the excitons, $\nabla \cdot \boldsymbol{i}$. Generally, the exciton current can be decomposed in a series in gradients of density. Within the minimal model, only the linear term is kept

$$\boldsymbol{i}(\boldsymbol{r}, t) = -D\nabla n(\boldsymbol{r}, t). \tag{2}$$

Here, $D \geqslant 0$ defines the diffusion coefficient. Eq. (2) is known as Fick's first law of diffusion. It plays a role of material relation linking current and density gradient similarly to Ohm's law in macroscopic electrodynamics relating electric current and field. In homogeneous equilibrium system D is a constant in space and time. In the presence of external perturbations, e.g., temperature or strain gradients, current $\boldsymbol{i}$ acquires additional contributions related to the drag forces acting on excitons, Section 4.3. By combining Eqs. (1) and (2) we arrive at the exciton diffusion equation

$$\frac{\partial n}{\partial t} - D\Delta n + R\{n\} = G, \tag{3}$$

with Δ being the Laplacian. Eqs. (1) – (3) are general and independent of specific microscopic models of exciton propagation. The only limitation of these equations is the "coarse grade" approach: function $n(\boldsymbol{r}, t)$ should be smooth on the scale of microscopic lengths and timescales underlying exciton dynamics. This means that $n(\boldsymbol{r}, t)$ should not vary significantly in space on the scale of excitonic de Broglie wavelength, its mean free path or hopping distance, as well as in time on the scales of scattering and hopping times.

Theoretical framework aimed at studies of exciton transport generally encompasses two key problems: (i) solution of diffusion Eq. (3) for appropriate experimental settings that include geometry of exciton propagation and specific forms of the generation and recombination terms, and (ii) microscopic calculations of the parameters entering Eq. (3). The latter concerns, in particular, the diffusion coefficient as function of system's parameters, justification of applicability of Eq. (3), and analysis of the propagation regimes beyond the diffusion model.

Before discussing these points in more detail, let us briefly address the solution of Eq. (3) in the simplest, yet arguably the most instructive case, of linear (monomolecular) recombination of excitons with $R\{n\} = n(\boldsymbol{r}, t)/\tau$; τ is the exciton lifetime that includes contributions of radiative and non-radiative recombination. General solution of Eq. (3) can be written in the form

$$n(\boldsymbol{r}, t) = \int_{-\infty}^{t} dt' \int d\boldsymbol{r}' g(\boldsymbol{r} - \boldsymbol{r}', t - t') e^{-\frac{t-t'}{\tau}} G(\boldsymbol{r}', t'),$$

$$g(\boldsymbol{r}, t) = \frac{1}{4\pi Dt} \exp\left(-\frac{r^2}{4Dt}\right). \tag{4}$$

Here the integration over $\boldsymbol{r}'$ is carried out over the sample area for a 2D system, and $g(\boldsymbol{r}, t)$ is the Greens function (fundamental solution) of the linear diffusion equation that describes diffusive spread of the excitons out of the spot where they were initially created. The mean squared displacement along a given axis x or y

$$\sigma^2(t) = \int d\boldsymbol{r} x^2 g(\boldsymbol{r}, t) = 2Dt, \tag{5}$$

increases linearly with time as a hallmark of diffusive behavior. Altogether, diffusive propagation can be considered as a random walk with uncorrelated directions of steps. Hence, as compared to ballistic regime where $\sigma \propto t$, the diffusion is characterized by slow expansion $\sigma \propto \sqrt{t}$. Eqs. (4) and (5) are widely used as a first step in the analysis of experimental data on exciton propagation, see Section 2.3. In some cases, however, especially if exciton-exciton interactions are significant, the model turns out to be insufficient, see Ref. de Quilettes et al. (2022) and Section 4.3.

2.2 Exciton propagation regimes

On a microscopic level, the exciton diffusion coefficient is given by the velocity autocorrelation function (Green, 1954; Kubo, 1957, 1966; Glazov et al., 2022)

$$D = \int_0^{\infty} \langle\langle \hat{v}_x(t) \hat{v}_x(0) \rangle\rangle \, dt, \tag{6}$$

where double angular brackets denote both quantum-mechanical and statistical averaging with appropriate symmetrization of non-commuting operators, and $\hat{v}_x(t)$ is the operator of the x component of exciton velocity taken in the Heisenberg representation (we assume in-plane isotropic system). Evaluation of D after Eq. (6) is highly involved and practically impossible for realistic systems since, generally, it requires the solution of the exciton density matrix equation with exciton-phonon and exciton-defects interactions taken into account. First theoretical works in this direction (Qiu et al., 2021; Cohen et al., 2023) demonstrate highly encouraging results (see Section 4.4). In many cases, however, it is still instructive to consider simplified, physically motivated models of exciton transport depending on the particular propagation regime that depends on relation between the parameters in the studied system.

Schematically, main regimes of exciton propagation are illustrated in Fig. 1. The hopping regime, Fig. 1A is typically realized in strongly disordered systems or in structures with external potential, e.g., in moiré superlattices, where excitons are localized in the potential minima. They can jump between these minima by direct tunneling or by emitting or absorbing phonons. In this regime, the exciton diffusion coefficient can be estimated as (Shklovskii and Efros, 1984; Baranovski, 2006; Akmaev et al., 2021)

$$D = \frac{d_{hop}^2}{\tau_{hop}} \propto \exp\left[-\left(\frac{T_0}{T}\right)^s\right]. \tag{7}$$

Here, d_{hop} is characteristic hopping distance roughly corresponding to the distance between potential minima, τ_{hop} is the waiting time, T is the temperature, and $k_B T_0$ is on the order of localization energy. The main feature of this regime is strong, exponential temperature dependence of the diffusion coefficient as seen from the right hand side of Eq. (7): Power s depends on the hopping regime, e.g., $s = 1/3$ for the variable range hopping in two dimensions. The hopping regime corresponds to the strong or Anderson localization of excitons.

Decrease in the disorder strength, i.e., effectively the ratio of T_0 to T, results in exciton delocalization so that the exciton propagation eventually becomes semiclassical, Fig. 1C. Here, excitons can be considered as wavepackets that freely propagate and experience relatively rare scattering

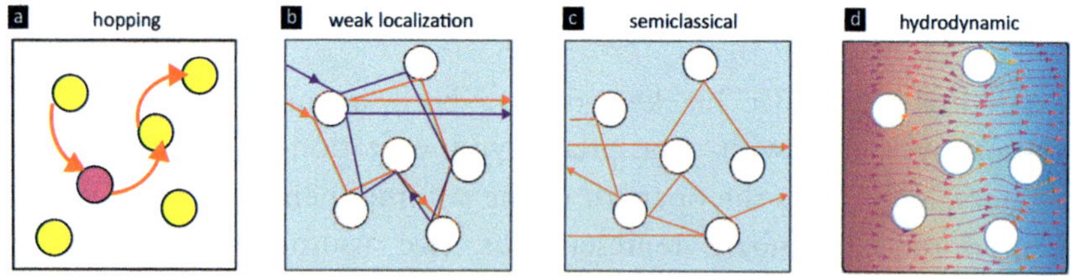

Fig. 1 Regimes of exciton propagation in 2D semiconductors: from the phonon-assisted hopping of localized excitons to the fluid-like flow. (A) Exciton hopping between localization sites. (B) Weak localization regime where interference of semiclassical trajectories is important. (C) Semiclassical regime consisting of free propagation interrupted by scattering. (D) Hydrodynamic, fluid-like propagation.

events with phonons and static defects, see Section 4.1 for detils. Between the hopping and semiclassical regimes is the region of weak localization, Fig. 1B, where interference of excitonic waves becomes important, as discussed in Section 4.2. Finally, in ultraclean systems and at elevated exciton densities excitons are expected to form fluids (Lozovik and Yudson, 1975; Fogler et al., 2014; Glazov and Suris, 2021) resulting in hydrodynamic transport, Fig. 1D, which qualitatively differs from the diffusive propagation, see Section 4.4 where hydrodynamic and other non-diffusive regimes are addressed.

2.3 Measuring exciton propagation

The general approach to measure diffusion of excitons in 2D TMDCs is essentially the same as for monitoring any type of an ensemble of diffusing particles, such as a drop of ink in water for example. It requires a local injection of spatially confined population followed by an observation of its spread over time. For excitons in TMDCs it is schematically illustrated in Fig. 2A.

Short laser pulses are typically focused by a high-aperture microscope objective to a spot with a micrometer-size diameter to create exciton population. If the laser beam and the optics are well aligned, the initial shape of the exciton distribution N_X is reasonably described by a Gaussian function

$$N_X(r, t = 0) \propto e^{-r^2/2\sigma_0^2}, \tag{8}$$

where r is the spatial coordinate and σ_0 the width of the distribution given as a standard deviation (the corresponding full-width-at-half-maximum equals $2\sqrt{2\ln 2}\ \sigma_0 \approx 2.35\ \sigma_0$). Measured σ_0 is a convolution of the exciton population profile and the resolution of the experimental setup in detection, e.g., point-spread-function for emission-type measurements. It is typically given as a variance, σ_0^2, in the literature. For a standard 100x objective with a numerical aperture of 0.9 and laser wavelengths in the visible spectral range, typical values for σ_0^2 are below 0.1 μm^2. Smaller initial widths can be obtained using lower wavelengths and objectives with higher apertures, e.g., with immersion oil for experiments at room temperature (Akselrod et al. 2014). For measurements through a glass of a cryostat, objectives with glass correction and larger working distances are required, resulting in smaller apertures on the order of 0.7, and σ_0^2 values of 0.2 μm^2 and above.

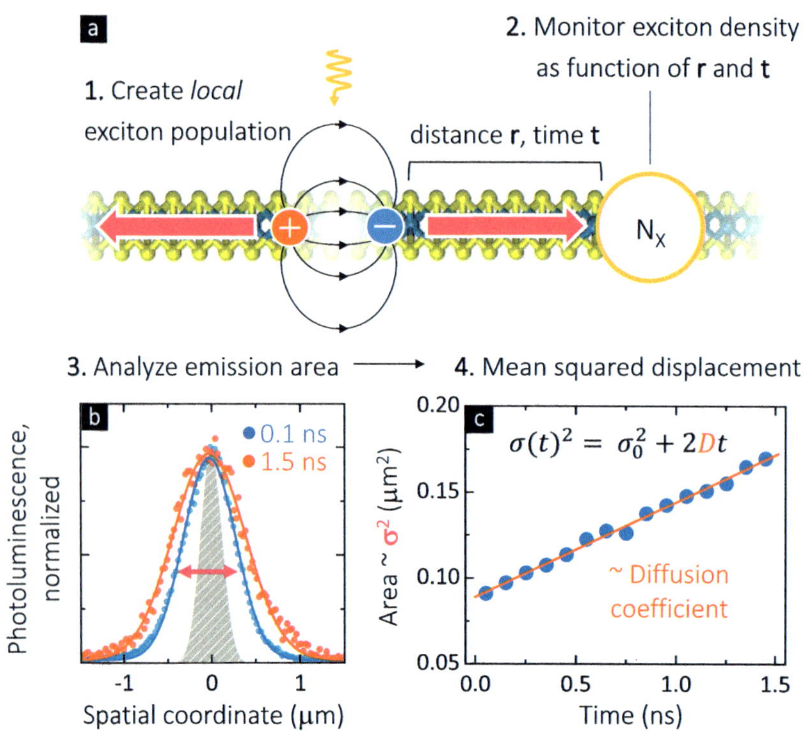

Fig. 2 Schematic illustration of typical experimental approaches to measure exciton diffusion in 2D materials: 1. Creation of local population and 2. Monitoring of the spatially- and time-resolved exciton density (A), 3. Analysis of the emission area (B), and 4. Extraction of the mean squared displacement $\sigma^2(t)$ (C). The diffusion coefficient D is extracted from the slope according $\sigma^2(t) = \sigma_0^2 + 2Dt$, Eq. (10). In (B) and (C), exemplary measurements of exciton diffusion in a freestanding WS_2 monolayer at room temperature and low excitation densities are used. The gray area represents the point-spread-function, i.e., the resolution of the setup. *Reprinted figure with permission from (Kulig et al. 2018). Copyright 2018 by the American Physical Society.*

After the injection, the excitons propagate as a function of time and spread according to (c.f. Eq. 4)

$$N_X(r, t) = A(t)e^{-r^2/2\sigma^2(t)} \tag{9}$$

with a mean squared displacement $\sigma^2(t) = \Delta\sigma^2(t) + \sigma_0^2$. It is a sum of the time-dependent increase of the distribution width $\Delta\sigma^2(t)$ and the time-independent component σ_0^2. The amplitude $A(t)$ typically decreases over time also due to the finite lifetime of the excitons. Most common techniques to detect the exciton distribution involve different types of transient microscopy based on spatially- and time-resolved optical measurements. For example, one can monitor changes in the non-linear optical susceptibility by using pump-probe type of geometries. This is achieved by scanning both the temporal and spatial separation between the pump and probe pulses (Kumar et al., 2014b; Yuan et al., 2017). In this manner, a series of pump-probe experiments is essentially performed for different spatial offsets between the pulses. One can also observe the scattering of a time-delayed probe pulse across a wide sample area, thus imaging propagation dynamics after the initial injection by a spatially localized pump pulse Delor et al. (2020). An alternative method involves imaging and detection of photoluminescence (PL) from the radiative recombination of excitons that often traces the dynamics of the total exciton population. Here, the time-resolution is provided by the detector itself that typically involves either an avalanche photo diode combined with time-correlated single photon counting (Goodman et al., 2020b; Akselrod et al., 2014; Seitz et al., 2020) or a streak camera (Kulig et al., 2018; Rosati et al., 2021b). The advantage of pump-probe detection is a higher time resolution limited by the duration of the laser pulses that is usually on the order of 100 fs in comparison to 1–10's of ps resolution of PL-based techniques. In addition, pump-probe signals are sensitive to both electronic and thermal excitations, thus allowing for detection of excitons, non-equilibrium phonons, or heat. Main advantages of PL-based methods include the possibility of performing measurements at extremely low excitation conditions (down to one exciton in the laser spot per pulse) as well as the sensitivity of the signals to the presence of photoexcited electron-hole pairs. Overall, these types of transient microscopy methods are largely complementary and were successfully employed to study exciton propagation in TMDCs. For a comprehensive overview we suggest a recent review by N. Ginsberg and W. Tisdale, see Ref. Ginsberg and Tisdale (2020).

Transient microscopy data is typically analyzed in terms of spatially-resolved profiles of the detected signals at different times after the excitation, as illustrated in Fig. 2B. It shows two representative cross sections of the PL signal from a freestanding WS_2 monolayer at room temperature (Kulig et al. (2018)). The profiles are fit to a Gaussian function using Eq. (9) and the time-dependent mean squared displacement $\sigma^2(t)$ is extracted, corresponding to the area of the exciton distribution (Fig. 2C). In case of the diffusion, see Eq. (5), this area increases linearly with time according to

$$\sigma^2(t) = \sigma_0^2 + \Delta\sigma^2(t) = \sigma_0^2 + 2Dt, \quad (10)$$

where D is the diffusion coefficient, introduced in Eq. (2). This representation further motivates the choice of $\sigma^2(t)$ and not $\sigma(t)$ for the analysis, since the time-independent initial broadening σ_0^2 enters Eq. (10) only as a constant offset. Thus, it is typically sufficient to consider only the time-dependent changes $\Delta\sigma^2(t)$. In general, however, the dependence of the mean squared displacement on time can be both superlinear, sublinear, or even have a negative slope. For the specific case of ballistic transport, as previously mentioned, it increases quadratically if the time-dependent changes are larger than the initial broadening, i.e., $\Delta\sigma^2(t) > \sigma_0^2$. Otherwise, for $\Delta\sigma^2(t) < \sigma_0^2$, even the ballistic transport can appear diffusive resulting in a linear increase of the mean squared displacement at early times. This can be illustrated by considering

$$\sigma^2(t) = (\sigma_0 + vt)^2 = \sigma_0^2 + 2\sigma_0 vt + (vt)^2 \approx \sigma_0^2 + 2\sigma_0 vt, \tag{11}$$

where v is the ballistic velocity and the approximation holds for $vt \ll \sigma_0$. In that case, other arguments such as considerations of the experimental conditions and the analysis of the value v can be required to distinguish diffusive and ballistic transport.

Overall, deviations of $\sigma^2(t)$ from the linear dependence signify a more complex behavior that can involve interactions, several types of propagating populations, time-dependent scattering, trapping and detrapping. In that case, it can be useful to analyze the time-dependent slope of $\sigma^2(t)$ that can be labeled as *effective* diffusivity extracted from $D_{\text{eff}} = 1/2 \cdot d\sigma^2(t)/dt$. For exciton propagation in semiconductors, including 2D TMDCs, observations of both linear and non-linear time-dependence of $\sigma^2(t)$ are common and even substantial deviations of the exciton distribution profile from the Gaussian shape can occur. For the latter case, the analysis in terms of the mean squared displacement should be understood only as a rough approximation. Finally, we note that one can also detect and analyze the spatial distribution of the excitons, i.e., their diffusion length L_D and their lifetime τ independently. These are connected to the diffusion coefficient by the relation $L_D = \sqrt{2D\tau}$ that allows for the estimation of the diffusion coefficient (Cadiz et al. 2018), while requiring accurate measurements of the absolute PL spot sizes. Moreover, the diffusion length can be obtained by employing sophisticated samples geometries based on monitoring the PL intensity as a function of distance from a region of efficient quenching, such as a metal contact (Kato and Kaneko, 2016). Altogether, the experimental results discussed in the following subsections are based on the methods for data acquisition and analysis presented above.

3. Experimental observations and main phenomena

3.1 Linear diffusion and the role of the environment

The exciton diffusion in 2D TMDCs was considered already at a rather early stage of the field in context of the excitons ability to reach grain boundaries (van der Zande et al. 2013) and the exciton-exciton annihilation (Mouri et al. 2014). First direct measurements, however, were reported for $MoSe_2$ using transient absorption microscopy (Kumar et al. 2014b), see Fig. 3A. The excitons in the monolayer $MoSe_2$ were found to diffuse only slightly slower compared to bulk and reach close to 0.5 μm diffusion length within their lifetime of about 100 ps. This finding of exciton propagation on a scale of a micrometer was confirmed in monolayer WS_2 by analyzing the dependence of PL intensity on the distance to the metal contacts (Kato and Kaneko 2016), see Fig. 3B. Interestingly, the spread of the measured diffusion coefficients was shown to be rather large in subsequent studies. For example, it could depend strongly on the choice of the preparation method, as demonstrated in Ref. Yuan et al. (2017), Fig. 3C using transient absorption microscopy. The exciton diffusion coefficient of a WS_2 monolayer was found to be $2.0 \pm 0.4\ cm^2/s$ for a mechanically exfoliated sample and only $0.1 \pm 0.04\ cm^2/s$ for a sample grown by chemical vapor deposition. This further contrasts diffusion coefficients of $12 \pm 3\ cm^2/s$ for $MoSe_2$ monolayer on SiO_2/Si substrate (Kumar et al. 2014b) and $0.3 \pm 0.1\ cm^2/s$ for freestanding and SiO_2/Si-supported WS_2 monolayer (Kulig et al. 2018). The data of a freestanding WS_2 monolayer from the latter study is shown as a streak camera image in Fig. 3D presenting as-measured and normalized PL intensity that illustrates the broadening of the profile over time.

This spread of values over almost two orders of magnitude cannot be explained by differences in intrinsic scattering channels alone. The strength of the exciton-phonon scattering in particular, which is dominant at room temperature, differs by less than a factor of two between WS_2 and $MoSe_2$ monolayers (Selig et al. 2016). Consequently, large fluctuations in measured diffusion coefficients are attributed to extrinsic factors, such as the presence of defects and inhomogeneous energy landscapes, i.e., disorder. In

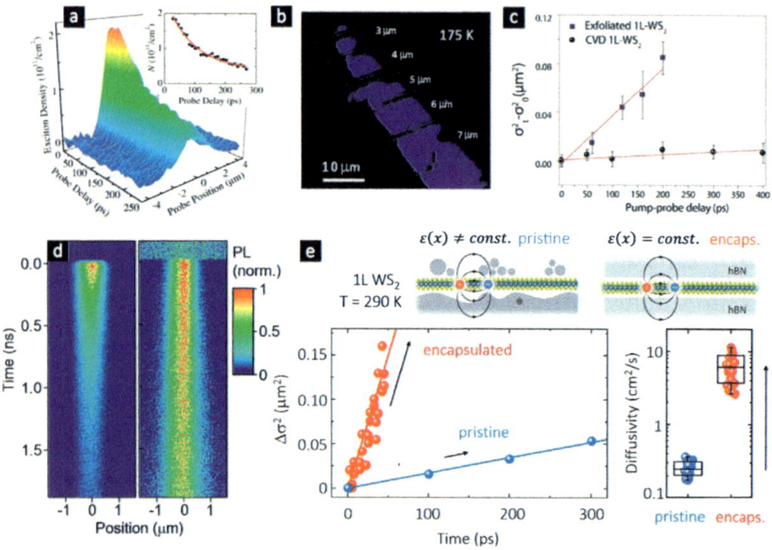

Fig. 3 (A) Time- and spatially-resolved profiles of the exciton density of a $MoSe_2$ monolayer sample obtained via transient absorption microscopy. (B) Map of PL intensity of a WS_2 monolayer sample placed between gold electrodes of different spacing. (C) Mean squared displacement of the exciton population measured by transient absorption microscopy in WS_2 monolayers fabricated by mechanical exfoliation and chemical vapor deposition. (D) Time- and spatially-resolved image of a PL cross section from a freestanding WS_2 monolayer measured by a streak camera and presented as-measured (left panel) and normalized at each time step (right panel). (E) Mean squared displacement and a summary of diffusion coefficients obtained from pristine and hBN-encapsulated WS_2 monolayers. The top panels show schematic illustrations of dielectric disorder from local fluctuations of the dielectric environment as well as its mitigation by encapsulation. The measurements in (A), (C–E) were performed at room temperature. *(A) Used with permission of Royal Society of Chemistry, from Kumar et al. (2014b), permission conveyed through Copyright Clearance Center, Inc.; (B) Reprinted with permission from Kato and Kaneko (2016). Copyright 2016 American Chemical Society; (C) Reprinted with permission from Yuan et al. (2017). Copyright 2017 American Chemical Society; (D) Reprinted figure with permission from Kulig et al. (2018). Copyright 2018 by the American Physical Society; (E) Data from Refs. Raja et al. (2019) and Zipfel et al. (2020).*

pristine samples on typical substrates, such as SiO_2 or glass, the energy scale of disorder estimated by the inhomogeneous spectral linewidth of the exciton resonances at low temperatures is on the order of 10–20 meV or even larger. These values are on the order of thermal fluctuations at room temperature ($k_BT = 26$ meV) and can thus substantially affect propagation of excitons in a similar manner as the mobility of free charges. The excitons can be effectively localized in areas with low potentials and scatter at the randomly distributed barriers within the disordered energy landscape. In addition to common sources of disorder in conventional semiconductors, such as the presence of structural defects and impurities, most important sources of disorder for 2D TMDCs are fluctuations of strain and dielectric environment. The former is related to local mechanical stress and deformations, for example caused by a rough substrate, and leads to energy shifts on the order of 10–20 meV per 0.1% strain. The latter stems from the sensitivity of the exciton energies to the dielectric screening of the monolayer surroundings (Wang et al. 2018). Local fluctuations of the dielectric constant of the environment, e.g., due to the substrate roughness and adsorbates, transfer directly into spatial fluctuations of the exciton energies (Raja et al. 2019). Characteristic energy scale of the dielectric disorder can reach values from 5 to 30 meV. In addition to strain and dielectric disorder, fluctuations of the local charge density and electric fields due to charged impurities in the substrate are further considered as a source of inhomogeneities.

Importantly, however, these sources of disorder can be mitigated and even almost completely suppressed by encapsulating monolayer TMDCs in high-quality hBN. hBN exhibits a smooth surface and provides homogeneous dielectric environment. As a consequence, encapsulation in hBN can lead to a strongly reduced inhomogeneous spectral broadening of the exciton transitions (Cadiz et al., 2017; Ajayi et al., 2017). It has also a substantial impact on the exciton diffusion, as shown in Fig. 3E (Cadiz et al., 2018; Raja et al., 2019). Room temperature diffusion coefficients increase from values on the order of 0.3 cm^2/s in pristine samples (both SiO_2/Si supported and freestanding) to 10 cm^2/s for the hBN-encapsulated ones. The latter can be also expressed as effective mobility on the order of 400 $cm^2/(Vs)$ according to $\mu = De/(k_B T)$ ($e > 0$ is the elementary charge). Despite the excitons not carrying an electric charge, this quantity can be useful for comparison to free carrier mobilities. In case of hBN-encapsulated TMDCs monolayers, the effective exciton mobility is thus comparable to the intrinsic limit given by carrier phonon scattering (Wang et al., 2021; Jin et al., 2014). We also note, however, that the exciton-phonon coupling that should determine the exciton diffusion is theoretically predicted to be substantially stronger than the scattering of free carriers with phonons. Thus, the exciton diffusion coefficients in the hBN-encapsulated samples measured at very low optical injection densities can be considered as too high, when compared with theoretical predictions. For example, a simple semiclassical model of $D = k_B T \times \tau_p / M$ (with τ_p being the scattering time and M exciton effective mass, see Eq. (12)) gives exciton diffusion coefficients on the order of $1\ \mathrm{cm}^2/\mathrm{s}$. This can mean that at very low electron-hole densities, there is a significant portion of unbound electron-hole pairs, as expected in thermal equilibrium by a process known as "entropy ionization" (Mock et al. 1978) or "exciton fission". In short, minimizing free energy requires a finite density of electrons and holes to stabilize bound states at finite temperatures. This is known as mass-action-law for chemical reactions or Saha equation for excitons (Kaindl et al. 2009). For a more detailed discussion of this process in 2D TMDCs and the underlying physics we refer the readers to Refs. (Zipfel et al., 2020; Steinhoff et al., 2017). We also note that the majority of propagating excitons are optically dark, either due to finite center-of-mass momentum outside of the radiative cone or due to the spin-selection rules of low lying states in W-based TMDCs. For these materials, the observed diffusion at room temperature should be understood as composite diffusion of several excitonic subspecies (Zipfel et al. 2020).

3.2 Non-linear diffusion

From the early studies of exciton dynamics in TMDCs it became clear that excitons interact efficiently with each other (Mouri et al., 2014; Kumar et al., 2014a; Sun et al., 2014). At elevated densities their recombination is strongly affected or even dominated by a process known as exciton-exciton annihilation. It is illustrated in Fig. 4A: if the exciton-exciton separation becomes comparable to the exciton diffusion length, the excitons can scatter inelastically with each other via Coulomb interaction. One exciton recombines non-radiatively by transferring the energy to the other that is excited to a higher energy state. Conceptually, this resembles a similar process for electrons in atoms and can thus be also referred to as Auger or Auger-Meitner recombination, see Section 4.3 and Eq. (18).

In the context of propagation exciton-exciton annihilation is known to affect the effectively measured broadening of the exciton distribution at elevated densities (Warren et al. 2000). For TMDCs this also leads to an apparent increase of the extracted diffusion coefficient with excitation density, as shown in Fig. 4B for SiO_2/Si-supported and freestanding WS_2 monolayers. The impact of exciton-exciton annihilation on the spatially- and time-dependent PL profiles is outlined in detail in Section 4 as a result of a bimolecular non-radiative process. It can be understood as follows: at sufficiently high densities the annihilation starts to contribute substantially to the exciton recombination. In the middle of the pump spot the excitons

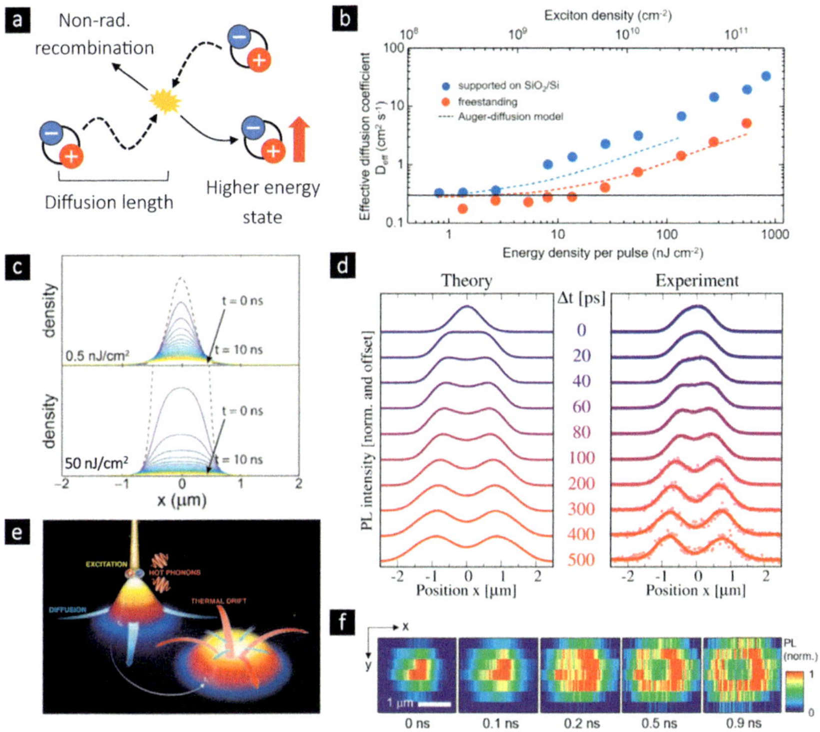

Fig. 4 (A) Schematic illustration of the exciton-exciton annihilation: one exciton recombines non-radiatively and the other is excited to a high-energy state. (B) Density-dependent effective diffusion coefficients in freestanding and SiO_2/Si supported WS_2 monolayers. Solid line corresponds to the linear diffusion coefficient that is constant at low densities. (C) Simulated profiles of the PL intensity in the presence of exciton-exciton annihilation at different times after the excitation for a low and high densities. (D) Theoretically calculated and experimentally measured profiles of the exciton emission in monolayer WS_2 for an estimated exciton density of 10^{12} cm^2/s. (E) Schematic illustration of the exciton halo formation from overheated excitons and Seebeck-like effect. (F) Series of images of normalized exciton PL forming a halo. All presented measurements were performed at room temperature. *(B, F) Reprinted figure with permission from Kulig et al. (2018). Copyright 2018 by the American Physical Society; (C) Reprinted with permission from Goodman et al. (2020b). Copyright 2020 American Chemical Society; (D, E) Reprinted with permission from Perea-Causín et al. (2019). Copyright 2019 American Chemical Society.*

thus decay faster than on the flanks where the density is smaller. As a consequence, the exciton distribution effectively broadens and acquires a super-Gaussian shape, i.e., a flat top. These changes of the profiles are illustrated in Fig. 4B by simulations from Ref. Goodman et al. (2020a), by comparing low and high excitation densities, where the exciton-exciton annihilation is either negligible or important. Moreover, the corresponding mean squared displacement becomes sub-linear. It increases rapidly at early times with the effective diffusion coefficient determined by the exciton-exciton annihilation rate and exciton density, Eq. (19), and more slowly at later times, converging to the slope given by linear diffusion, see Section 4.3. The impact of exciton-exciton annihilation is thus highly relevant to interpret observed broadening in the experiments as well as to identify the actual diffusion coefficient, either at low excitation densities or at later times after the excitation (assuming absence of trapping and detrapping (Seitz et al., 2021; Kurilovich et al., 2022)).

In addition, in the regime of sufficiently high excitation densities on the order of 100's of nJ/cm^2 in WS_2 and WSe_2 monolayers (Kulig et al., 2018; Rosati et al., 2021b), the exciton distribution can evolve into a ring-shaped profile. It corresponds to a double-peak structure in the PL cross section, see Fig. 4C, and can be also imaged in the 2D plane, see Fig. 4E. This halo is reversible with respect to the pump power and resembles similar observations in quantum well structures at cryogenic temperatures (Butov et al. 2004), albeit it appears at room temperature in 2D TMDCs. Importantly, a purely density-dependent non-linearity in the diffusion equation (see Section 4.3) does not lead to the appearance of a halo. As soon as the density distribution acquires a flat-top profile, the diffusion current disappears due to the lack of a population gradient. This requires a memory type of effect instead, i.e., the dynamics of excitons should be determined by events at earlier times. Here we also note that exciton halos and more complex shapes can also appear as a consequence of non-parabolic dispersion Qiu et al. (2021) in the regime of ballistic wave packet propagation. Since non-resonant excitation conditions are often used in experiments, photoexcited electron-hole pairs relax to the lowest energy state under the emission of phonons. These phonons can be reabsorbed by excitons and lead to a local heating of the exciton distribution, a phenomenon known as "hot-phonon effect" (Shah et al. 1970). Moreover, an additional key process that is independent from the excitation conditions is the exciton-exciton annihilation discussed above. Since one of the excitons gains an energy corresponding to the exciton transition energy on the order of 1.5–2 eV, depending on the material, the exciton-exciton annihilation can lead to a substantial heating of the exciton distribution by many 100's of K. The temperature gradient due to the overheated excitons in the middle of the excitation spot then leads to a Seebeck-like drift term in the diffusion equation, as schematically illustrated in Fig. 4D and outlined in detail in Ref. Perea-Causín et al. (2019). The resulting theoretically calculated profiles closely resemble experimental observation, c.f., Fig. 4C. Moreover, the expansion of the hot phonon cloud can push the excitons from the center and lead to the appearance of the halos (Glazov, 2019). This type of phenomena are known as "phonon-wind" (Smith et al. 1989) and "phonon-drag" effects corresponding to ballistic and diffusion regimes of phonon propagation, respectively. Finally we note that exciton-exciton annihilation is strongly reduced in hBN-encapsulated TMDCs (Zipfel et al., 2020; Hoshi et al., 2017); the microscopic origin of this effect may be related to the suppression of disorder-assisted scattering, reduced Coulomb-interaction strength or shifts of the higher energy levels out of resonance conditions (Han et al., 2018; Erkensten et al., 2021a). As a consequence, non-linear propagation phenomena related to exciton-exciton annihilation, such as density-dependent increase of the effective diffusion and halo formation are less pronounced or even suppressed in hBN-encapsulated samples (Zipfel et al. 2020).

3.3 Impact of free carriers and temperature dependence

In 2D TMDCs, the presence of free carriers is shown to strongly affect the exciton properties and the resulting optical response (Mak et al., 2013; Ross et al., 2013). It leads to the formation of composite states as a consequence of attractive exciton-carrier interaction. These can be either described in the molecular picture of a three-particle trion (Lampert, 1958; Kheng et al., 1993; Courtade et al., 2017) or as an exciton dressed by the electron-hole excitations of the Fermi sea, known as tetron or attractive Fermi-polaron (Suris, 2003; Koudinov et al., 2014; Sidler et al., 2017). The two descriptions are essentially equivalent at sufficiently low free carrier densities (Glazov, 2020b). Of particular importance for the propagation of excitonic complexes is their scattering with free charge carriers and the resulting changes of the center-of-mass momentum. The scattering is rather efficient for neutral excitons and increases linearly with the carrier density, as shown in Fig. 5A by calculations from Ref. Efimkin and MacDonald (2017). Understanding this increase in an analytical model requires description of the exciton-carrier scattering in the presence of additional dissipation (Wagner et al. 2023), for example by phonons. This is further illustrated in Fig. 5B for two different temperatures: the scattering rate with phonons is faster at 30 K compared to 5 K, leading to a higher slope of the exciton-carrier scattering rate (observed as an increase of

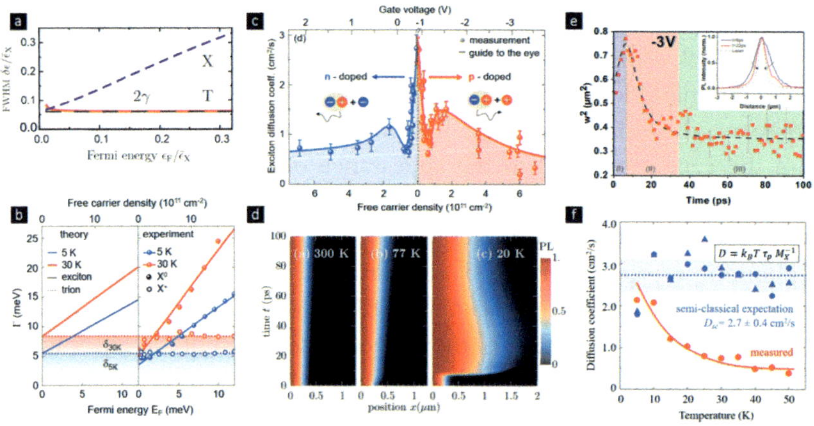

Fig. 5 (A) Calculated linewidths of excitons (X, repulsive Fermi polarons) and trions (T, attractive Fermi polarons) as function of the free carrier density as Fermi energy in units of hydrogen-like exciton binding energy. (B) Calculated (left panel) and measured (right panel) exciton-electron scattering rates and non-radiative spectral linewidths, respectively. The data is taken from bright exciton and trion states measured in reflectance on hBN-encapsulated monolayer WSe_2. (C) Measured diffusion coefficient in the spectral range of neutral and charged dark excitons in hBN-encapsulated monolayer WSe_2 as function of free carrier densities in a gate-tunable sample. (D) Calculated PL signal of a WS_2 monolayer, normalized at each time step, at different temperature as a function of spatial position and time. A narrowing of the spatial distribution is observed at the lowest temperature indicating the region of negative effective diffusion. (E) Squared width ($w^2 = 2\sigma^2$) of the measured bright trion PL in hBN-encapsulated WSe_2 as a function of time. (F) Temperature dependence of exciton diffusion in hBN-encapsulated WSe_2 monolayer (red points) and semiclassical expectation (blue points) from temperature-dependent linewidths. Lines are the guides for eye. *(A) Reprinted figure with permission from Efimkin and MacDonald (2017). Copyright 2017 by the American Physical Society; (B, C) Reprinted with permission from Wagner et al. (2023). Copyright 2023 American Chemical Society; (D) Used with permission of Royal Society of Chemistry, from Rosati et al. (2020b), permission conveyed through Copyright Clearance Center, Inc.; (E) Reprinted figure with permission from Beret et al. (2023). Copyright 2023 by the American Physical Society; (F) Reprinted figure with permission from Wagner et al. (2021). Copyright 2021 by the American Physical Society.*

spectral linewidth in the experiment). Importantly, in contrast to excitons, the scattering of trions or attractive Fermi polarons with free carriers is very weak, as also reported for ZnSe quantum wells (Astakhov et al. 2000) and monolayer TMDCs (Carbone et al., 2020; Wagner et al., 2020).

The different scattering rates of excitons and trions with free carriers are evident in diffusion. Fig. 5C shows the dependence of the diffusion coefficient in the spectral range of dark exciton complexes in hBN-encapsulated monolayer WSe_2 (Wagner et al. 2023). The data are obtained in the linear optical excitation regime in a gate-tunable sample at the lattice temperature of 5 K. As a consequence of efficient exciton-carrier scattering the diffusion coefficient decreases substantially with the initial increase of the free carrier density. At higher densities, as the majority of the excitons are bound to trions (attractive Fermi polarons), the scattering with free carriers becomes less efficient. The diffusion coefficient increases again and is followed by only a weak decrease with the free carrier density, both in *n*- and *p*-doped regimes. Here we also note that trions, formed at elevated carrier densities, propagate freely and no evidence of localization is observed (Wagner et al. 2023). Interestingly, in a material such as WSe_2, the interplay between bright and dark exciton states can even lead to an effectively negative diffusion during initial relaxation of the excitons, as theoretically predicted in Ref. Rosati et al. (2020a) and shown in Fig. 5D. This phenomenon arises from a rapid propagation of hot excitons that are short lived and cool down towards lower lying dark states with slower diffusion. This type of behavior is indeed observed for the first 20 ps after the excitation by detecting spatially-resolved distribution of bright trions in WSe_2 (Beret et al. 2023), see Fig. 5E.

While the impact of free carriers can be readily understood using semiclassical description of the diffusion, a question arises whether this description is generally valid for excitons in 2D TMDCs. As discussed in Ref. Glazov (2020a), the estimated mean free path of the excitons at low temperatures is comparable or smaller than the size of the semiclassical exciton wave packet given by the de Broglie wavelength. This roughly defines the applicability limit of the semiclassical description of quasiparticle transport (Mikhnenko et al. 2015), known as Mott-Ioffe-Regel criterion. For excitons in 2D TMDCs, quantum interference effects (i.e., weak localization) are predicted to provide corrections to the exciton diffusion coefficient at low temperatures, where the interaction with low-frequency acoustic phonons determines diffusion, see Section 4.2. In the experiment, the exciton diffusion coefficient is also found to deviate substantially from the expected semiclassical behavior at elevated temperatures in a similar manner (Wagner et al. 2021), as shown in Fig. 5F. This behavior is observed over 100's of ps after the excitation that allows the exciton system to cool down and not be affected by propagation of non-equilibrium and overheated excitons (Rosati et al., 2021b; Cordovilla Leon et al., 2019). The absolute change of the diffusion coefficient, however, is much larger than theoretically predicted, implying that other phenomena such as the interplay between different valleys can contribute to the observations. At the current stage, the microscopic description of this process in a comprehensive quantum mechanical picture including both realistic exciton bandstructures and quantum interference phenomena remains an open question.

3.4 Propagation of interlayer excitons in heterobilayers

When two TMDCs monolayers are combined, their respective valence and conduction bands are typically offset in the same direction with respect to energy, resulting in the type-II band alignment (Kang et al. 2013), see Fig. 6A. Upon optical excitation this leads to rapid charge separation (Hong et al. 2014) and formation of the interlayer excitons with spatially-separated electrons and holes located in different layers (Yu et al., 2015; Rivera et al., 2015). The interest in TMDCs heterobilayers and interlayer excitons is motivated by the possibility to tune their properties through angular and lattice mismatch (Yu et al., 2017; Wu et al., 2017; Zhang et al., 2018; Jin et al., 2019; Seyler et al., 2019; Alexeev et al., 2019; Tran et al., 2019) leading to observations of a variety of many-body phenomena (Xu et al., 2020; Regan et al., 2020; Tang et al., 2020) and offering additional pathways to manipulate exciton currents (Unuchek et al., 2018; Sun et al., 2022). These effects are intimately connected to both the nature of the excitonic states and their propagation dynamics in the heterostructures that is currently an active field of research (Rivera et al., 2016; Calman et al., 2018; Jauregui et al., 2019; Yuan et al., 2020; Choi et al., 2020; Li et al., 2021; Wang et al., 2021). In the following we summarize some of the key observations related to transport of interlayer excitons. We note, however, that these should strongly depend on the type and shape of the heterostructure in question, presence of moiré -type states (Yu et al., 2017; Wu et al., 2017) as well as local and mesoscopic atomic reconstruction (Weston et al., 2020; Rosenberger et al., 2020; Zhao et al., 2023).

Among initial observations for interlayer excitons were very long population and valley-polarization lifetimes of at least several 10's of nanoseconds (Rivera et al. 2016). During that time the PL of the interlayer excitons spreads to a region to a few micrometer diameter for a $MoSe_2/WSe_2$ heterobilayer at $T = 30$ K. Spin-valley polarization of the excitons, initially induced by a circularly polarized optical excitation, is largely conserved as the excitons propagate. Moreover, it shows a ring-shaped profile resembling the exciton population rings in monolayer materials evidencing valley-dependent interactions (Rivera et al. 2016). One type of the interaction between interlayer excitons is the

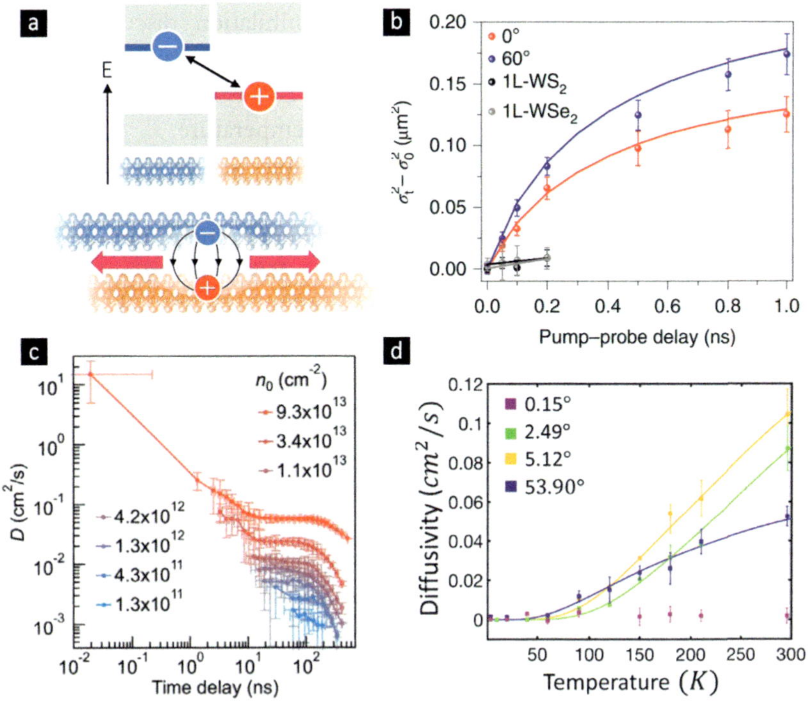

Fig. 6 (A) Schematic illustration of interlayer excitons formed at type-II transitions in TMDC heterobilayers. (B) Mean squared displacement of the exciton diffusion in CVD-grown WS_2/WSe_2 heterobilayer at T = 295 K for two different angular alignments and in comparison to the individual monolayer results. (C) Effective diffusion coefficients of interlayer excitons in mechanically exfoliated, hBN-encapsulated $MoSe_2$/WSe_2 heterobilayer at T = 4 K for a range of excitation densities up to the Mott transition. (D) Temperature dependent diffusion coefficients of interlayer excitons in mechanically exfoliated $MoSe_2$/WSe_2 heterobilayers for different twist angles. *(B) Reprinted with permission from Yuan et al. (2020); (C) Reprinted figure with permission from Wang et al. (2021). Copyright 2021 by the American Physical Society; (D) Reprinted with permission from Li et al. (2021). Copyright 2021 American Chemical Society.*

dipole-dipole repulsion, discussed extensively in the context of coupled quantum wells (Rapaport et al. 2006). In a simplified picture it arises due to the repulsion of the carriers of the same sign located in the same layer for two interlayer excitons in a type II heterostructures. We note, however, that additional contributions to the exciton-exciton interaction that are attractive were recently predicted as well (Erkensten et al. 2022). A consequence of the exciton-exciton repulsion is a faster initial broadening of the spatial distribution of excitons followed by slower dynamics as the excitons recombine and their density decreases. In that regard this is similar to the consequences of exciton-exciton annihilation discussed in the Sections 3.2 and 4.3, and Eq. (22), albeit it leads to an actual expansion of the exciton distribution. This type of behavior is illustrated in Fig. 6B for a WS_2/WSe_2 heterobilayer measured at room temperature.

In addition to interactions, interlayer excitons are often found to be localized at low densities and temperatures in TMDCs heterobilayers. As shown in Fig. 6C and D, the corresponding diffusion coefficients are on the order of 10^{-2} to 10^{-3} cm^2/s in this regime. With increasing excitation densities the role of the localization becomes less pronounced due to the interactions, state filling and additional many-body effects up to the density range of the Mott transitions, where excitons cease to be bound (Haug and Koch, 2009; Wang et al., 2019, 2021; Choi et al., 2023). Exciton localization is accompanied by a characteristic temperature dependence, demonstrated in Fig. 6D for $MoSe_2/WSe_2$ heterobilayers (Li et al. 2021). The exciton propagation is thermally activated and the diffusion coefficient increases strongly with temperature up to the values on the order of $0.1\,cm^2/s$ at room temperature. Combined with weak light-mater coupling, exciton localization naturally explains very long population lifetimes observed at low temperatures in TMDCs heterobilayers, preventing the excitons to quickly reach non-radiative traps and recombine (Zipfel et al. 2020).

4. Theory of exciton transport

4.1 Semiclassical model of exciton propagation

In the semiclassical regime excitons can be considered as wavepackets in the momentum and real spaces that propagate ballistically between the scattering events, Fig. 1C. In that case the velocity autocorrelation function decays exponentially with time and the diffusion coefficient takes the simple form

$$\langle\langle \hat{v}_x(t)\hat{v}_x(0)\rangle\rangle = \langle v_x^2\rangle e^{-|t|/\tau_p} \quad \Rightarrow \quad D = \frac{k_B T}{M}\tau_p. \tag{12}$$

Here, τ_p is the momentum relaxation time, i.e., the time between the scattering events where the exciton momentum or velocity significantly change and M is the exciton translational motion mass. One also takes into account that for non-degenerate excitons (those obeying Boltzmann distribution) statistical average of the velocity squared can be readily evaluated using the equipartition theorem. The semiclassical description is applicable provided that exciton kinetic energy, i.e., $k_B T$, exceeds the collisional broadening of excitonic states, $\hbar/\tau_p$. Equivalently, the mean free path $\ell = v_T \tau_p$, with $v_T = \sqrt{2k_B T/M}$ being the thermal velocity, should be ideally by far larger than the de Broglie wavelength $\lambda = 2\pi\hbar/\sqrt{2Mk_B T}$:

$$\frac{k_B T \tau_p}{\hbar} \gg 1 \quad \Leftrightarrow \quad \ell \gg \lambda. \tag{13}$$

As mentioned in Section 3.3, Eq. (13) is known as Mott-Ioffe-Regel criterion (Mikhnenko et al. 2015).

Consistent semiclassical description of the exciton propagation can be carried out using kinetic Boltzmann equation for the distribution function $f_{m,\boldsymbol{k}}(\boldsymbol{r}, t)$ where m enumerates exciton bands and spin states and $\boldsymbol{k}$ is the exciton momentum (Zipfel et al., 2020; Glazov, 2019; Perea-Causín et al., 2019; Rosati et al., 2021a, 2021b; Choi et al., 2023)

$$\frac{\partial f_{m,\boldsymbol{k}}}{\partial t} + \boldsymbol{v}_{m,\boldsymbol{k}} \cdot \frac{\partial f_{m,\boldsymbol{k}}}{\partial \boldsymbol{r}} + \frac{\boldsymbol{F}}{\hbar} \cdot \frac{\partial f_{m,\boldsymbol{k}}}{\partial \boldsymbol{k}} + \hat{Q}\{f_{m,\boldsymbol{k}}\} + \hat{R}\{f_{m,\boldsymbol{k}}\}$$
$$= G_{m,\boldsymbol{k}}(\boldsymbol{r}, t). \tag{14}$$

Here $\boldsymbol{v}_{m,\boldsymbol{k}} = \hbar^{-1}\partial E_{m,\boldsymbol{k}}/\partial \boldsymbol{k}$ is the exciton velocity, $E_{m,\boldsymbol{k}}$ is its dispersion, $G_{m,\boldsymbol{k}}(\boldsymbol{r}, t)$ stands for the generation rate, the operator $\hat{R}\{f_{m,\boldsymbol{k}}\}$ describes the recombination processes, and

$$\hat{Q}\{f_{m,\boldsymbol{k}}\} = \sum_{m',\boldsymbol{k}'} W_{m,\boldsymbol{k};m',\boldsymbol{k}'} f_{m,\boldsymbol{k}} - W_{m',\boldsymbol{k}';m,\boldsymbol{k}} f_{m',\boldsymbol{k}'}, \tag{15}$$

is the collision integral. It describes exciton transitions between different states due to the scattering by static defects, phonons, electrons, other excitons, etc., with $W_{m,\boldsymbol{k};m',\boldsymbol{k}'}$ being corresponding transition rates. For generality we have also included in Eq. (14) the external drag force $\boldsymbol{F}$ acting on excitons. Eqs. (14) and (15) together with either microscopically calculated or approximate, parameterized bandstructure and appropriately calculated scattering rates provide general framework for describing exciton propagation. Indeed, summation of Eq. (14) over $\boldsymbol{k}$ and m yields continuity Eq. (1), while microscopic expressions for the exciton flux and diffusion coefficient (2) can be derived by perturbative solution of Eq. (14) assuming that the spatial gradient term is small. Particularly, within a single band model and the effective mass approximation, where $E_{\boldsymbol{k}} = \hbar^2 k^2/2M$ and subscript m is omitted, we obtain Eq. (12) with

$$\frac{1}{\tau_p} = \sum_{\mathbf{k}'} W_{\mathbf{k},\mathbf{k}'}(1 - \cos\vartheta_{\mathbf{k},\mathbf{k}'}). \tag{16}$$

This microscopic expression for τ_p is derived under approximation of elastic scattering $E_{\mathbf{k}} = E_{\mathbf{k}'}$; $\vartheta_{\mathbf{k},\mathbf{k}'} = \angle\mathbf{k},\ \mathbf{k}'$ is the scattering angle. The same expression can be derived using the Greens function approach to evaluate the velocity autocorrelation function (6) if one keeps only ladder diagrams, see Ref. Glazov et al. (2022) for details.

An important consequence of the semiclassical model is the temperature dependence of the exciton diffusion coefficient. A simple, experimentally accessible scenario is the exciton diffusion in a monolayer semiconductor with suppressed static disorder. In this case, the exciton-phonon scattering provides the dominant channel for the momentum relaxation over wide temperature range. In particular, for sufficiently low temperatures the acoustic phonons with linear dispersion are the main source of scattering. In that case $\tau_p \propto T^{-1}$ and the expected semiclassical value of diffusivity becomes temperature independent (Wagner et al., 2021; Glazov, 2020a; Glazov et al., 2022), see Fig. 5F. With increase in the temperature the scattering by the zone-edge acoustic phonons and optical phonons come into play that should result in decreasing D with increasing temperature. The analysis of $D(T)$ for various scattering mechanisms for mono- and bilayer transition metal dichalcogenides is presented in Ref. Glazov et al. (2022). In any case, for semiclassical propagation the temperature dependence of the exciton diffusion coefficient is relatively weak and usually opposite to the exponential one in the hopping regime with thermally activated diffusion, Eq. (7).

4.2 Quantum effects

Interestingly, in 2D TMDCs the Mott-Ioffe-Regel condition (13) for the semiclassical propagation can be readily violated: in the relevant case of long-wavelength acoustic phonon scattering $\hbar/\tau_p = ck_BT$ with the material-dependent coefficient c on the order of unity. Thus, even in this case the semiclassical description becomes insufficient and quantum effects should be taken into account. The analysis shows that the main contribution beyond semiclassical Eq. (12) results from the interference of exciton de Broglie waves at the close-loop trajectories as illustrated in Fig. 1B: The phases of the waves propagating the same path co- and counter-clockwise are the same, giving rise to the constructive interference and, consequently, to an effectively reduced speed of propagation. This behavior is known as weak localization that can play a crucial role in the electron transport and is related to the coherent backscattering of light and matter waves propagating in disordered media (Gnedin and Dolginov, 1964; Ivchenko et al., 1977; Gor'kov et al., 1979; Golubentsev, 1984; Arseev and Dzyubenko, 1998). Weak localization serves as a precursor for strong, Anderson localization and takes an intermediate place in the transport regimes between the hopping, Fig. 1A, and semiclassical, Fig. 1C, propagation.

For 2D TMDCs the weak localization results in temperature-dependent correction to the classical diffusion coefficient (12) in the form (Glazov, 2020a; Glazov et al., 2022)

$$\delta D = -\frac{\hbar}{2\pi M}\ln\left(\frac{\tau_\phi}{\tau_p}\right) < 0, \tag{17}$$

where $\tau_\phi \gg \tau_p$ is the phase relaxation rate with respect to the complete loop between clock- and counter-clock-wise directions, and $|\delta D| \ll D$. The phase relaxation is related to the "inelastisity" of the exciton-acoustic phonon scattering. If the longitudinal acoustic phonon scattering dominates then $\tau_\phi/\tau_p \propto T^{-1/3}$ leading to a decrease of the diffusion coefficient $D + \delta D$ with increasing temperature (Glazov, 2020a; Glazov et al., 2022). At higher temperatures, this effect disappears due to the thermal activation of additional phonon modes with flat dispersion that results in strong inelasticity of the scattering and suppression of δD. Noteworthy, the weak localization affects the diffusion coefficient while keeping the exciton scattering time τ_p and, accordingly, the linewidth intact. As outlined in Section 3.3, it provides a partial, albeit quantitatively insufficient explanations for the deviations of the measured diffusion coefficients from the semiclassical behavior (Wagner et al. 2021). Altogether, it allows one to disentangle semiclassical and quantum contributions to the diffusion.

4.3 Nonlinear transport: exciton-exciton annihilation and repulsion

In the above section, we considered exciton diffusion as a single-particle effect. Such an approach is valid for low exciton densities. Increase in the exciton density, however, gives rise to exciton-exciton interactions which affect (i) exciton energies via repulsive Coulomb interactions (Erkensten et al., 2022; Shahnazaryan et al., 2017) and (ii) exciton population lifetimes via exciton-exciton annihilation (Han et al., 2018; Danovich et al., 2016; Erkensten et al., 2021b). From the formal viewpoint, exciton-exciton interactions affect both real and imaginary parts of the exciton self-energy.

Estimates show that main effect is on the exciton lifetime for the case of TMDCs monolayers, and $R\{n\}$ in Eq. (3) can be presented in effectively bimolecular approximation as

$$R\{n\} = \frac{n}{\tau} + R_A n^2, \tag{18}$$

where, as above, τ accounts for the single-exciton lifetime and R_A is the exciton-exciton annihilation coefficient. The term $R_A n^2$ describes the process where two excitons collide and one of them non-radiatively recombines, while another one takes the remaining energy and the momentum. Numerical and analytical solutions of diffusion Eq. (3) with the recombination in the form of Eq. (18) shows that exciton distribution becomes flat-top as compared to Gaussian (4) (Kulig et al., 2018; Perea-Causín et al., 2019; de Quilettes et al., 2022; Goodman et al., 2020a). The mean square displacement $\sigma^2(t)$ then deviates from the linear dependence on time. On very short time scales compared to $1/(R_A n^2)$, however, it can be described by an effective diffusion coefficient

$$D_{\text{eff}} = D + \frac{R_A}{8\pi} \int n(\boldsymbol{r}, 0)\, d\boldsymbol{r}. \tag{19}$$

As shown in the context of Fig. 4 in Section 3.2, such an increase is readily observed in TMDCs monolayers. Fast bimolecular recombination results over time in a decrease of the exciton density and, in turn, diminishes the role of the exciton-exciton annihilation effect on the effective diffusion. Hence, the D_{eff} decreases as a function of time down to the initial value D yielding overall subdiffusive behavior of $\sigma^2(t)$.

Let us now turn to the effect of the exciton energy variation as a function of density that can be represented as

$$\delta E(\boldsymbol{r}, t) = U_0 n(\boldsymbol{r}, t), \tag{20}$$

with the coefficient U_0. The energy variation is more prominent in heterostructures where interlayer excitons possess static dipole moments. Exciton-exciton interactions are typically considered as repulsive (Shahnazaryan et al., 2017; Erkensten et al., 2022) with $U_0 > 0$, albeit attractive contributions from exchange can lead to decrease of the overall interaction strength (Erkensten et al. 2022). Spatial variation of density then gives rise to the variation of energy and, accordingly, to the force $\boldsymbol{F} = -\partial \delta E/\partial \boldsymbol{r}$ acting on the excitons. This force is directed in the direction of decreasing density and thus effectively increases he rate of the exciton propagation. In the presence of such a force, the exciton current $\boldsymbol{i}$ in Eq. (2) acquires the drift contribution $\propto \boldsymbol{F}$ [cf. Eq. (14)]:

$$\boldsymbol{i} = -D\nabla n + \frac{D}{k_B T}\boldsymbol{F}n. \tag{21}$$

The relation of the coefficients for the diffusive and drift contributions is given by the Einstein relation stating that at the equilibrium conditions and fixed number of excitons the diffusive and drift currents should compensate each other. The analysis shows that the effective diffusion coefficient can be approximated as (Erkensten et al., 2022; Wietek et al., 2023)

$$D_{\text{eff}} = D + \frac{U_0 D}{k_B T} n(\mathbf{0}, 0). \tag{22}$$

The Auger and repulsive contributions to D_{eff}, Eqs. (19) and (22), can be largely considered as additive. However, they can be in principle separated based on the analysis of kinetics of excitons and their spatial profiles $n(\boldsymbol{r}, t)$.

It is worth stressing that exciton-exciton collisions do not directly contribute to τ_p and do not directly affect the actual, linear diffusion coefficient in Eq. (12) owing to momentum conservation (Gantmakher and Levinson, 1987; Wietek et al., 2023) unless there are several exciton ensembles with different dispersions and scattering mechanisms. It leads to a faster spread of the mean squared displacement that can be approximated by an *effective* diffusion coefficient, extracted from the slope. However, also the bare diffusion coefficient D can in principle become a function of density, e.g., due to the screening of the long-range disorder potential by excitons (Ivanov, 2002). Such propagation scenario deserves further studies especially for heterostructures with long-range moiré -type potentials. Finally, accounting for the external force $\boldsymbol{F}$ in the expression for the exciton flux density (21) makes it possible to theoretically study a range of situations where excitons are subject to external forces. These include exciton drift in strain-induced potentials (Rosati et al., 2021a; Moon et al., 2020), thermal drift (Seebeck effect) with $\boldsymbol{F} \propto \nabla T$ (Perea-Causín et al. 2019) as well as the phonon wind and drag effects (Glazov, 2019), eventually leading to the halo formation, see Section 3.2 and Fig. 4D–F.

4.4 Beyond diffusion

Despite its convenience, the model of diffusive propagation of excitons (3) has a number of limitations. They are related, on the one hand, to the assumption of small density gradients in Eqs. (2), (21), and, on the other, to perturbative treatment of interactions where the recombination rate (18) and energy shift (20) are decomposed in a density series.

Beyond these approximations, the regime of strong, non-perturbative, exciton-exciton interactions presents significant complexity from the theoretical viewpoint because of formation of collective condensate-like many-body states (Glazov and Suris, 2021). Without going into details of this growing field of research, we note that due to bosonic nature of excitons and repulsive interaction (at least for certain conditions), the fate of the excitons at low temperatures and elevated densities could be either a superfluid or a Wigner-crystal-like state (Lozovik and Yudson, 1975; Fogler et al., 2014; Glazov and Suris, 2021; Laikhtman and Rapaport, 2009). In the superfluid case, excitonic liquid can flow past obstacles with no friction, Fig. 1D. As a result, the exciton propagation is no longer diffusive. Rather, excitons expand quasi-ballistically driven by interaction-induced pressure with $\sigma(t) \sim c_s t$ where c_s is the speed of sound in the condensate (Kuznetsov and Kagan, 2020). An interesting scenario for future studies is the viscous hydrodynamics of excitons in clean systems where the exciton-exciton scattering rate exceeds their scattering rate by static defects and phonons, but the excitons still remain in an uncondensed, non-superfluid state.

Importantly, the ongoing development of high-level many-body perturbation theories based on *ab initio* approaches to bandstructure calculations naturally calls for fully microscopic calculations of the exciton transport. Such calculations are naturally highly involved since they require combining (i) the atomistic computation of the single particle states and phonons in the crystal from, e.g., density functional theory, ideally including many-body corrections, (ii) solution of the Bethe-Salpeter equation to obtain the excitons, and (iii) numerical calculation of dynamics of many-body wavefunction Ψ or density matrix ρ of the system. Particular complexity arises from the fact that, generally, phonon and exciton degrees of freedom are coupled. Still, recently Refs. (Qiu et al., 2021; Cohen et al., 2023) reported first results in this direction. Fig. 7 illustrates exciton propagation from the simulation of the wavefunction dynamics in a single layer MoS_2 [panel A] and from the density distribution dynamics in a model system of pentacene crystal [panel B]. Interestingly, exciton expansion shown in Fig. 7B directly demonstrates a crossover from ballistic propagation, $\sigma^2 \propto t^2$, at short time scales to diffusive propagation with $\sigma^2 \propto t$ at longer times when scattering processes come into play. The main qualitative features of exciton propagation are

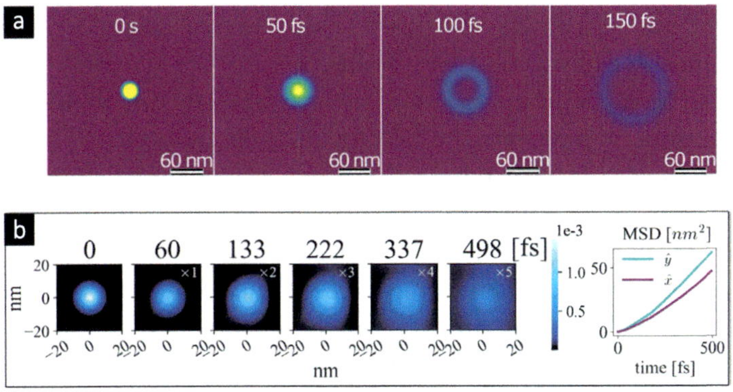

Fig. 7 *Ab initio* approaches to exciton propagation. (A) Time evolution of the amplitude squared ($|\Psi(\boldsymbol{r}, t)|^2$) of an initial Gaussian exciton wave packet for monolayer MoS_2. (B) Spatial propagation of exciton wavepacket in pentancene crystal during the first 500 fs. Blue distribution shows the bright state occupation, normalized and amplified with respect to the initial step. The mean squared displacement along the *x* (purple) and *y* (teal) are shown on the right. *(A) Reprinted from (Qiu et al. 2021) under the Creative Commons License. Copyright 2021 American Chemical Society; (B) Reprinted with permission from (Cohen et al. 2023).*

directly captured and there is no doubt that such methods will quickly develop and gain the predictive power to address a variety of complex systems and intriguing scenarios.

5. Conclusion

To conclude we presented an overview of basic effects encompassing transport of excitons in two-dimensional transition metal dichalcogenides. We have briefly addressed the methodological approaches both to measurements and theoretical analysis of exciton propagation, focusing on linear and non-linear diffusion as key phenomena. Several experimental manifestations of the exciton diffusion are presented including the impact of relevant conditions including temperature, disorder, and presence of free carriers. From the theoretical standpoint, we have outlined basic approached to describing exciton diffusion including semiclassical approach and its extensions. The role of nonlinearities, particularly, exciton-exciton annihilation and repulsion are discussed. The consequences of these processes for exciton propagation, including enhancement of the observed diffusion coefficient and generally subdiffusive behavior are addressed both from experimental and theoretical viewpoints.

While basic features of exciton propagation in transition-metal dichalcogenide monolayers and heterostructures are captured by state-of-the-art models, the research in this field is far from being complete. Possible avenues for future studies include development of deep understanding of the origins of non-classical exciton propagation and the transition between localization and free propagation, as well asstudies, both theoretical and experimental, of the exciton transport beyond the diffusive regime. Of particular importance are consequences of complex bandstructures, excitons with non-parabolic dispersion as well as hybridized excitons in heterostructures. Moreover, the impact of correlated states of free carriers such as the Wigner crystals observed both in monolayer materials and in moiré-type structures on the exciton propagation as well as the correlated states of the excitons themselves represent intriguing avenues of research. Propagation of spin-valley degrees of freedom offers equally exciting directions and interesting phenomena, including exciton valley-Hall effect and directional transport of valley pseudospins. One can also consider the interplay of different dimensionalities, from effectively zero- to one- and two-dimensions for excitons in TMDCs, that can be realized in this class of materials. Naturally, this offers only a brief glimpse into possible perspectives and directions currently pursued and those motivating future research in a rich and vibrant field of excitonic physics in 2D materials.

References

Ajayi, O.A., Ardelean, J.V., Shepard, G.D., Wang, J., Antony, A., Taniguchi, T., Watanabe, K., Heinz, T.F., Strauf, S., Zhu, X.-Y., Hone, J.C., 2017. Approaching the intrinsic photoluminescence linewidth in transition metal dichalcogenide monolayers. 2D Mater. 4 (3), 031011. https://doi.org/10.1088/2053-1583/aa6aa1.

Akmaev, M.A., Glazov, M.M., Kochiev, M.V., Vinokurov, P.V., Smagulova, S.A., Belykh, V.V., 2021. Spatiotemporal dynamics of free and bound excitons in CVD-grown MoS_2 monolayer. Appl. Phys. Lett. 119 (11), 113102. https://doi.org/10.1063/5.0063372.

Akselrod, G.M., Deotare, P.B., Thompson, N.J., Lee, J., Tisdale, W.A., Baldo, M.A., Menon, V.M., Bulović, V., 2014. Visualization of exciton transport in ordered and disordered molecular solids. Nat. Commun. 5, 3646. https://doi.org/10.1038/ncomms4646.

Alexeev, E.M., Ruiz-Tijerina, D.A., Danovich, M., Hamer, M.J., Terry, D.J., Nayak, P.K., Ahn, S., Pak, S., Lee, J., Sohn, J.I., Molas, M.R., Koperski, M., Watanabe, K., Taniguchi, T., Novoselov, K.S., Gorbachev, R.V., Shin, H.S., Fal'ko, V.I., Tartakovskii, A.I., 2019. Resonantly hybridized excitons in moiré superlattices in van der Waals heterostructures. Nature 567 (7746), 81–86. https://doi.org/10.1038/s41586-019-0986-9.

Arora, A., 2021. Magneto-optics of layered two-dimensional semiconductors and heterostructures: progress and prospects. J. Appl. Phys. 129 (12), 120902. https://doi.org/10.1063/5.0042683.

Arseev, P.I., Dzyubenko, A.B., 1998. Exciton magnetotransport in two-dimensional systems: weak-localization effects. J. Exp. Theor. Phys. 87 (1), 200–209. https://doi.org/10.1134/1.558641.

Astakhov, G.V., Kochereshko, V.P., Yakovlev, D.R., Ossau, W., Nürnberger, J., Faschinger, W., Landwehr, G., 2000. Oscillator strength of trion states in ZnSe-based quantum wells. Phys. Rev. B 62 (15), 10345. https://doi.org/10.1103/PhysRevB.62.10345.

Baranovski, S., 2006. (Ed.). Charge Transport in Disordered Solids with Applications in Electronics. Wiley.

Basov, D.N., Fogler, M.M., García de Abajo, F.J., 2016. Polaritons in van der Waals materials. Science 354 (6309), 195. https://doi.org/10.1126/science.aag1992.

Beret, D., Ren, L., Robert, C., Foussat, L., Renucci, P., Lagarde, D., Balocchi, A., Amand, T., Urbaszek, B., Watanabe, K., Taniguchi, T., Marie, X., Lombez, L., 2023. Nonlinear diffusion of negatively charged excitons in monolayer WSe_2. Phys. Rev. B 107 (4), 045420. https://doi.org/10.1103/PhysRevB.107.045420.

Blundo, E., Cappelluti, E., Felici, M., Pettinari, G., Polimeni, A., 2021. Strain-tuning of the electronic, optical, and vibrational properties of two-dimensional crystals. Appl. Phys. Rev. 8 (2), 021318. https://doi.org/10.1063/5.0037852. ⟨https://pubs.aip.org/aip/apr/article/1056778⟩.

Butov, L.V., Zrenner, A., Abstreiter, G., Böhm, G., Weimann, G., 1994. Condensation of indirect excitons in coupled AlAs/GaAs quantum wells. Phys. Rev. Lett. 73 (2), 304–307. https://doi.org/10.1103/PhysRevLett.73.304.

Butov, L.V., Levitov, L.S., Mintsev, A.V., Simons, B.D., Gossard, A.C., Chemla, D.S., 2004. Formation mechanism and low-temperature instability of exciton rings. Phys. Rev. Lett. 92 (11), 117404. https://doi.org/10.1103/PhysRevLett.92.117404.

Cadiz, F., Courtade, E., Robert, C., Wang, G., Shen, Y., Cai, H., Taniguchi, T., Watanabe, K., Carrere, H., Lagarde, D., Manca, M., Amand, T., Renucci, P., Tongay, S., Marie, X., Urbaszek, B., 2017. Excitonic linewidth approaching the homogeneous limit in MoS_2-based van der Waals heterostructures. Phys. Rev. X 7 (2), 021026. https://doi.org/10.1103/PhysRevX.7.021026.

Cadiz, F., Robert, C., Courtade, E., Manca, M., Martinelli, L., Taniguchi, T., Watanabe, K., Amand, T., Rowe, A.C.H., Paget, D., Urbaszek, B., Marie, X., 2018. Exciton diffusion in WSe_2 monolayers embedded in a van der Waals heterostructure. Appl. Phys. Lett. 112 (15), 152106. https://doi.org/10.1063/1.5026478.

Calman, E.V., Fogler, M.M., Butov, L.V., Hu, S., Mishchenko, A., Geim, A.K., 2018. Indirect excitons in van der Waals heterostructures at room temperature. Nat. Commun. 9 (1), 1895. https://doi.org/10.1038/s41467-018-04293-7.

Carbone, M.R., Mayers, M.Z., Reichman, D.R., 2020. Microscopic model of the doping dependence of linewidths in monolayer transition metal dichalcogenides. J. Chem. Phys. 152 (19), 194705. https://doi.org/10.1063/5.0008730.

Cheng, G., Li, B., Jin, Z., Zhang, M., Wang, J., 2021. Observation of diffusion and drift of the negative trions in monolayer WS_2. Nano Lett. 21 (14), 6314–6320. https://doi.org/10.1021/acs.nanolett.1c02351.

Choi, J., Hsu, W.-T., Lu, L.-S., Sun, L., Cheng, H.-Y., Lee, M.-H., Quan, J., Tran, K., Wang, C.-Y., Staab, M., Jones, K., Taniguchi, T., Watanabe, K., Chu, M.-W., Gwo, S., Kim, S., Shih, C.-K., Li, X., Chang, W.-H., 2020. Moiré potential impedes interlayer exciton diffusion in van der Waals heterostructures. Sci. Adv. 6 (39). https://doi.org/10.1126/sciadv.aba8866.

Choi, J., Embley, J., Blach, D.D., Perea-Causín, R., Erkensten, D., Kim, D.S., Yuan, L., Yoon, W.Y., Taniguchi, T., Watanabe, K., Ueno, K., Tutuc, E., Brem, S., Malic, E., Li, X., Huang, L., 2023. Fermi pressure and Coulomb repulsion driven rapid hot plasma expansion in a van der Waals heterostructure. Nano Lett. 23 (10), 4399–4405. https://doi.org/10.1021/acs.nanolett.3c00678.

Cohen, G., Haber, J.B., Neaton, J.B., Qiu, D.Y., Refaely-Abramson, S., 2023. Phonon-driven femtosecond dynamics of excitons in crystalline pentacene from first principles. https://doi.org/10.48550/arXiv.2305.04223.

Cordovilla Leon, D.F., Li, Z., Jang, S.W., Cheng, C.-H., Deotare, P.B., 2018. Exciton transport in strained monolayer WSe_2. Appl. Phys. Lett. 113 (25), 252101. https://doi.org/10.1063/1.5063263.

Cordovilla Leon, D.F., Li, Z., Jang, S.W., Deotare, P.B., 2019. Hot exciton transport in WSe_2 monolayers. Phys. Rev. B 100 (24), 241401. https://doi.org/10.1103/PhysRevB.100.241401.

Courtade, E., Semina, M., Manca, M., Glazov, M.M., Robert, C., Cadiz, F., Wang, G., Taniguchi, T., Watanabe, K., Pierre, M., Escoffier, W., Ivchenko, E.L., Renucci, P., Marie, X., Amand, T., Urbaszek, B., 2017. Charged excitons in monolayer WSe_2: experiment and theory. Phys. Rev. B 96. https://doi.org/10.1103/PhysRevB.96.085302.

Danovich, M., Zólyomi, V., Fal'ko, V.I., Aleiner, I.L., 2016. Auger recombination of dark excitons in WS_2 and WSe_2 monolayers. 2D Mater. 3 (3), 035011. https://iopscience.iop.org/article/10.1088/2053-1583/3/3/035011.

Datta, K., Lyu, Z., Li, Z., Taniguchi, T., Watanabe, K., Deotare, P.B., 2022. Spatiotemporally controlled room-temperature exciton transport under dynamic strain. Nat. Photon. 16 (3), 242–247. https://doi.org/10.1038/s41566-021-00951-3.

de Quilettes, D.W., Brenes, R., Laitz, M., Motes, B.T., Glazov, M.M., Bulović, V., 2022. Impact of photon recycling, grain boundaries, and nonlinear recombination on energy transport in semiconductors. ACS Photon. 9 (1), 110–122. https://doi.org/10.1021/acsphotonics.1c01186.

Delor, M., Weaver, H.L., Yu, Q., Ginsberg, N.S., 2020. Imaging material functionality through three-dimensional nanoscale tracking of energy flow. Nat. Mater. 19 (1), 56–62. https://doi.org/10.1038/s41563-019-0498-x.

Deng, S., Shi, E., Yuan, L., Jin, L., Dou, L., Huang, L., 2020. Long-range exciton transport and slow annihilation in two-dimensional hybrid perovskites. Nat. Commun. 11 (1), 664. https://doi.org/10.1038/s41467-020-14403-z.

Dirnberger, F., Ziegler, J.D., Faria Junior, P.E., Bushati, R., Taniguchi, T., Watanabe, K., Fabian, J., Bougeard, D., Chernikov, A., Menon, V.M., 2021. Quasi-1D exciton channels in strain-engineered 2D materials. Sci. Adv. 7 (44). https://doi.org/10.1126/sciadv.abj3066.

Efimkin, D.K., MacDonald, A.H., 2017. Many-body theory of trion absorption features in two-dimensional semiconductors. Phys. Rev. B 95 (3), 035417. https://doi.org/10.1103/PhysRevB.95.035417.

Erkensten, D., Brem, S., Wagner, K., Gillen, R., Perea-Causín, R., Ziegler, J.D., Taniguchi, T., Watanabe, K., Maultzsch, J., Chernikov, A., Malic, E., 2021a. Dark exciton-exciton annihilation in monolayer WSe_2. Phys. Rev. B 104 (24), L241406. https://doi.org/10.1103/PhysRevB.104.L241406.

Erkensten, D., Brem, S., Wagner, K., Gillen, R., Perea-Causín, R., Ziegler, J.D., Taniguchi, T., Watanabe, K., Maultzsch, J., Chernikov, A., Malic, E., 2021b. Dark exciton-exciton annihilation in monolayer WSe_2. Phys. Rev. B 104, L241406. https://doi.org/10.1103/PhysRevB.104.L241406.

Erkensten, D., Brem, S., Perea-Causín, R., Malic, E., 2022. Microscopic origin of anomalous interlayer exciton transport in van der Waals heterostructures. Phys. Rev. Mater. 6, 094006. https://doi.org/10.1103/PhysRevMaterials.6.094006.

Erland, J., Razbirin, B.S., Pantke, K.-H., Lyssenko, V.G., Hvam, J.M., 1993. Exciton diffusion in CdSe. Phys. Rev. B 47 (7), 3582–3587. https://doi.org/10.1103/PhysRevB.47.3582.

Fogler, M.M., Butov, L.V., Novoselov, K.S., 2014. Indirect excitons in van der Waals heterostructures. Nat. Commun. 5, 4555. https://doi.org/10.1038/ncomms5555.

Frenkel, J., 1931. On the transformation of light into heat in solids. Phys. Rev. 37 (1), 17–44. https://doi.org/10.1103/PhysRev.37.17.

Fu, Y., He, D., He, J., Bian, A., Zhang, L., Liu, S., Wang, Y., Zhao, H., 2019. Effect of dielectric environment on excitonic dynamics in monolayer WS_2. Adv. Mater. Interfaces, 1901307. https://doi.org/10.1002/admi.201901307.

Gantmakher, V.F., Levinson, Y.B., 1987. Carrier Scattering in Metals and Semiconductors. North-Holland Physics Publishing, Amsterdam.

Ginsberg, N.S., Tisdale, W.A., 2020. Spatially resolved photogenerated exciton and charge transport in emerging semiconductors. Ann. Rev. Phys. Chem. 71 (1), 1–30. https://doi.org/10.1146/annurev-physchem-052516-050703.

Glazov, M.M., Suris, R.A., 2021. Collective states of excitons in semiconductors. Phys. Uspekhi 63 (11), 1051. https://doi.org/10.3367/UFNe.2019.10.038663.

Glazov, M.M., Iakovlev, Z.A., Refaely-Abramson, S., 2022. Phonon-induced exciton weak localization in two-dimensional semiconductors. Appl. Phys. Lett. 121 (19), 192106. https://doi.org/10.1063/5.0122633.

Glazov, M.M., 2019. Phonon wind and drag of excitons in monolayer semiconductors. Phys. Rev. B 100 (4), 045426. https://doi.org/10.1103/PhysRevB.100.045426.

Glazov, M.M., 2020a. Quantum interference effect on exciton transport in monolayer semiconductors. Phys. Rev. Lett. 124 (16), 166802. https://doi.org/10.1103/PhysRevLett.124.166802.

Glazov, M.M., 2020b. Optical properties of charged excitons in two-dimensional semiconductors. J. Chem. Phys. 153 (3), 034703. https://doi.org/10.1063/5.0012475.

Gnedin, Y.N., Dolginov, A., 1964. Theory of multiple scattering. JETP 18, 784. http://www.jetp.ras.ru/cgi-bin/dn/e_018_03_0784.pdf.

Golubentsev, A.A., 1984. Suppression of interference effects in multiple scattering of light. JETP 59, 26. http://www.jetp.ras.ru/cgi-bin/dn/e_059_01_0026.pdf.

Goodman, A.J., Lien, D.-H., Ahn, G.H., Spiegel, L.L., Amani, M., Willard, A.P., Javey, A., Tisdale, W.A., 2020a. Substrate-dependent exciton diffusion and annihilation in chemically treated MoS_2 and WS_2. J. Phys. Chem. C 124 (22), 12175–12184. https://doi.org/10.1021/acs.jpcc.0c04000.

Goodman, A.J., Lien, D.-H., Ahn, G.H., Spiegel, L.L., Amani, M., Willard, A.P., Javey, A., Tisdale, W.A., 2020b. Substrate-dependent exciton diffusion and annihilation in chemically treated MoS_2 and WS_2. J. Phys. Chem. C 124 (22), 12175–12184. https://doi.org/10.1021/acs.jpcc.0c04000.

Gor'kov, L., Larkin, A., Khmelnitskii, D., 1979. Particle conductivity in a two-dimensional random potential. JETP Lett. 30, 228. https://doi.org/10.1142/9789814317344_0022.

Green, M.S., 1954. Markoff random processes and the statistical mechanics of time-dependent phenomena. II. Irreversible processes in fluids. J. Chem. Phys. 22 (3), 398–413. https://doi.org/10.1063/1.1740082.

Gross, E.F., Karrjew, N.A., 1952. Light absorption by cuprous oxide crystal in infrared and visible part of the spectrum. Dokl. Akad. Nauk SSSR 84, 471.

Han, B., Robert, C., Courtade, E., Manca, M., Shree, S., Amand, T., Renucci, P., Taniguchi, T., Watanabe, K., Marie, X., Golub, L.E., Glazov, M.M., Urbaszek, B., 2018. Exciton states in monolayer $MoSe_2$ and $MoTe_2$ probed by upconversion spectroscopy. Phys. Rev. X 8 (3), 031073. https://doi.org/10.1103/PhysRevX.8.031073.

Hao, S., Bellus, M.Z., He, D., Wang, Y., Zhao, H., 2020. Controlling exciton transport in monolayer $MoSe_2$ by dielectric screening. Nanoscale Horizons 5 (1), 139–143. https://doi.org/10.1039/C9NH00462A.

Haug, H., Koch, S.W., 2009. Quantum Theory of the Optical and Electronic Properties of Semiconductors, fifth ed. World Scientific, Singapore.

Hillmer, H., Hansmann, S., Forchel, A., Morohashi, M., Lopez, E., Meier, H.P., Ploog, K., 1988. Two-dimensional exciton transport in GaAs/GaAlAs quantum wells. Appl. Phys. Lett. 53 (20), 1937–1939. https://doi.org/10.1063/1.100328.

Hong, X., Kim, J., Shi, S.-F., Zhang, Y., Jin, C., Sun, Y., Tongay, S., Wu, J., Zhang, Y., Wang, F., 2014. Ultrafast charge transfer in atomically thin MoS_2/WS_2 heterostructures. Nat. Nanotechnol. 9 (9), 682–686. https://doi.org/10.1038/nnano.2014.167. ⟨https://www.nature.com/articles/nnano.2014.167⟩.

Hoshi, Y., Kuroda, T., Okada, M., Moriya, R., Masubuchi, S., Watanabe, K., Taniguchi, T., Kitaura, R., Machida, T., 2017. Suppression of exciton-exciton annihilation in tungsten disulfide monolayers encapsulated by hexagonal boron nitrides. Phys. Rev. B 95 (24), 241403. https://doi.org/10.1103/PhysRevB.95.241403.

Hotta, T., Higuchi, S., Ueda, A., Shinokita, K., Miyauchi, Y., Matsuda, K., Ueno, K., Taniguchi, T., Watanabe, K., Kitaura, R., 2020. Exciton diffusion in hBN-encapsulated monolayer $MoSe_2$. Phys. Rev. B 102 (11), 115424. https://doi.org/10.1103/PhysRevB.102.115424.

Ivanov, A.L., 2002. Quantum diffusion of dipole-oriented indirect excitons in coupled quantum wells. Europhys. Lett. (EPL) 59 (4), 586–591. https://doi.org/10.1209/epl/i2002-00144-3.

Ivchenko, E.L., Pikus, G.E., Razbirin, B.S., Starukhin, A.I., 1977. Optical orientation and alignment of free excitons in GaSe under resonant excitation theory. JETP 45, 1172. https://ui.adsabs.harvard.edu/abs/1977ZhETF..72.2230I/abstract.

Ivchenko, E.L., 2005. Optical Spectroscopy of Semiconductor Nanostructures. Alpha Science, Harrow UK.

Jauregui, L.A., Joe, A.Y., Pistunova, K., Wild, D.S., High, A.A., Zhou, Y., Scuri, G., De Greve, K., Sushko, A., Yu, C.-H., Taniguchi, T., Watanabe, K., Needleman, D.J., Lukin, M.D., Park, H., Kim, P., 2019. Electrical control of interlayer exciton dynamics in atomically thin heterostructures. Science 366 (6467), 870–875. https://doi.org/10.1126/science.aaw4194.

Jin, Z., Li, X., Mullen, J.T., Kim, K.W., 2014. Intrinsic transport properties of electrons and holes in monolayer transition-metal dichalcogenides. Phys. Rev. B 90 (4), 045422. https://doi.org/10.1103/PhysRevB.90.045422.

Jin, C., Regan, E.C., Yan, A., Iqbal Bakti Utama, M., Wang, D., Zhao, S., Qin, Y., Yang, S., Zheng, Z., Shi, S., Watanabe, K., Taniguchi, T., Tongay, S., Zettl, A., Wang, F., 2019. Observation of moiré excitons in WSe_2/WS_2 heterostructure superlattices. Nature 567 (7746), 76–80. https://doi.org/10.1038/s41586-019-0976-y.

Kaindl, R.A., Hägele, D., Carnahan, M.A., Chemla, D.S., 2009. Transient terahertz spectroscopy of excitons and unbound carriers in quasi-two-dimensional electron-hole gases. Phys. Rev. B. 79 (4), 1–13. https://doi.org/10.1103/PhysRevB.79.045320.

Kang, J., Tongay, S., Zhou, J., Li, J., Wu, J., 2013. Band offsets and heterostructures of two-dimensional semiconductors. Appl. Phys. Lett. 102 (1), 012111. https://doi.org/10.1063/1.4774090.

Kato, T., Kaneko, T., 2016. Transport dynamics of neutral excitons and trions in monolayer WS_2. ACS Nano 10 (10), 9687–9694. https://doi.org/10.1021/acsnano.6b05580.

Kheng, K., Cox, R., d'Aubigné, M., Bassani, F., Saminadayar, K., Tatarenko, S., 1993. Observation of negatively charged excitons X_ in semiconductor quantum wells. Phys. Rev. Lett. 71 (11), 1752–1755. https://doi.org/10.1103/PhysRevLett.71.1752.

Koudinov, A.V., Kehl, C., Rodina, A.V., Geurts, J., Wolverson, D., Karczewski, G., 2014. Suris tetrons: possible spectroscopic evidence for four-particle optical excitations of a two-dimensional electron gas. Phys. Rev. Lett. 112 (14), 147402. https://doi.org/10.1103/PhysRevLett.112.147402.

Kubo, R., 1957. Statistical-mechanical theory of irreversible processes. I. General theory and simple applications to magnetic and conduction problems. J. Phys. Soc. Japan 12 (6), 570–586. https://doi.org/10.1143/JPSJ.12.570.

Kubo, R., 1966. The fluctuation-dissipation theorem. Rep. Prog. Phys. 29 (1), 255. ⟨http://stacks.iop.org/0034-4885/29/i=1/a=306⟩.

Kulig, M., Zipfel, J., Nagler, P., Blanter, S., Schüller, C., Korn, T., Paradiso, N., Glazov, M.M., Chernikov, A., 2018. Exciton diffusion and halo effects in monolayer semiconductors. Phys. Rev. Lett. 120 (20), 207401. https://doi.org/10.1103/PhysRevLett.120.207401.

Kumar, N., Cui, Q., Ceballos, F., He, D., Wang, Y., Zhao, H., 2014a. Exciton-exciton annihilation in $MoSe_2$ monolayers. Phys. Rev. B 89 (12), 125427. https://doi.org/10.1103/PhysRevB.89.125427.

Kumar, N., Cui, Q., Ceballos, F., He, D., Wang, Y., Zhao, H., 2014b. Exciton diffusion in monolayer and bulk $MoSe_2$. Nanoscale 6, 4915–4919. https://doi.org/10.1039/c3nr06863c.

Kurilovich, A.A., Mantsevich, V.N., Mardoukhi, Y., Stevenson, K.J., Chechkin, A.V., Palyulin, V.V., 2022. Non-Markovian diffusion of excitons in layered perovskites and transition metal dichalcogenides. Phys. Chem. Chem. Phys. 24 (22), 13941–13950. https://doi.org/10.1039/D2CP00557C.

Kuznetsov, E.A., Kagan, M.Y., 2020. Semiclassical expansion of quantum gases into a vacuum. Theor. Math. Phys. 202 (3), 399–411. https://doi.org/10.1134/S0040577920030125.

Laikhtman, B., Rapaport, R., 2009. Exciton correlations in coupled quantum wells and their luminescence blue shift. Phys. Rev. B 80, 195313. https://doi.org/10.1103/PhysRevB.80.195313.

Lampert, M.A., 1958. Mobile and immobile effective-mass-particle complexes in nonmetallic solids. Phys. Rev. Lett. 1 (12), 450–453. https://doi.org/10.1103/PhysRevLett.1.450.

Li, Z., Lu, X., CordovillaLeon, D.F., Lyu, Z., Xie, H., Hou, J., Lu, Y., Guo, X., Kaczmarek, A., Taniguchi, T., Watanabe, K., Zhao, L., Yang, L., Deotare, P.B., 2021. Interlayer exciton transport in $MoSe_2/WSe_2$ heterostructures. ACS Nano 15 (1), 1539–1547. https://doi.org/10.1021/acsnano.0c08981.

Lozovik, Y.E., Yudson, V.I., 1975. Feasibility of superfluidity of paired spatially separated electrons and holes; a new superconductivity mechanism. JETP Lett. 22, 274. https://www.osti.gov/biblio/7285279.

Mak, K.F., Lee, C., Hone, J., Shan, J., Heinz, T.F., 2010. Atomically thin MoS_2: a new direct-gap semiconductor. Phys. Rev. Lett. 105 (13), 136805. https://doi.org/10.1103/PhysRevLett.105.136805.

Mak, K.F., He, K., Lee, C., Lee, G.H., Hone, J., Heinz, T.F., Shan, J., 2013. Tightly bound trions in monolayer MoS_2. Nat. Mater. 12 (12), 207–211. https://doi.org/10.1038/nmat3505.

Mikhnenko, O.V., Blom, P.W.M., Nguyen, T.-Q., 2015. Exciton diffusion in organic semiconductors. Energy Environ. Sci. 8 (7), 1867–1888. https://doi.org/10.1039/C5EE00925A.

Mock, J., Thomas, G., Combescot, M., 1978. Entropy ionization of an exciton gas. Solid State Commun. 25 (5), 279–282. https://doi.org/10.1016/0038-1098(78)90958-4.

Moon, H., Grosso, G., Chakraborty, C., Peng, C., Taniguchi, T., Watanabe, K., Englund, D., 2020. Dynamic exciton funneling by local strain control in a monolayer semiconductor. Nano Lett. 20 (9), 6791–6797. https://doi.org/10.1021/acs.nanolett.0c02757.

Mouri, S., Miyauchi, Y., Toh, M., Zhao, W., Eda, G., Matsuda, K., 2014. Nonlinear photoluminescence in atomically thin layered WSe_2 arising from diffusion-assisted exciton-exciton annihilation. Phys. Rev. B 90 (15), 155449. https://doi.org/10.1103/PhysRevB.90.155449.

Mueller, T., Malic, E., 2018. Exciton physics and device application of two-dimensional transition metal dichalcogenide semiconductors. npj 2D Mater. Appl. 2 (1), 29. https://doi.org/10.1038/s41699-018-0074-2.

Novoselov, K.S., Jiang, D., Schedin, F., Booth, T.J., Khotkevich, V.V., Morozov, S.V., Geim, A.K., 2005. Two-dimensional atomic crystals. Proc. Natl. Acad. Sci. U. S. A. 102 (30), 10451–10453. https://doi.org/10.1073/pnas.0502848102.

Perea-Causín, R., Brem, S., Rosati, R., Jago, R., Kulig, M., Ziegler, J.D., Zipfel, J., Chernikov, A., Malic, E., 2019. Exciton propagation and halo formation in two-dimensional materials. Nano Lett. 19 (10), 7317–7323. https://doi.org/10.1021/acs.nanolett.9b02948.

Pitaevskii, L., Stringari, S., 2003. Bose-Einstein Condensation. Clarendon Press, Oxford.

Qiu, D.Y., Cohen, G., Novichkova, D., Refaely-Abramson, S., 2021. Signatures of dimensionality and symmetry in exciton band structure: consequences for exciton dynamics and transport. Nano Lett. 21 (18), 7644–7650. https://doi.org/10.1021/acs.nanolett.1c02352.

Raja, A., Chaves, A., Yu, J., Arefe, G., Hill, H.M., Rigosi, A.F., Berkelbach, T.C., Nagler, P., Schüller, C., Korn, T., Nuckolls, C., Hone, J., Brus, L.E., Heinz, T.F., Reichman, D.R., Chernikov, A., 2017. Coulomb engineering of the bandgap and excitons in two-dimensional materials. Nat. Commun. 8, 15251. https://doi.org/10.1038/ncomms15251.

Raja, A., Waldecker, L., Zipfel, J., Cho, Y., Brem, S., Ziegler, J.D., Kulig, M., Taniguchi, T., Watanabe, K., Malic, E., Heinz, T.F., Berkelbach, T.C., Chernikov, A., 2019. Dielectric disorder in two-dimensional materials. Nat. Nanotechnol. 14 (9), 832–837. https://doi.org/10.1038/s41565-019-0520-0.

Rapaport, R., Chen, G., Snoke, D., Simon, S.H., Pfeiffer, L., West, K., Liu, Y., Denev, S., 2004. Charge separation of dense two-dimensional electron-hole gases: mechanism for exciton ring pattern formation. Phys. Rev. Lett. 92 (11), 117405. https://doi.org/10.1103/PhysRevLett.92.117405.

Rapaport, R., Chen, G., Simon, S.H., 2006. Nonlinear dynamics of a dense two-dimensional dipolar exciton gas. Phys. Rev. B 73 (3), 033319. https://doi.org/10.1103/PhysRevB.73.033319.

Regan, E.C., Wang, D., Jin, C., Bakti Utama, M.I., Gao, B., Wei, X., Zhao, S., Zhao, W., Zhang, Z., Yumigeta, K., Blei, M., Carlström, J.D., Watanabe, K., Taniguchi, T., Tongay, S., Crommie, M., Zettl, A., Wang, F., 2020. Mott and generalized Wigner crystal states in WSe_2/WS_2 moiré superlattices. Nature 579 (7799), 359–363. https://doi.org/10.1038/s41586-020-2092-4.

Rivera, P., Schaibley, J.R., Jones, A.M., Ross, J.S., Wu, S., Aivazian, G., Klement, P., Seyler, K., Clark, G., Ghimire, N.J., Yan, J., Mandrus, D.G., Yao, W., Xu, X., 2015. Observation of long-lived interlayer excitons in monolayer $MoSe_2$-WSe_2 heterostructures. Nat. Commun. 6, 6242. https://doi.org/10.1038/ncomms7242.

Rivera, P., Seyler, K.L., Yu, H., Schaibley, J.R., Yan, J., Mandrus, D.G., Yao, W., Xu, X., 2016. Valley-polarized exciton dynamics in a 2D semiconductor heterostructure. Science 351 (6274), 688–691. https://doi.org/10.1126/science.aac7820.

Rosati, R., Wagner, K., Brem, S., Perea-Causín, R., Wietek, E., Zipfel, J., Ziegler, J.D., Selig, M., Taniguchi, T., Watanabe, K., Knorr, A., Chernikov, A., Malic, E., 2020a. Temporal evolution of low-temperature phonon sidebands in transition metal dichalcogenides. ACS Photonics 7 (10), 2756–2764. https://doi.org/10.1021/acsphotonics.0c00866.

Rosati, R., Perea-Causín, R., Brem, S., Malic, E., 2020b. Negative effective excitonic diffusion in monolayer transition metal dichalcogenides. Nanoscale 12 (1), 356–363. https://doi.org/10.1039/C9NR07056G. ⟨http://xlink.rsc.org/?DOI=C9NR07056G⟩.

Rosati, R., Schmidt, R., Brem, S., Perea-Causín, R., Niehues, I., Kern, J., Preuß, J.A., Schneider, R., Michaelis de Vasconcellos, S., Bratschitsch, R., Malic, E., 2021a. Dark exciton anti-funneling in atomically thin semiconductors. Nat. Commun. 12 (1), 7221. https://doi.org/10.1038/s41467-021-27425-y.

Rosati, R., Brem, S., Perea-Causín, R., Schmidt, R., Niehues, I., Michaelis de Vasconcellos, S., Bratschitsch, R., Malic, E., 2021b. Strain-dependent exciton diffusion in transition metal dichalcogenides. 2D Mater. 8 (1), 015030. https://doi.org/10.1088/2053-1583/abbd51.

Rosenberger, M.R., Chuang, H.-J., Phillips, M., Oleshko, V.P., McCreary, K.M., Sivaram, S.V., Hellberg, C.S., Jonker, B.T., 2020. Twist angle-dependent atomic reconstruction and Moiré patterns in transition metal dichalcogenide heterostructures. ACS Nano 14 (4), 4550–4558. https://doi.org/10.1021/acsnano.0c00088.

Ross, J.S., Wu, S., Yu, H., Ghimire, N.J., Jones, A.M., Aivazian, G., Yan, J., Mandrus, D.G., Xiao, D., Yao, W., Xu, X., 2013. Electrical control of neutral and charged excitons in a monolayer semiconductor. Nat. Commun. 4, 1474. https://doi.org/10.1038/ncomms2498.

Seitz, M., Magdaleno, A.J., Alcázar-Cano, N., Meléndez, M., Lubbers, T.J., Walraven, S.W., Pakdel, S., Prada, E., Delgado-Buscalioni, R., Prins, F., 2020. Exciton diffusion in two-dimensional metal-halide perovskites. Nat. Commun. 11 (1), 2035. https://doi.org/10.1038/s41467-020-15882-w.

Seitz, M., Meléndez, M., Alcázar-Cano, N., Congreve, D.N., Delgado-Buscalioni, R., Prins, F., 2021. Mapping the trap-state landscape in 2D metal-halide perovskites using transient photoluminescence microscopy. Optical Mater. 9 (18), 2001875. https://doi.org/10.1002/adom.202001875.

Selig, M., Berghäuser, G., Raja, A., Nagler, P., Schüller, C., Heinz, T.F., Korn, T., Chernikov, A., Malic, E., Knorr, A., 2016. Excitonic linewidth and coherence lifetime in monolayer transition metal dichalcogenides. Nat. Commun. 7, 13279. https://doi.org/10.1038/ncomms13279.

Seyler, K.L., Rivera, P., Yu, H., Wilson, N.P., Ray, E.L., Mandrus, D.G., Yan, J., Yao, W., Xu, X., 2019. Signatures of moiré-trapped valley excitons in $MoSe_2/WSe_2$ heterobilayers. Nature 567 (7746), 66–70. https://doi.org/10.1038/s41586-019-0957-1.

Shah, J., Leite, R., Scott, J., 1970. Photoexcited hot LO phonons in GaAs. Solid State Commun. 8 (14), 1089–1093. https://doi.org/10.1016/0038-1098(70)90002-5.

Shahnazaryan, V., Iorsh, I., Shelykh, I.A., Kyriienko, O., 2017. Exciton-exciton interaction in transition-metal dichalcogenide monolayers. Phys. Rev. B 96, 115409. https://doi.org/10.1103/PhysRevB.96.115409.

Shahnazaryan, V., Kyriienko, O., Rostami, H., 2019. Exciton routing in the heterostructure of a transition metal dichalcogenide monolayer on a paraelectric substrate. Phys. Rev. B 100 (16), 165303. https://doi.org/10.1103/PhysRevB.100.165303.

Shklovskii, B.I., Efros, A.L., 1984. Electronic Properties of Doped Semiconductors, Vol. 45 of Springer Series in Solid-State Sciences. Springer Berlin Heidelberg, Berlin, Heidelberg. https://doi.org/10.1007/978-3-662-02403-4.

Sidler, M., Back, P., Cotlet, O., Srivastava, A., Fink, T., Kroner, M., Demler, E., Imamoglu, A., 2017. Fermi polaron-polaritons in charge-tunable atomically thin semiconductors. Nat. Phys. 13 (3), 255–261. https://doi.org/10.1038/nphys3949. ⟨http://www.nature.com/articles/nphys3949⟩.

Smith, L.M., Wake, D.R., Wolfe, J.P., Levi, D., Klein, M.V., Klem, J., Henderson, T., Morkoç, H., 1988. Picosecond imaging of photoexcited carriers in quantum wells: anomalous lateral confinement at high densities. Phys. Rev. B 38 (8), 5788–5791. https://doi.org/10.1103/PhysRevB.38.5788.

Smith, L.M., Preston, J.S., Wolfe, J.P., Wake, D.R., Klem, J., Henderson, T., Morkoç, H., 1989. Phonon-wind-driven transport of photoexcited carriers in a semiconductor quantum well. Phys. Rev. B 39 (3), 1862–1870. https://doi.org/10.1103/PhysRevB.39.1862.

Soubelet, P., Klein, J., Wierzbowski, J., Silvioli, R., Sigger, F., Stier, A.V., Gallo, K., Finley, J.J., 2021. Charged exciton kinetics in monolayer $MoSe_2$ near ferroelectric domain walls in periodically poled $LiNbO_3$. Nano Lett. 21 (2), 959–966. https://doi.org/10.1021/acs.nanolett.0c03810.

Splendiani, A., Sun, L., Zhang, Y., Li, T., Kim, J., Chim, C.-Y., Galli, G., Wang, F., 2010. Emerging photoluminescence in monolayer MoS_2. Nano Lett. 10 (4), 1271–1275. https://doi.org/10.1021/nl903868w.

Steinhoff, A., Florian, M., Rösner, M., Schönhoff, G., Wehling, T.O., Jahnke, F., 2017. Exciton fission in monolayer transition metal dichalcogenide semiconductors. Nat. Commun. 8 (1), 1166. https://doi.org/10.1038/s41467-017-01298-6.

Steininger, F., Knorr, A., Stroucken, T., Thomas, P., Koch, S.W., 1996. Dynamic evolution of spatiotemporally localized electronic wave packets in semiconductor quantum wells. Phys. Rev. Lett. 77 (3), 550–553. https://doi.org/10.1103/PhysRevLett.77.550.

Sun, D., Rao, Y., Reider, G.A., Chen, G., You, Y., Brézin, L., Harutyunyan, A.R., Heinz, T.F., 2014. Observation of rapid exciton-exciton annihilation in monolayer molybdenum disulfide. Nano Lett. 14 (10), 5625–5629. https://doi.org/10.1021/nl5021975.

Sun, Z., Ciarrocchi, A., Tagarelli, F., GonzalezMarin, J.F., Watanabe, K., Taniguchi, T., Kis, A., 2022. Excitonic transport driven by repulsive dipolar interaction in a van der Waals heterostructure. Nat. Photon. 16 (1), 79–85. https://doi.org/10.1038/s41566-021-00908-6.

Suris, R.A., 2003. Correlation between trion and hole in Fermi distribution in process of trion photo-excitation in doped QWs. In: Optical Properties of 2D Systems with Interacting Electrons. Springer, Netherlands, Dordrecht, pp. 111–124. https://doi.org/10.1007/978-94-010-0078-9_9.

Tang, Y., Li, L., Li, T., Xu, Y., Liu, S., Barmak, K., Watanabe, K., Taniguchi, T., MacDonald, A.H., Shan, J., Mak, K.F., 2020. Simulation of Hubbard model physics in WSe_2/WS_2 moiré superlattices. Nature 579 (7799), 353–358. https://doi.org/10.1038/s41586-020-2085-3.

Tikhodeev, S., Kopelevich, G., Gippius, N., 1998. Exciton transport in Cu_2O: phonon wind versus superfluidity. Phys. Status Solidi (b) 206 (1). https://doi.org/10.1002/(SICI)1521-3951(199803)206:1<45::AID-PSSB45>3.0.CO;2-5.

Tran, K., Moody, G., Wu, F., Lu, X., Choi, J., Kim, K., Rai, A., Sanchez, D.A., Quan, J., Singh, A., Embley, J., Zepeda, A., Campbell, M., Autry, T., Taniguchi, T., Watanabe, K., Lu, N., Banerjee, S.K., Silverman, K.L., Kim, S., Tutuc, E., Yang, L., MacDonald, A.H., Li, X., 2019. Evidence for moiré excitons in van der Waals heterostructures. Nature 567 (7746), 71–75. https://doi.org/10.1038/s41586-019-0975-z.

Uddin, S.Z., Kim, H., Lorenzon, M., Yeh, M., Lien, D.-H., Barnard, E.S., Htoon, H., Weber-Bargioni, A., Javey, A., 2020. Neutral exciton diffusion in monolayer MoS_2. ACS Nano 14 (10), 13433–13440. https://doi.org/10.1021/acsnano.0c05305.

Unuchek, D., Ciarrocchi, A., Avsar, A., Watanabe, K., Taniguchi, T., Kis, A., 2018. Room-temperature electrical control of exciton flux in a van der Waals heterostructure. Nature 560 (7718), 340–344. https://doi.org/10.1038/s41586-018-0357-y.

van der Zande, A.M., Huang, P.Y., Chenet, D. a., Berkelbach, T.C., You, Y., Lee, G.-H., Heinz, T.F., Reichman, D.R., Muller, D. a., Hone, J.C., 2013. Grains and grain boundaries in highly crystalline monolayer molybdenum disulphide. Nat. Mater. 12 (6), 554–561. https://doi.org/10.1038/nmat3633.

Wagner, K., Wietek, E., Ziegler, J.D., Semina, M.A., Taniguchi, T., Watanabe, K., Zipfel, J., Glazov, M.M., Chernikov, A., 2020. Autoionization and dressing of excited excitons by free carriers in monolayer WSe_2. Phys. Rev. Lett. 125 (26), 267401. https://doi.org/10.1103/PhysRevLett.125.267401.

Wagner, K., Zipfel, J., Rosati, R., Wietek, E., Ziegler, J.D., Brem, S., Perea-Causín, R., Taniguchi, T., Watanabe, K., Glazov, M.M., Malic, E., Chernikov, A., 2021. Nonclassical exciton diffusion in monolayer WSe_2. Phys. Rev. Lett. 127 (7), 076801. https://doi.org/10.1103/PhysRevLett.127.076801.

Wagner, K., Iakovlev, Z.A., Ziegler, J.D., Cuccu, M., Taniguchi, T., Watanabe, K., Glazov, M.M., Chernikov, A., 2023. Diffusion of excitons in a two-dimensional Fermi sea of free charges. Nano Lett. 23 (11), 4708–4715. https://doi.org/10.1021/acs.nanolett.2c03796.

Wang, G., Chernikov, A., Glazov, M.M., Heinz, T.F., Marie, X., Amand, T., Urbaszek, B., 2018. Colloquium: excitons in atomically thin transition metal dichalcogenides. Rev. Modern Phys. 90 (2), 021001. https://doi.org/10.1103/RevModPhys.90.021001.

Wang, J., Guo, Y., Huang, Y., Luo, H., Zhou, X., Gu, C., Liu, B., 2019. Diffusion dynamics of valley excitons by transient grating spectroscopy in monolayer WSe_2. Appl. Phys. Lett. 115 (13), 131902. https://doi.org/10.1063/1.5116263.

Wang, J., Shi, Q., Shih, E.-M., Zhou, L., Wu, W., Bai, Y., Rhodes, D., Barmak, K., Hone, J., Dean, C.R., Zhu, X.-Y., 2021. Diffusivity reveals three distinct phases of interlayer excitons in $MoSe_2/WSe_2$ heterobilayers. Phys. Rev. Lett. 126 (10), 106804. https://doi.org/10.1103/PhysRevLett.126.106804.

Warren, J.T., O'Hara, K.E., Wolfe, J.P., 2000. Two-body decay of thermalized excitons in CuO_2. Phys. Rev. B 61 (12), 8215–8223. https://doi.org/10.1103/PhysRevB.61.8215.

Weston, A., Zou, Y., Enaldiev, V., Summerfield, A., Clark, N., Zólyomi, V., Graham, A., Yelgel, C., Magorrian, S., Zhou, M., Zultak, J., Hopkinson, D., Barinov, A., Bointon, T.H., Kretinin, A., Wilson, N.R., Beton, P.H., Fal'ko, V.I., Haigh, S.J., Gorbachev, R., 2020. Atomic reconstruction in twisted bilayers of transition metal dichalcogenides. Nat. Nanotechnol. 15 (7), 592–597. https://doi.org/10.1038/s41565-020-0682-9.

Wietek, E., Florian, M., Göser, J.M., Taniguchi, T., Watanabe, K., Högele, A., Glazov, M.M., Steinhoff, A., Chernikov, A., 2023, Non-linear and negative effective diffusivity of optical excitations in moiré-free heterobilayers.

Wilson, J., Yoffe, A., 1969. The transition metal dichalcogenides discussion and interpretation of the observed optical, electrical and structural properties. Adv. Phys. 18 (73), 193–335. https://doi.org/10.1080/00018736900101307.

Wu, F., Lovorn, T., MacDonald, A.H., 2017. Topological exciton bands in Moiré Heterojunctions. Phys. Rev. Lett. 118 (14), 147401. https://doi.org/10.1103/PhysRevLett.118.147401.

Xiao, J., Zhao, M., Wang, Y., Zhang, X., 2017. Excitons in atomically thin 2D semiconductors and their applications. Nanophotonics 6, 1309. https://doi.org/10.1515/nanoph-2016-0160.

Xu, X., Yao, W., Xiao, D., Heinz, T.F., 2014. Spin and pseudospins in layered transition metal dichalcogenides. Nat. Phys. 10 (5), 343–350. https://doi.org/10.1038/nphys2942.

Xu, Y., Liu, S., Rhodes, D.A., Watanabe, K., Taniguchi, T., Hone, J., Elser, V., Mak, K.F., Shan, J., 2020. Correlated insulating states at fractional fillings of moiré superlattices. Nature 587 (7833), 214–218. https://doi.org/10.1038/s41586-020-2868-6.

Yu, H., Cui, X., Xu, X., Yao, W., 2015. Valley excitons in two-dimensional semiconductors. Natl. Sci. Rev. 2 (1), 57–70. https://doi.org/10.1093/nsr/nwu078.

Yu, H., Liu, G.-B., Tang, J., Xu, X., Yao, W., 2017. Moiré excitons: from programmable quantum emitter arrays to spin-orbit-coupled artificial lattices. Sci. Adv. 3 (11), e1701696. https://doi.org/10.1126/sciadv.1701696.

Yuan, L., Wang, T., Zhu, T., Zhou, M., Huang, L., 2017. Exciton dynamics, transport, and annihilation in atomically thin two-dimensional semiconductors. J. Phys. Chem. Lett. 8 (14), 3371–3379. https://doi.org/10.1021/acs.jpclett.7b00885.

Yuan, L., Zheng, B., Kunstmann, J., Brumme, T., Kuc, A.B., Ma, C., Deng, S., Blach, D., Pan, A., Huang, L., 2020. Twist-angle-dependent interlayer exciton diffusion in WS_2/WSe_2 heterobilayers. Nat. Mater. 19 (6), 617–623. https://doi.org/10.1038/s41563-020-0670-3.

Zhang, N., Surrente, A., Baranowski, M., Maude, D.K., Gant, P., Castellanos-Gomez, A., Plochocka, P., 2018. MoiréIntralayer excitons in a $MoSe_2$/MoS_2 heterostructure. Nano Lett. 18 (12), 7651–7657. https://doi.org/10.1021/acs.nanolett.8b03266.

Zhao, S., Li, Z., Huang, X., Rupp, A., Göser, J., Vovk, I.A., Kruchinin, S.Y., Watanabe, K., Taniguchi, T., Bilgin, I., Baimuratov, A.S., Högele, A., 2023. Excitons in mesoscopically reconstructed moiré heterostructures. Nat. Nanotechnol. https://doi.org/10.1038/s41565-023-01356-9.

Ziman, J., 2001. Electrons and Phonons. Oxford University Press. https://doi.org/10.1093/acprof:oso/9780198507796.001.0001.

Zipfel, J., Kulig, M., Perea-Causín, R., Brem, S., Ziegler, J.D., Rosati, R., Taniguchi, T., Watanabe, K., Glazov, M.M., Malic, E., Chernikov, A., 2020. Exciton diffusion in monolayer semiconductors with suppressed disorder. Phys. Rev. B 101 (11), 115430. https://doi.org/10.1103/PhysRevB.101.115430.

CHAPTER FOUR

Control of exciton transport/dynamics in 2D materials using surface acoustic waves

Kanak Datta[a,b,*] and Parag B. Deotare[a,b]
[a]Department of Electrical Engineering and Computer Science, University of Michigan, Ann Arbor, MI, United States
[b]Department of Materials Science and Engineering, University of Michigan, Ann Arbor, MI, United States
*Corresponding author. e-mail address: kanak@umich.edu

Contents

Abstract

Monolayer 2D semiconductors are highly suitable for realizing ultrafast, energy efficient, room temperature excitonic devices for information transfer, encoding and processing. This is primarily due to the existence of high binding energy excitons in a material system in which transport is primarily governed by bands. The atomic level thickness further enables vertical scaling beyond the limits of bulk 3D semiconductors. More importantly, the vast and enriched excitonic properties of 2D materials are extremely sensitive to external stimuli and therefore can be successfully probed as well as manipulated using electric field, magnetic field, doping, mechanical forces, dielectric environment etc. Propagating surface acoustic wave (SAW) is one approach that can utilize the mechanical flexibility of the 2D material system to efficiently control the exciton photophysics. Over the past few years, many exciting studies have been reported on SAW interaction with charge carriers as well as excitons in 2D materials. In this chapter, we discuss the results of excitonic interactions with the traveling strain and electric field associated with the propagating SAW.

Semiconductors and Semimetals, Volume 112
ISSN 0080-8784, https://doi.org/10.1016/bs.semsem.2023.09.005

In particular, we focus on room temperature experimental demonstration of directional exciton transport under traveling strain wave and exciton modulation due to the traveling electric field.

1. Introduction

Exciton, an excited state of matter that is capable of transporting energy, plays an important role in natural processes such as photosynthesis as well as energy converting semiconductor devices. While excitonic devices have been successfully commercialized for display and lighting applications, further applications are currently hindered due to limited diffusive transport that is mostly undirected. Providing directionality to energy transport adds a new dimensionality that can not only bolster the performance of the current excitonic devices but also introduces a new platform for next generation data processing and communication technology (Baldo and Stojanovi, 2009; Grosso et al., 2009; Butov, 2017; Unuchek et al., 2018). Such optoexcitonic circuits would be extremely energy efficient due to the charge neutrality that makes it immune to lossy stray and parasitic capacitance. At the same time, it will utilize the seamless transition with photons for long distance data communication as well as generation/detection.

Exciton transport is an important property that is challenging to control, yet critical for achieving data processing and communication devices. Strongly bound Frenkel excitons in organic semiconductors have very poor mobility. Such excitons are traditionally used in flexible, low-cost, high performance optoelectronic applications such as photovoltaics, organic light emitting diodes (OLEDs), thin film transistors (TFTs) and displays. On the other hand, weakly bound Wannier–Mott excitons in inorganic semiconductors are well known for their large mobility but, are seldom stable at room temperature. Excitons of this type are traditionally found in low dimensional semiconducting media such as III–V semiconductor quantum wells, quantum dots, quantum wires etc. However, monolayer Transition Metal Dichalcogenides (TMDCs), that have recently gathered a lot of interest in the research community, support excitons even at room temperature due to the reduced dielectric screening, large binding energy and strong in-plane coulomb interaction (Liu et al., 2015; Chernikov et al., 2014; Novoselov et al., 2016; Yu et al., 2022), thus making it a potential material system for excitonic research.

Transport of excitons is primarily density driven and thus isotropic in nature. However, a non-uniform potential can create an energy gradient to preferentially direct the exciton flux. This can be achieved in the presence of a spatially varying external stimulus such as an electric field (High et al., 2008; Kuznetsova et al., 2010; Jauregui et al., 2019; Liu et al., 2020) or strain (Fu et al., 2014b; Cordovilla Leon et al., 2018; Harats et al., 2020; Moon et al., 2020), which can create a spatial gradient in the energy potential. The use of strain engineering is more versatile as it does not require electrical contacts and is compatible with highly flexible, atomically thin TMDCs material system (Berkelbach et al., 2013; Chernikov et al., 2014). One approach to utilize the mechanical flexibility of 2D TMDCs to achieve directional exciton transport is by coupling with acoustic waves. Surface acoustic waves (SAW) are shear waves that propagate along a solid surface with the wave amplitude decaying exponentially into the solid half space (Morgan, 2007). Typical velocity of such waves are in the region of a few thousands of meter per second (Morgan, 2007). SAW devices find applications in many branches of modern science and technology including wireless communication, sensing chips, filters, actuators, resonators, life science, and microfluid applications (Campbell, 1989; Hashimoto, 2000; Mandal and Banerjee, 2022). The tightly confined nature of the SAW phonons makes them ideally suited for interfacing with, probing, and controlling elementary species in condensed matter physics ranging from quantum limit (Manenti et al., 2016; Schütz, 2017; Moores et al., 2018) to collective coherent excitations (Cerda-Méndez et al., 2017; Delsing et al., 2019).

In this chapter, we briefly discuss the recent progress on excitons and acoustic wave interaction in 2D materials. We begin by discussing the prospect of TMDCs as excitonic media, effect of strain on optical and exciton transport properties in TMDCs, and possible exciton modulation schemes under surface acoustic waves. We then discuss room temperature exciton transport in TMDCs under traveling strain wave. Finally, we discuss the photophysical modulation by dielectric engineering and carrier screening on exciton interaction with SAW and provide outlook for future directions.

2. The exciting prospect of 2D TMDCs as robust excitonic media

Gradual scaling of TMDCs from bulk to monolayer results in out-of-plane quantum confinement and therefore exciting changes in the electronic bandstructure (Pettenkofer et al., 2001; Ellis et al., 2011; Jing et al., 2020). In monolayer form, TMDCs generally contain three atomic planes where chalcogen atoms are contained in the top and bottom planes in a triangular lattice configuration. The middle plane contains the transition metal atom in a different triangular lattice configuration (Liu et al., 2015). The monolayer band extrema (conduction band minima and valance band maxima) are typically located at K points in the hexagonal Brillouin zone. The valance band show strong spin orbit splitting due to the presence of heavy metal transition atom (Zhu et al., 2011; Komider et al., 2013; Wang et al., 2018). The monolayer electronic bandstructure is further complicated by variation of the mixture of the atomic orbitals at different symmetry points in the Brillouin zone—mostly d orbitals from the transition metal atoms and the p orbitals from the chalcogen atoms (Shi et al., 2013; Roldán et al., 2014; Zhao et al., 2015). As multiple monolayers of these materials are stacked together, this complex orbital hybridization further alters the energy of conduction of valance band at different symmetry points, rendering direct to indirect bandgap transition (Kuc et al., 2011; Komsa and Krasheninnikov, 2012). Broken inversion symmetry in the monolayer lattice, orbital hybridization, strong spin-orbit interaction further gives rise to exciting spin-valley properties and optical selection rules (Yao et al., 2008; Cao et al., 2012; Mak et al., 2012). Strong out-of-plane confinement, significantly weakened dielectric screening resulting from the monolayer nature, strong in-plane coulomb interaction allows exciton binding energy in the range of 100 s of meV. Contrary to Frenkel excitons in disordered organic semiconductors, excitons in 2D materials show superior transport properties due to well-defined bandstructure (Cadiz et al., 2018; Kang et al., 2019; Goodman et al., 2020). For example, exciton diffusivity in TMDCs is more than two orders of magnitude higher than that reported for organic semiconductors (Mikhnenko et al., 2015; Cordovilla Leon et al., 2018; Kulig et al., 2018). The overall transport behaviors, however, is strongly influenced by the surrounding environment (Lin et al., 2014; Raja et al., 2017), due to the 2D nature and factors, such as impurity and defects (Hwang et al., 2007; Morozov et al., 2008), energetic disorders (Wierzbowski et al., 2017), surface roughness (Dean et al., 2010), and dielectric environment (Li et al., 2021) etc. The excitonic behavior can be further tuned by generation densities, doping that gives access to exotic many body interactions (Chernikov et al., 2015; Pei et al., 2019; Qiu et al., 2019), and mechanical deformation (Yang et al., 2021; Yu et al., 2022). As discussed earlier, excitons in 2D TMDCs are influenced by the strong spin–orbit coupling (Zhu et al., 2011; Premasiri and Gao,

2019), valley degeneracy and valley polarization (Rivera et al., 2016; Schaibley et al., 2016) in the bandstructure that results in coexisting optically bright and dark (excitonic) states (Zhang et al., 2015). Such enriched excitonic properties coupled to a wide range of control knobs, make 2D material a robust platform for storing and encoding optical information, as well as converting to other resonant material excitation modes such as plasmons, polaritons and cavities (Flatten et al., 2018; Gu et al., 2019; Latini et al., 2019; Azzam et al., 2021; Yu et al., 2021), thereby providing a potential pathway to realize high performance optical, optoelectronic and information processing devices. Furthermore, when multiple monolayers of these materials are vertically stacked, indirect excitons are formed where electrons and holes are spatially localized at different monolayers (Novoselov et al., 2016; Liu and Hersam, 2018; Cheng et al., 2019). Such indirect excitons further enhance the application bandwidth of TMDCs (Novoselov et al., 2016; Liu et al., 2019b; Ciarrocchi et al., 2022).

Owing to the atomic level thickness (monolayer thickness ~0.7 nm), TMDCs can also be seamlessly transferred on and integrated to many commercially available and artificially synthesized substrates, allowing exciton manipulation by underlying substrate functionalization (Akinwande et al., 2017; Bae et al., 2019; Goodman et al., 2020). More importantly in the current context, the mechanical flexibility of TMDCs can be utilized to control transport of various excitonic species by coupling with acoustic waves.

3. Strain effects on excitonic properties in TMDCs

The ability to control mechanical deformation under external forces makes strain engineering a robust platform for manipulating material, electronic and photophysical properties and thereby realizing high performing electronic, optoelectronic and excitonic devices (Yu et al., 2017, 2022; Peng et al., 2020). This is further exemplified in 2D materials (Yu et al., 2017, 2022; Peng et al., 2020) such as TMDCs that have been reported to withstand mechanical deformation of more than 10% in monolayer form (Scalise et al., 2011; Ghorbani-Asl et al., 2013). The optical and electronic properties of such monolayers show strong sensitivity to mechanical deformations when compared to their bulk multilayer counterparts and other semiconducting media. For instance, the absorption spectra of monolayer tungsten diselenide (WSe_2) reportedly show uniaxial strain tunability of about 47 mev/%–60 meV/% (Desai et al., 2014; Aslan et al., 2018).

Theoretical bandstructure calculation shows a proportional lowering of bandgap under tensile strain and semiconductor to metal transition at larger strain values (Scalise et al., 2011; Ghorbani-Asl et al., 2013). The modulation of individual band extrema and hence the quasiparticle properties at such symmetry positions in momentum space is rather non-trivial. External strain generally produces unequal shifts in the conduction and valence band energies, results in direct to indirect bandgap transition and vice versa, modifies band curvature at different symmetry points in the energy-momentum space. The resulting band curvature modification results in a change in the exciton effective mass (Shi et al., 2013; Rasmussen and Thygesen, 2015; Wang et al., 2015; Song and Yang, 2017) that directly influences transport. Alongside bandstructure modification, strain also modulates exciton (bound electron-hole pair) population in different valleys, and affects inter and intra-valley scattering rates (Niehues et al., 2018; Aslan et al., 2020; Moon et al., 2020). Therefore, static strain can be used as an efficient control knob for controlling resonant exciton emission in monolayer TMDCs. Traditional schemes of static strain application in TMDCs include transferring such layered materials on patterned nanostructures (Branny et al., 2017; Mangu et al., 2017; Palacios-Berraquero et al., 2017; Cordovilla Leon et al., 2018; Luo et al., 2018), controlled deformation using micro-mechanics (Christopher et al., 2019) or nanoscale AFM tips (Rosenberger et al., 2019; Koo et al., 2021), controlled bending of flexible substrates (Castellanos-Gomez et al., 2013; Desai et al., 2014; Island et al., 2016; Dai et al., 2019) etc.

Since external strain modulates the band structure of TMDCs monolayers and therefore, the exciton potential, directional exciton transport can be achieved by applying non-uniform strain gradient in an excitonic semiconducting medium. A spatial modulation in exciton potential would exert a directional force on an exciton resulting in drift velocity, $v = \mu \frac{dU}{dx}$; where v, μ, $\frac{dU}{dx}$ refer to the exciton velocity, mobility, and exciton potential gradient respectively. Fu and co-workers reported experimental observations on the dynamics of excitons in bent ZnO microwires (Fu et al., 2014a) using picosecond resolved cathodoluminescence. Due to their giant bandgap tunability, flexible nature and ability to withstand large mechanical deformation and strong light–matter interaction, TMDCs are also an attractive platform for studying directional exciton transport under a non-uniform strain gradient—a phenomenon commonly referred to as *exciton funneling* (Wang et al., 2019; Moon et al., 2020; Lee et al., 2022; Qi et al., 2022). Room temperature exciton funneling in TMDCs monolayers has been experimentally reported using patterned substrates (Cordovilla Leon et al., 2018), AFM tips (Moon et al., 2020) and nano-scale tips (Koo et al., 2021). Under high strain, such funneling can also lead to quasiparticle inter-conversion (Harats et al., 2020). However, the efficiency of such funneling process is limited by the balance between drift and diffusion processes at thermal equilibrium, exciton lifetime in the medium under study, non-radiative recombination processes that dominate at high excitation density (Harats and Bolotin, 2021). At room temperature, the theoretical limit of such funneling process has been reported to be 50% which can be further improved at low temperature where exciton drift dominated over diffusive transport (Harats and Bolotin, 2021). The funneling efficiency and the transport distance can be largely improved by using TMDCs hetero bilayers with long radiative lifetime (Kim et al., 2017; Jiang et al., 2021).

4. Acoustic wave mediated exciton manipulation in excitonic media

Dynamic strain is necessary to achieve exciton transport beyond funneling. This can be achieved using traveling acoustic waves. Coherent acoustic phonon oscillations tightly confined to the solid surface have been widely used to study, control, and interact with elementary particle species in semiconductor nanostructures (Delsing et al., 2019). Of the various types of surface acoustic modes in solids, Rayleigh mode is typically of interest to research communities for studies in condensed matter physics (Biryukov et al., 1995). In Rayleigh type SAW waves, the surface particles oscillate normal and parallel to the solid surface and the propagating wave adopts an elliptical polarization along normal to the substrate. The wave energy remains tightly confined to the solid surface and the wave amplitude decays exponentially into the solid (effective penetration depth ≈ SAW wavelength) (Morgan, 2007). In condensed matter physics, SAW transducers have been successfully used to study indirect exciton transport in III–V quantum wells (Rudolph et al., 2007a; Lazić et al., 2014), on-chip transfer and control of electron gas (Hermelin et al., 2011; McNeil et al., 2011; Bertrand et al., 2016), accessing individual defect centers in solids (Arcizet et al., 2011; Golter et al., 2016; Schütz, 2017), precision modulation of single quantum dot emission (Gell et al., 2008; Couto et al., 2009; Metcalfe et al., 2010), charge transport (Rocke et al., 1997; Hernández-Mínguez et al., 2012), control cavity dynamics and optomechanics (De Lima and Santos, 2005; Fuhrmann et al., 2011; Tadesse and Li, 2015; Cerda-Méndez et al., 2017;

Villa et al., 2017; Okada et al., 2018). SAW is typically generated by patterning metal interdigitated electrodes (IDTs) on piezoelectric substrate like lithium niobate, lithium titanate, barium titanate, gallium arsenide, quartz etc. (Morgan, 2007). As such electrodes are excited using radio frequency (RF) source, the inverse piezoelectric effect converts the electrical excitation to propagating strain wave. Since the wave propagates along the surface, it generates compressive and tensile strain in the 2D semiconductor transferred on the piezoelectric substrate. Such strain periodically modulates the semiconductor bandgap, resulting in out-of-phase bandgap modulation which is known as Type-I band edge modulation. The potential pockets (separated by SAW wavelength and traveling at the speed of the acoustic wave in the material) created by the SAW have been shown to efficiently capture and transport indirect excitons in III–V semiconductor quantum wells at cryogenic temperatures (Rudolph et al., 2007b, 2007a) up to several hundreds of micrometers. Indirect excitons are ideally suited for such application due to their long lifetime (several hundreds of ns) (Golub et al., 1990; Zrenner et al., 1992) and high mobility in III–V quantum wells (Hillmer et al., 1989; Leonard et al., 2012; Dorow et al., 2018). A second modulation scheme is also observed in the acoustic wave assisted particle transport domain which is known as Type-II band edge modulation, originating from the piezoelectric nature of the traveling SAW. The electric field of the traveling SAW creates in-phase modulation of the semiconductor energy bands, creating potential pockets separated by half the wavelength of the SAW (Rocke et al., 1997). This type of modulation causes excitonic dissociation that results in ambipolar transport of electrons and holes and their possible remote recombination (Völk et al., 2010; Kinzel et al., 2011; Hernández-Mínguez et al., 2012; Büyükköse et al., 2014; Lazić et al., 2015). A simple schematic representation of these two types of modulation is shown in Fig. 1.

In 2D materials, surface acoustic wave has been successfully used to realize acoustic charge transport in graphene (Miseikis et al., 2012; Bandhu et al., 2013; Zheng et al., 2016; Hernández-Minguez et al., 2018), high-performance scalable optical devices (Preciado et al., 2015; Zheng et al., 2018), modulating exciton photoluminescence (Rezk et al., 2015, 2016), emission from defect sites (Lazić et al., 2019), cavity optomechanics (Patel et al., 2023). TMDCs are an attractive platform for studying traveling strain induced long range room temperature exciton transport because of the large excitonic binding energy. In addition, the indirect excitons in 2D heterobilayers show lifetimes ranging hundreds of ns at low temperatures (Jauregui et al., 2019; Ciarrocchi et al., 2022). Moreover, recent

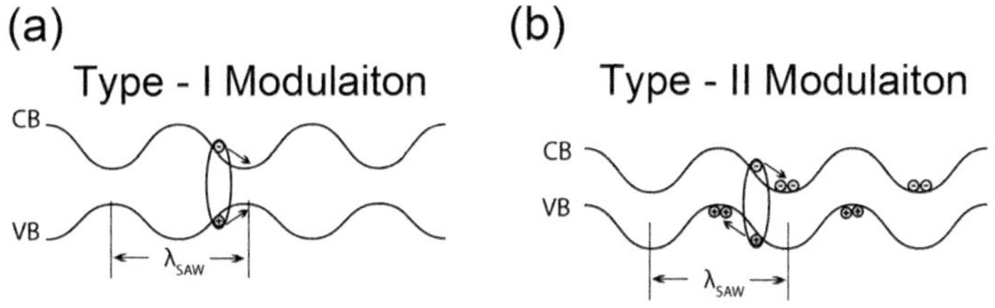

Fig. 1 Effects of band edge modulation on excitons under surface acoustic wave in a semiconducting media: (a) Type-I modulation: photoexcited excitons travel to the lowest potential location separated by the SAW wavelength and are swept away at the SAW velocity. (b) Type-II modulation: photoexcited excitons dissociate under the SAW piezoelectric field, electrons and holes are transported to the conduction band minima and valance band maxima respectively; separated by half the SAW wavelength. The charge carriers are then transported away at the SAW velocity.

improvements in growth and transfer techniques of TMDCs provides reliable interface and seamless integration with piezoelectric substrates (Jia et al., 2022; Wang et al., 2022). Recently, the traveling acoustic field has been successfully used to achieve long-range exciton transport in TMDCs monolayers (Datta et al., 2022) and hetero-bilayers (Peng et al., 2022) and will be discussed in further details in the chapter. The in-plane electric field of the SAW wave can also be beneficial to control exciton ionization, and acoustic stark effect (Datta et al., 2021; Scolfaro et al., 2021).

4.1 Device architecture for selecting between type-I and type-II modulation

The relative strength of type-I and type-II modulation can be controlled by the dielectric environment of the monolayer. We fabricated SAW electrodes on a 128^0 Y-cut lithium niobate substrate ($LiNbO_3$) using optical lithography followed by low pressure e-beam evaporation of Pt/Au and lift-off. Detailed fabrication parameters and process description can be found in Datta et al. (2021, 2022). Following device fabrication, monolayer WSe_2 flakes were transferred on the substrate using a dry transfer technique (Datta et al., 2022). To study the effect of type-I modulation, we utilized hexagonal boron nitride (h-BN) as the dielectric moderator. Such moderator allows the preservation of large binding energy and reduces exciton dissociation and ionization under SAW electric field. The h-BN encapsulation of the TMDCs monolayer also improves the transport properties of excitons (Hoshi et al., 2017; Hotta et al., 2020). In the absence of h-BN, strong dielectric screening from the piezoelectric substrate lowers exciton binding energy in the monolayer and triggers strong dissociation and ionization under SAW electric field (Datta et al., 2021).

5. Exciton transport under traveling strain wave in monolayer WSe_2

Traveling strain assisted long range exciton transport has been widely investigated in III–V semiconductor quantum wells at cryogenic temperature (Rudolph et al., 2007b, 2007a). Large radiative lifetime and low-temperature mobility makes indirect excitons in III–V quantum well system ideal media for realizing μm scale energy transport. However, generation of such species requires application of external electric field. Direct excitons in 2D materials are advantageous in this regard as they can be generated simply under optical excitation and are stable at room temperature due to their large binding energy.

Fig. 2a shows a simple schematic representation of the device geometry used in the experiment as described in the previous section. A brightfield image of the h-BN encapsulated monolayer WSe_2 transferred in front of the SAW interdigitated electrodes is shown in Fig. 2b. A magnified version of the sample is also shown in the inset. Fig. 2c–d show the results of exciton transport measurement using scanning SPAD technique (Cordovilla Leon et al., 2018; Datta et al., 2022). The technique enables visualization of exciton density as it evolves in space and time. Fig. 2e and f show the spatiotemporal exciton density (normalized along space at each time instance) distributions at two representative RF input excitations: −60 dBm (~1 nW) and 13 dBm (~20 mW) respectively at an optical fluence of 1.2 $\mu J/cm^2$. The near-symmetric exciton density at −60 dBm depicts diffusion transport while a clear non-symmetric distribution is observed at 13 dBm RF input power. The exciton density shows a gradual shift with time in the direction of SAW propagation. This is also visible line profiles of the measured exciton density at two-time instances: 0 ns and 3.5 ns, at two RF powers: Fig. 2e −60 dBm and Fig. 2f 13 dBm. Similar results based on acoustic wave assisted exciton transport were reported by Peng et al. (2022) in bilayer WSe_2 as shown in Fig. 2g–i. As seen in the figure, PL emission from the edges of the bilayer flakes increases proportionally with RF power.

In our experiments, we observed signature of exciton transport in the weak coupling regime (García-Cristóbal et al., 2004). In this regime the

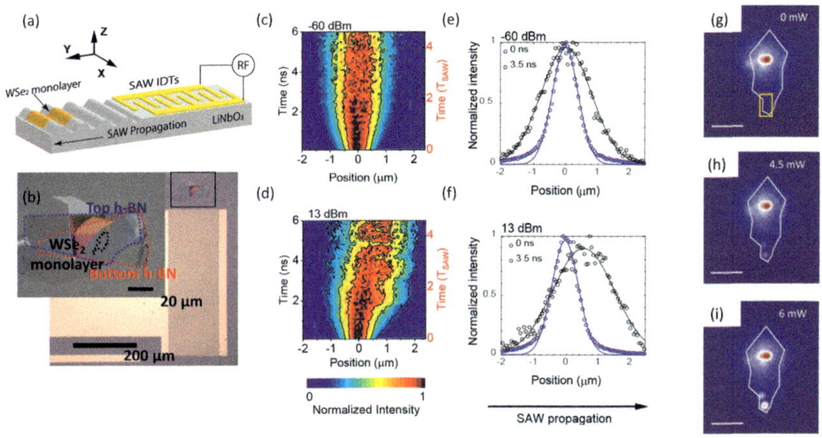

Fig. 2 Exciton transport under type-I band edge modulation in monolayer WSe_2. (a) Simple schematic representation of the device geometry used in this work. (b) the h-BN encapsulated monolayer WSe_2 sample transferred in front the IDTs along the SAW propagation direction obtained using a bright field microscope A magnified image of the monolayer flake sandwiched between h-BN flakes is also shown. The top and bottom h-BN flakes are also highlighted. Spatiotemporal exciton density profile (linear scale) from phase synchronized TCSPC measurement for (c) −60 dBm; (d) 13 dBm RF excitation input at optical fluence of 1.2 μJ/cm^2. A spatial shift in exciton density can be observed under RF excitation. Normalized exciton density profiles at different time instances as a function of space along with the Gaussian fit under (e) −60 dBm and (f) 13 dBm RF input power. At minimum RF input power (−60 dBm), the exciton density broadens symmetrically in space due to exciton diffusion in the monolayer. At high RF input, an asymmetric spatial shift of the exciton density in the direction of the acoustic wave propagation is observed with the observed shift increasing with time. To improve the signal to noise ratio, the raw data was binned using a 100 ps window. PL images at different SAW power of (g) 0 mW, (h) 4.5 mW, (i) 6 mW from Peng et al. (2022). The solid white line outlines the WSe_2 flake. The two bright emission spots at the edge of the WSe_2 flake highlight the SAW-driven transport of IXs. Scale bar, 10 μm. *(a), (c)–(f) Reproduced with permission from Datta et al. (2022). (g)–(i) Reproduced with permission from Peng et al. (2022).*

exciton flux is unable to keep pace with the traveling strain wave. This is seen in Fig. 3a as the periodic oscillation of the fitted gaussian peak evolution with time. The evolution of the gaussian peak can be fitted using a simplified linear relationship:

$$h(t) = v_{avg}t + h_{offset} \tag{1}$$

where, $h(t)$ refers to the net displacement of the generated exciton density, v_{avg} refers to the average drift velocity of the exciton density under applied dynamic strain, and h_{offset} is a constant to accommodate the fitting error (Datta et al., 2022). This allows the extraction of the average velocity

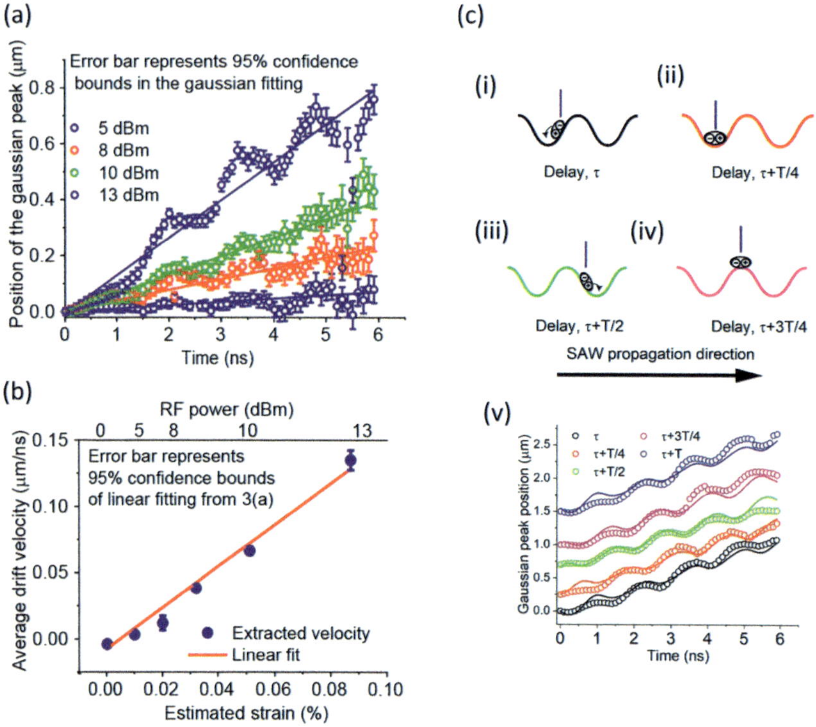

Fig. 3 Exciton transport in monolayer WSe_2 under RF excitation at 1.2 μJ/cm^2 optical fluence. (a) Evolution of the exciton density peak extracted using gaussian fits (circles) and the approximate linear fit (solid lines) at various RF powers. The error bars represent the uncertainties in the gaussian fit. (b) Average drift velocity extracted from the effective shift in exciton density that shows a clear increase with estimated strain i.e. applied RF power. (c) Spatiotemporally controlled energy transport under phase modulated optical excitation for 15 dBm RF excitation. Schematic presentation of photogenerated excitons at different SAW phases along with exciton potential: (i) τ; (ii) $\tau + T/4$; (iii) $\tau + T/2$; (iv) $\tau + 3T/4$; T refers to the SAW period. The vertical blue line represents the position of excitation pulse on SAW wave. Detailed description can be found in the text. (v) Phase synchronized evolution of the gaussian peak position (circles) extracted by fitting the normalized exciton density measured using TCSPC technique. The gaussian peak position oscillates at the period of the SAW wave. A progressive delay in the Gaussian peak is also observed as the time delay is increased at an increment of $T/4$. Alongside oscillations, a net drift in the Gaussian peak is observed in the direction of acoustic wave propagation. *(a)–(c) Reproduced with permission from Datta et al. (2022).*

of the photogenerated excitons that shows a linear increase with the volumetric strain as shown in Fig. 3b. The average velocity can be further improved by increasing the RF excitation power, reducing the SAW wavelength to increase strain, or by improving intrinsic exciton transport properties by suppressed scattering from impurity states or surface roughness (Hotta et al., 2020). From our measurement, we extract the maximum drift velocity under 13 dBm RF excitation to be 600 m/s, much smaller than the SAW velocity in $LiNbO_3$ [3979 m/s (Morgan, 2007)]. From the maximum exciton velocity, the room temperature mobility can be estimated using the following equation (Datta et al., 2022):

$$\begin{aligned} v_{\max} &= \mu \left| \frac{\partial E_g}{\partial \varepsilon} \right| \frac{\partial \varepsilon}{\partial x} \Big|_{\max} \\ &= \mu \left| \frac{\partial E_g}{\partial \varepsilon} \right| k\varepsilon_0 \end{aligned} \tag{2}$$

where, $\mu \frac{\partial E_g}{\partial \varepsilon}$, $k = \frac{2\pi}{\lambda_{SAW}}$, and ε_0 refer to the exciton mobility, strain sensitivity of monolayer bandgap (Aas and Bulutay, 2018), acoustic wave momentum and maximum dynamic strain in the monolayer, respectively. λ_{SAW} refers to the wavelength of the acoustic wave ($\lambda_{SAW} = 4.7\ \mu m$). Using the extracted maximum drift velocity at 13 dBm RF input power, we estimate the exciton mobility in the monolayer to be 900 $cm^2/eV/s$ which is nearly two orders lower [$\sim 10^4\ cm^2/eV/s$ (Gärtner et al., 2006; Violante et al., 2014; Dorow et al., 2018)] compared to indirect excitons in III-V materials at cryogenic temperatures (Rudolph et al., 2007b; Violante et al., 2014). This further corroborates the weak coupling of photogenerated excitons to the traveling strain field. In one SAW period, only a fraction of the photogenerated excitons get coupled to the traveling potential pocket, resulting in asymmetric funneling of the exciton density and a net spatial shift is observed in the direction of the SAW over many periods. Under such a transport regime, the transport distance becomes primarily limited by the radiative lifetime of the generated species and sample size (Datta et al., 2022). The small intrinsic exciton mobility, resulting from impurities, defects, and phonons at room temperature, along with short radiative lifetime results in smaller transport distance (~μm) under dynamic strain in monolayer TMDCs. However, the results show the potential of the 2D material system for future room temperature excitonic devices, especially as defect densities get lowered with rapid progress in material growth.

The use of SAW also enables precise acoustic steering of the exciton density (Datta et al., 2022). The dynamic acoustic strain field generated by a SAW wave can be represented by:

$$\varepsilon(x, t) = \varepsilon_0 \cos(\omega t - kx + \varphi) \tag{3}$$

where, ε_0 refers to the amplitude of the strain field, ω refers to the angular frequency ($\omega = 2\pi f_{SAW}$; f_{SAW} refers to the acoustic resonance frequency), and φ refers to the instantaneous phase of the acoustic wave. Therefore, for a given optical excitation position from the IDT, the generated exciton density interacts with a certain phase of the traveling strain field. Using Eq. (3), the dynamic strain induced drift velocity of the exciton flux can be written as (Cordovilla Leon et al., 2018):

$$v_{ex}(x, t) = \mu_\varepsilon \frac{\partial \varepsilon(x, t)}{\partial x} = \mu_\varepsilon k \sin(\omega t - kx + \varphi) \tag{4}$$

where $\mu_\varepsilon = \mu \frac{\partial E_g}{\partial \varepsilon}$ refers to the strain mobility of the excitons (Cordovilla Leon et al., 2018). At photoexcitation ($t=0$), the instantaneous velocity of the exciton flux is a function of the acoustic phase, $v_{ex}(x, 0) = \mu_\varepsilon k \sin(-kx + \varphi)$. Therefore, precise control over the initial direction of photogenerated exciton flux can be achieved by controlling the relative phases as shown in Fig. 3c. We achieve such control by introducing a delay, τ ($\varphi = \frac{2\pi}{T}\tau$; $T = \frac{1}{f_{SAW}}$) in the laser trigger signal (refer Datta et al., 2022 for schematics of the characterization setup). Fig. 3c plots the spatiotemporal evolution of the center of the exciton density distribution as a function of the time delay (τ) with increments of $T/4$ (refer Datta et al., 2022 for the corresponding spatiotemporal exciton density maps), which were further numerically modeled using the following modified drift–diffusion equation (Cordovilla Leon et al., 2018):

$$\frac{\partial n_{ex}(x, t)}{\partial t} = D_{ex} \frac{\partial^2 n_{ex}(x, t)}{\partial x^2} + \frac{\partial \left(v_{ex} n_{ex}(x, t) \right)}{\partial x} + G - \frac{n_{ex}(x, t)}{\tau_{ex}} - \frac{n_{ex}(x, t)}{\tau_{ion}} \tag{5}$$

Here, n_{ex}, D_{ex}, v_{ex}, τ_{ex} and τ_{ion} refer to the population, diffusion coefficient, velocity, recombination time and ionization time of the neutral exciton (García-Cristóbal et al., 2004) respectively. G refers to the photogenerated exciton density under optical excitation. The solid lines in Fig. 3c–v are fits from the numerical simulation that successfully captures the observed dynamics. Consistently, we observed similar progressive evolution with instantaneous phase in the transient photoluminescence (TRPL). Detailed discussion on the fitting parameters used in our study and modeling can be found in Datta et al. (2022).

6. Engineering exciton photophysics under traveling electric field

In the last section we discussed exciton transport in TMDCs monolayer under type-I band edge modulation. h-BN encapsulation was used to moderate the dielectric environment and preserve the large room temperature exciton binding energy of monolayer WSe_2. However, when a monolayer WSe_2 is directly transferred on piezoelectric $LiNbO_3$ substrate, the exciton binding energy is lowered due to strong dielectric surrounding (Komsa and Krasheninnikov, 2012; Berkelbach et al., 2013; Lin et al., 2014; Stier et al., 2016). As a result, strong dissociation is observed under SAW electric field. The dielectric surrounding also increases the exciton polarizability resulting in a redshift in the monolayer PL spectra. At the same time, strong exciton ionization by the in-plane electric field of the SAW causes broadening in the exciton PL.

To investigate the room temperature exciton photophysics under surface acoustic wave, we transferred a monolayer WSe_2 using mechanical exfoliation technique on 128° Y-cut piezoelectric $LiNbO_3$ substrate. SAW is generated by patterning interdigitated electrodes (IDTs) with a 6 μm period. As the RF excitation is turned on, the electric field from the SAW results in strong dissociation of the photogenerated excitons. The traveling electric field of the SAW creates type-II band edge modulation—same phase modulation of the conduction and valance band of the semiconducting media. This scheme creates traveling potential pockets (separated by half of the SAW wavelength) where dissociated electrons (holes) accumulate at the conduction band minima (valance band maxima) and are moved from the position of photogeneration at the SAW velocity. The transferred charges can later be made to recombine at a remote location as the SAW electric field is screened by external means (Rocke et al., 1997). Fig. 4 shows the monolayer WSe_2 PL spectra measured under a set of RF excitation at 0.5 μW optical power. The PL spectra shows a clear quenching of monolayer WSe_2 emission and a redshift as the RF excitation is increased. The observed PL quenching is weaker than other low dimensional excitonic

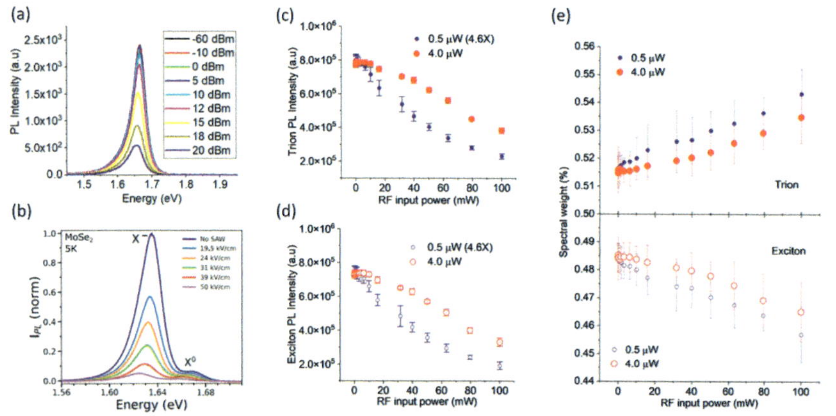

Fig. 4 Monolayer TMDCs exciton PL modulation under piezoelectric modulation. (a) Monolayer WSe_2 PL at a set of RF powers measured at 0.5 μW optical power. Strong PL quenching along with a gradual redshift under RF excitation is observed from the PL spectra. (b) Acoustic modulation of monolayer $MoSe_2$ PL (in normalized scale) from Scolfaro et al. (2021). Strong quenching in monolayer $MoSe_2$ PL is observed as the RF power is increased. Integrated exciton PL intensity as a function of RF input power at two optical power—0.5 μW and 4 μW for (c) trion and (d) neutral excitons. (e) Trion and neutral exciton PL spectral weight at two optical excitations as a function of RF power. As observed from our experiments, trion spectral weight increases with input RF power while neutral exciton spectral weight decreases with input RF power. *(a), (c)–(e) Reproduced with permission from Datta et al. (2021). (b) Reproduced with permission from Scolfaro et al. (2021) under creative commons.*

media like GaAs (Rocke et al., 1997; Santos et al., 1998) and ZnCdSe (Fuhrmann et al., 2009) quantum wells which is attributed to large exciton binding energy in monolayer semiconductors. This is found to be in agreement with the theoretical work from Huang et al. (2017) and experimental work from Scolfaro et al. (2021) on monolayer $MoSe_2$ (shown in Fig. 4b).

We extracted the integrated PL intensities of exciton ($\int_\lambda I_{PL,ex}(\lambda)d\lambda$) and trion ($\int_\lambda I_{PL,tr}(\lambda)d\lambda$) by fitting the PL spectra at different RF excitation power using two gaussian functions [corresponding to neutral A-excitons and trions respectively (Aivazian et al., 2015; Arora et al., 2015; You et al., 2015; Huang et al., 2016; He et al., 2020; Rivera et al., 2021)]. Here, $I_{PL,ex}(\lambda)$, $I_{PL,tr}(\lambda)$ refer to the estimated PL spectrum for the excitons and trions. The integrated intensities for trion and neutral excitons at 0.5 μW and 4 μW optical input power for a set of RF excitation is shown in Fig. 4(c) and Fig. 4(d) respectively. We estimated the exciton and trion spectral weight using $\eta_{PL,tr} = \frac{\int_\lambda I_{PL,tr}(\lambda)d\lambda}{\int_\lambda I_{PL}(\lambda)d\lambda}$ and exciton, $\eta_{PL,ex} = \frac{\int_\lambda I_{PL,ex}(\lambda)d\lambda}{\int_\lambda I_{PL}(\lambda)d\lambda}$; where, $\int_\lambda I_{PL}(\lambda)d\lambda = \int_\lambda I_{PL,ex}(\lambda)d\lambda + \int_\lambda I_{PL,tr}(\lambda)d\lambda$. The extracted exciton (trion) spectral weight shows a gradual decrease (increase) with RF applied power. This indicates an increase in trion formation rate as more free carriers become available due to piezoelectric dissociation of excitons in monolayer WSe_2. The estimated trion and exciton PL spectral weight with RF input power is shown in Fig. 4e. This is further supported by the observed increase in trion binding energy with applied RF excitation (Datta et al., 2021).

The observed PL quenching is accompanied by a strong redshift in the PL spectra under RF power. SAW electric field induced Stark shift has been previously reported in low dimensional semiconductors (Choy et al., 1998; Ivanov and Littlewood, 2001; Santos et al., 2004). Under an electric field, the spectral shift of excitons is generally presented by the quadratic relationship: $\Delta E = -\frac{1}{2}\alpha {E_X}^2$; where α and E_X refer to the in-plane neutral exciton polarizability and the applied electric field respectively. Upon fitting the exciton PL redshift, we extract the monolayer neutral exciton polarizability to be $8.43 \pm 0.18 \times 10^{-6}$ Dm/V. The measured neutral exciton PL redshift and the quadratic fit is shown in Fig. 5a. This strong redshift is attributed to the in-plane electric field of the SAW as the out-plane polarizability of monolayer TMDCs has been reported to be at least few orders of magnitude smaller (Klein et al., 2016; Roch et al., 2018). The observed large in-plane polarizability is due to strong dielectric screening from the piezoelectric $LiNbO_3$ substrate that weakens the excitonic electron–hole interaction and making the bound state more susceptible to applied electric field (Rezk et al., 2015, 2016).

The observed strong PL quenching and redshift were accompanied by a broadening of the exciton PL spectra. Various physical mechanisms have been attributed to PL broadening in low dimensional semiconductor exciton PL such as exciton ionization under electric field, scattering with free carriers at high excitation densities, phonons, strain induced bandgap modulation. From our experiments, we ruled out possible scattering induced PL broadening at high optical excitation (such broadening generally takes place at higher densities $>10^{12}$ cm^{-2} (Chernikov et al., 2015), one-two order of magnitude higher than our experiments). We also observed no linewidth broadening due to optical excitation induced lattice heating (Datta et al., 2021). Therefore, we attribute the linewidth broadening of the exciton PL to dynamic strain (Violante et al., 2014) (type-I modulation) in the monolayer and piezoelectric field induced

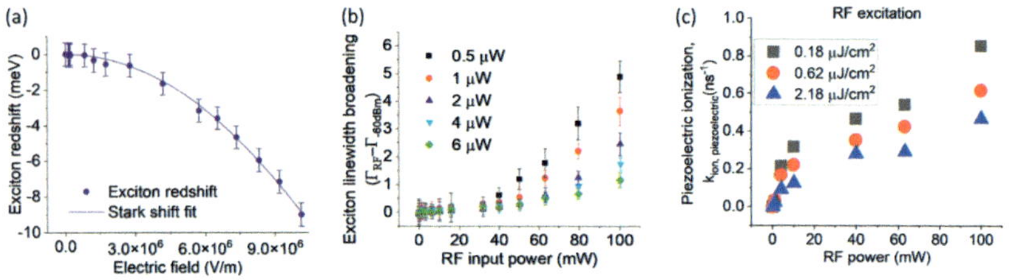

Fig. 5 Exciton PL spectral shift and ionization under piezoelectric modulation: (a) Measured neutral exciton PL redshift under applied RF power and the quadratic fit from the Stark shift equation discussed in the text at 0.5 μW optical input power. (b) Exciton linewidth broadening at various RF input power that shows a clear inverse dependence of the optical excitation that refers to stronger ionization at lower optical excitation. (c) Extracted piezoelectric ionization rate as a function of input RF power that shows an inverse relationship with optical excitation density. *(a)–(c) Reproduced with permission from Datta et al. (2021).*

ionization (type-II modulation). The piezoelectric field induced broadening can be estimated using a two-dimensional exciton ionization model (Miller et al., 1985):

$$\Gamma_{ion}(E) = \beta E_b \sqrt{\frac{\gamma F_0}{F}} \exp\left(- \frac{32}{3} \frac{\gamma F_0}{F} \right) \tag{6}$$

where β and γ are dimensionless parameters that account for substrate induced dielectric screening that modifies the coulomb interaction in the monolayer. A smaller value of the scaling factor ($\gamma < 1$) refers to a stronger screening and hence, a larger linewidth broadening change under electric field and vice versa. F_0 represents the field required for exciton ionization under substrate induced screening: $F_0 = \frac{E_b^*}{e a_b^*}$; where, E_b^* and a_b^* are the screened exciton binding energy and Bohr radius respectively and e is the electronic charge.

The total linewidth broadening under RF excitation can now be written as:

$$\begin{aligned} \Gamma_{RF} - \Gamma_{-60dBm} &= \Gamma_{ion}(E) + \Gamma_\varepsilon \\ &= \beta E_b \sqrt{\frac{\gamma F_0}{F}} \exp\left(- \frac{32}{3} \frac{\gamma F_0}{F} \right) + \Gamma_\varepsilon \end{aligned} \tag{7}$$

Where Γ_ε refers to the dynamic strain induced broadening which has been extracted in our work (Datta et al., 2021) using the formulation from Weiß and Krenner (2018).

As seen from Fig. 5b, the neutral exciton peak broadens with RF power, with reduced broadening as the optical power increases. The observed broadening of the neutral exciton state shows a stronger dependence on the SAW electric field—a clear signature of substrate induced coulomb screening. The measured exciton linewidth broadening at different optical input power and fitting using Eq. (7) and the fitting parameters can be found in our work (Datta et al., 2021). Therefore, it can be concluded that the observed excitonic linewidth broadening is contributed by the in-plane component of the traveling SAW electric field and modulation of the monolayer bandgap by the dynamic strain field. The contribution due to the field induced ionization is significantly larger than the strain induced broadening. This is further supported by TRPL measurements performed under RF power at different optical excitation densities (Datta et al., 2021). The transient photoluminescence from monolayer WSe_2 can be fitted using a biexponential function:

$$I(t) = A\exp(-k_f t) + B\exp(-k_s t) \tag{8}$$

where the fast (k_f) component is attributed to the decay (radiative and nonradiative) from the excitonic state and slow (k_s) component is typically attributed to the emission from defect bound states (Sim et al., 2013; Mouri et al., 2014; Yan et al., 2014; Liu et al., 2019a). Using $k_{eff} = k_f$, the decay rate at different RF excitation for a fixed optical excitation and subtracting the excitation decay rate at the lowest RF excitation (−60 dBm), we can extract the ionization rate due to piezoelectric field:

$$k_{ion,piezoelectric} = k_{eff} - k_{decay,-60dBm} \tag{9}$$

The extracted ionization due to piezoelectric field is shown in Fig. 5. The piezoelectric ionization also shows an inverse optical excitation density dependence. The piezoelectric ionization trend observed in TRPL measurements supports the results obtained from PL measurements—a stronger dissociation of excitons in monolayer WSe_2 by the in-plane component of the SAW piezoelectric field under strong piezoelectric substrate induced screening at lower optical excitation and such interactions can be further modulated by free carrier screening.

7. Controlling type-I and type-II band edge modulation in monolayer semiconductors

So far, we have discussed signatures of both types of band edge modulation schemes: type-I and type-II. While type-I modulation scheme is suitable for directional exciton transport under strain from SAWs, type-II modulation leads to free carrier transport following exciton dissociation under SAW electric field. For a non-encapsulated monolayer, due to strong dielectric screening from the piezoelectric $LiNbO_3$ substrate (Lin et al., 2014), excitons dissociate into free carriers under SAW electric field. The free carriers are spatially separated and remain trapped in the energetically stable positions of the conduction (minima) and valance (maxima) bands and are transferred at the SAW velocity (Rocke et al., 1997). Thus, SAW induced dissociation results in an additional non-radiative decay channel and a faster decay is observed in transient photoluminescence. In addition, as excitons are dissociated into free carriers that remain spatially separated, no dynamic oscillations are observed in the spatiotemporal evolution of the exciton density. Therefore, type-II band edge modulation dominates in this regime. The generated exciton density does not show any directional transport feature as the SAW is turned *ON*. On the contrary, incorporating h-BN encapsulation reduces the screening effect from the $LiNbO_3$ substrate. The reduced screening increases the binding energy such that the SAW piezoelectric field is unable to dissociate the exciton. This suppresses the extent of type-II band edge modulation and allows the observation of exciton dynamics under type-I band edge modulation. The spatiotemporal exciton density shows strong oscillation in the order of SAW period due to exciton coupling to the traveling strain wave (Datta et al., 2022). We show the difference in type-I and type-II band edge modulation on a non-encapsulated monolayer WSe_2 and h-BN encapsulated monolayer WSe_2 under same SAW frequency in Fig. 6. Under the same RF power and optical excitation, excitons in an encapsulated monolayer show clear modulation at SAW period in gaussian peak evolution with time and transient photoluminescence (Fig. 6a and b respectively). Such periodic modulation is not observed for a non-encapsulated monolayer WSe_2 (Fig. 6c and d). Detailed information can be found in Datta et al. (2022).

The relative extent of type-I and type-II modulation can be further tuned in an encapsulated monolayer by optical fluence. As the SAW excitation is turned *ON*, exciton dissociation takes place under SAW electric field. At high optical fluence, strong free carrier screening acts as

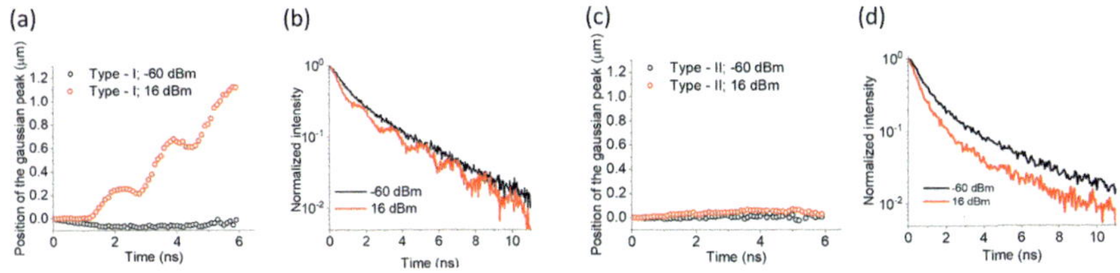

Fig. 6 Comparison of exciton dynamics under RF excitation for h-BN encapsulated monolayer WSe_2 (type-I band edge modulation) and a non-encapsulated WSe_2 (type-II band edge modulation) sample. Under the same RF power, (a) h-BN encapsulated monolayer shows modulation at SAW period. (b) Transient photoluminescence under RF excitation for h-BN encapsulated monolayer that shows oscillations at the SAW period. (c) Temporal evolution of exciton density peak extracted from single gaussian fitting under −60 dBm and 16 dBm RF power for non-encapsulated monolayer WSe_2. (d) TRPL under RF excitation for non-encapsulated monolayer. The non-encapsulated sample show a strong reduction in TRPL due to increased exciton dissociation under strong dielectric screening. Contrary to the encapsulated sample, the non-encapsulated sample neither shows any spatial drift nor shows any dynamic modulation. The data has been binned with 100 ps window to improve the signal to noise ratio following acquisition. *Reproduced with permission from Datta et al. (2022).*

moderator to reduce the effect of SAW electric field and lowers the exciton decay. On the contrary, under low optical fluence, exciton dissociation and ionization is enhanced as the screening from free carriers is reduced. This is observed from the TRPL measurements performed at two optical fluences for a set of RF powers. The results are shown in Fig. 7a and b. While TRPL performed at high optical fluence show oscillations in exciton decay at the SAW period due to inter-valley scattering, the TRPL at lower optical fluence does not show such oscillatory behavior. This observation is shown by a simple schematic in Fig. 7c. The extracted exciton decay rate change also shows a stronger increase at lower optical fluence as RF excitation is ramped up.

Optical excitation can be further utilized to as a possible tuning knob for exciton transport under SAW which occurs from density dependent many-body phenomenon such as non-radiative Auger scattering (Mouri et al., 2014; Yuan et al., 2017), halo formation (Kulig et al., 2018), hot exciton effects (Song and Dery, 2013; Cordovilla Leon et al., 2019), non-equilibrium phonon drag effects (Glazov, 2019). Strong non-linear interactions increases the exciton kinetic energy and density dependent diffusive transport (Kulig et al., 2018; Perea-Causín et al., 2019). Such high kinetic energy at higher optical fluence allows stronger exciton coupling to the traveling wave as the RF excitation is increased to 13 dBm. This is

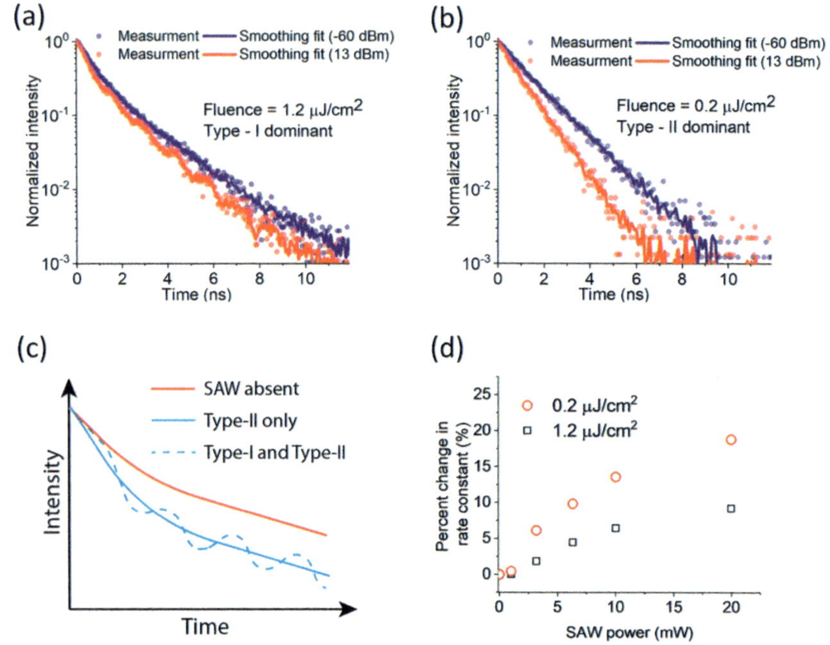

Fig. 7 Optical fluence dependent time resolved photoluminescence under RF excitations (−60 dBm and 13 dBm) for (a) 1.2 μJ/cm^2 and (b) 0.2 μJ/cm^2 optical fluences with smoothing fit. At lower optical fluence, we observe a sharp increase in the exciton decay rate due to an increase in exciton dissociation process at lower optical fluence. On the contrary, at a high optical fluence, the TRPL shows oscillations that recovers to the values at −60 dBm. (c) Schematic representation of TRPL under different band edge modulations. The figure is not drawn to scale. (d) Percent increase in the exciton decay rate as a function of RF excitation power. A clear increase in the decay rate is observed at low optical fluence due to lowered free carrier screening. *Reproduced with permission from Datta et al. (2022).*

clearly seen from the temporal evolution of the gaussian peak at two optical fluence (0.2 μJ/cm^2 and 1.2 μJ/cm^2) under the same RF excitation (13 dBm) as shown in Fig. 8.

8. Concluding remarks and outlook

To summarize, we have discussed SAW induced exciton control in monolayer TMDCs at room temperature, primarily focusing on directional transport under traveling strain and piezoelectric modulation of exciton photophysics. Although SAW technology and devices are routinely used to

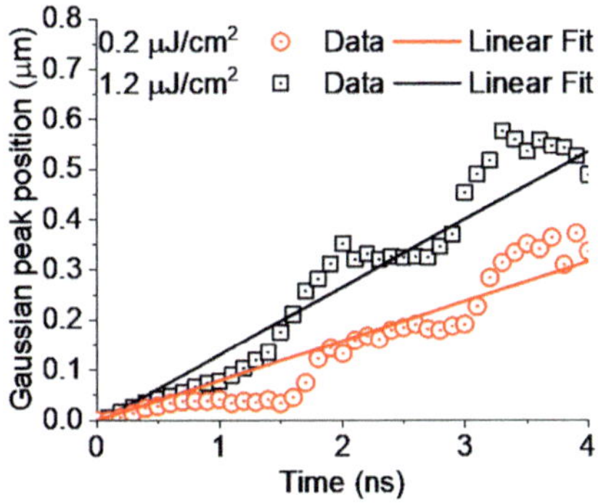

Fig. 8 Optical excitation dependent exciton diffusion in h-BN/WSe_2/h-BN sample. Peak position of the exciton density extracted from the gaussian fitting for two exciton densities under 13 dBm RF excitation. The average drift of exciton density under RF excitation is higher for higher exciton density. *Reproduced with permission from Datta et al. (2022).*

study particle and quasiparticle species in III–V semiconductor quantum well and other nanostructures, their integration to 2D semiconductors is still an emerging field. While phenomena such as directional exciton transport, acoustic stark shift, acoustic steering, acousto-galvanic effect etc. have been experimentally verified, demonstration of valley spin-acoustic effect (Sonowal et al., 2022), valley acousto-electric effect (Kalameitsev et al., 2019), valley acousto-electric Hall effect (Sonowal et al., 2020) is still missing and only theoretically proposed (Nie et al., 2023). As the field develops, further advancement in 2D materials growth, transfer and fabrication techniques will enable longer transport distances. Exfoliated monolayers typically show the best intrinsic transport properties and therefore the highest drift velocity under traveling strain. Therefore, large area monolayers with good intrinsic transport properties remain the key to achieve long range directional transport using traveling strain wave at room temperature. Improvement in transfer techniques can also help in achieving devices with better 2D material/piezoelectric substrate interface with lower defect densities that hinder energy transfer from the traveling strain wave to excitons, act as scattering centers for the traveling strain as well as excitons. So, the advancement of our fundamental understandings as well as applications of acousto-excitonics in 2D materials will require a multidisciplinary effort encompassing the field of chemistry, physics, material science and engineering.

Acknowledgment

This book chapter writing was supported by the Army Research Office under Grant Number W911NF-21-1-0207.

References

Aas, S., Bulutay, C., 2018. Strain dependence of photoluminescence and circular dichroism in transition metal dichalcogenides: a *k.p* analysis. Opt. Express 26 (22), 28672. https://doi.org/10.1364/oe.26.028672.

Aivazian, G., et al., 2015. Magnetic control of valley pseudospin in monolayer WSe_2. Nat. Phys. 11 (2), 148–152. https://doi.org/10.1038/nphys3201.

Akinwande, D., et al., 2017. A review on mechanics and mechanical properties of 2D materials—Graphene and beyond. Extreme Mech. Lett. 13, 42–77. https://doi.org/10.1016/J.EML.2017.01.008.

Arcizet, O., et al., 2011. A single nitrogen-vacancy defect coupled to a nanomechanical oscillator. Nat. Phys. 7 (11), 879–883. https://doi.org/10.1038/nphys2070.

Arora, A., et al., 2015. Excitonic resonances in thin films of WSe_2: From monolayer to bulk material. Nanoscale 7 (23), 10421–10429. https://doi.org/10.1039/c5nr01536g.

Aslan, O.B., et al., 2020. Strained bilayer WSe_2 with reduced exciton-phonon coupling. Phys. Rev. B 101 (11), 115305. https://doi.org/10.1103/PHYSREVB.101.115305/FIGURES/4/MEDIUM.

Aslan, O.B., Deng, M., Heinz, T.F., 2018. Strain tuning of excitons in monolayer WSe_2. Phys. Rev. B 98 (11), 115308. https://doi.org/10.1103/PhysRevB.98.115308.

Azzam, S.I., Parto, K., Moody, G., 2021. Prospects and challenges of quantum emitters in 2D materials. Appl. Phys. Lett. 118 (24), 240502. https://doi.org/10.1063/5.0054116/14549116/240502_1_ACCEPTED_MANUSCRIPT.PDF.

Bae, S.H., et al., 2019. Integration of bulk materials with two-dimensional materials for physical coupling and applications. Nat. Mater. 18 (6), 550–560. https://doi.org/10.1038/s41563-019-0335-2.

Baldo, M., Stojanovi, V., 2009. Optical switching: Excitonic interconnects. Nat. Photonics 3 (10), 558–560. https://doi.org/10.1038/nphoton.2009.178.

Bandhu, L., Lawton, L.M., Nash, G.R., 2013. Macroscopic acoustoelectric charge transport in graphene. Appl. Phys. Lett. 103 (13), 133101. https://doi.org/10.1063/1.4822121.

Berkelbach, T.C., Hybertsen, M.S., Reichman, D.R., 2013. Theory of neutral and charged excitons in monolayer transition metal dichalcogenides. Phys. Rev. B - Condens. Matter Mater. Phys. 88 (4), 045318. https://doi.org/10.1103/PhysRevB.88.045318.

Bertrand, B., et al., 2016. Fast spin information transfer between distant quantum dots using individual electrons. Nat. Nanotechnol. 11 (8), 672–676. https://doi.org/10.1038/nnano.2016.82.

Biryukov, S.V., et al., 1995. Surface Acoustic Waves in Inhomogeneous Media. https://doi.org/10.1007/978-3-642-57767-3.

Branny, A., et al., 2017. Deterministic strain-induced arrays of quantum emitters in a two-dimensional semiconductor. Nat. Commun. 8 (1), 1–7. https://doi.org/10.1038/ncomms15053.

Butov, L.V., 2017. Excitonic devices. Superlattices Microstruct. 108, 2–26. https://doi.org/10.1016/j.spmi.2016.12.035.

Büyükköse, S., et al., 2014. High-frequency acoustic charge transport in GaAs nanowires. Nanotechnology 25, 135204. https://doi.org/10.1088/0957-4484/25/13/135204.

Cadiz, F., et al., 2018. Exciton diffusion in WSe_2 monolayers embedded in a van der Waals heterostructure. Appl. Phys. Lett. 112 (15), 152106. https://doi.org/10.1063/1.5026478.

Campbell, C.1989. Surface Acoustic Wave Devices and Their Signal Processing Applications, p. 485.

Cao, T., et al., 2012. Valley-selective circular dichroism of monolayer molybdenum disulphide. Nat. Commun. 3 (1), 1–5. https://doi.org/10.1038/ncomms1882.

Castellanos-Gomez, A., et al., 2013. Local strain engineering in atomically thin MoS_2. Nano Lett. 13 (11), 5361–5366. https://doi.org/10.1021/nl402875m.

Cerda-Méndez, E.A., et al., 2017. Quantum fluids of light in acoustic lattices. J. Phys. D: Appl. Phys. 51 (3), 033001. https://doi.org/10.1088/1361-6463/AA9EC7.

Cheng, J., et al., 2019. Recent advances in optoelectronic devices based on 2D materials and their heterostructures. Adv. Optical Mater. 7 (1), 1800441. https://doi.org/10.1002/ADOM.201800441.

Chernikov, A., et al., 2014. Exciton binding energy and nonhydrogenic Rydberg series in monolayer WS_2. Phys. Rev. Lett. 113 (7), 076802. https://doi.org/10.1103/PhysRevLett.113.076802.

Chernikov, A., et al., 2015. Electrical tuning of exciton binding energies in monolayer WS_2. Phys. Rev. Lett. 115 (12), 126802. https://doi.org/10.1103/PhysRevLett.115.126802.

Choy, W.C.H., Li, E.H., Weiss, B.L., 1998. Electro-optic and electro-absorptive modulations of AlGaAs/GaAs quantum well using surface acoustic wave. J. Appl. Phys. 83 (2), 858–866. https://doi.org/10.1063/1.366768.

Christopher, J.W., et al., 2019. Monolayer MoS_2 Strained to 1.3% with a microelectromechanical system. J. Microelectromech. Syst. 28 (2), 254–263. https://doi.org/10.1109/JMEMS.2018.2877983.

Ciarrocchi, A., et al., 2022. Excitonic devices with van der Waals heterostructures: valleytronics meets twistronics. Nat. Rev. Mater. 7 (6), 449–464. https://doi.org/10.1038/s41578-021-00408-7.

Cordovilla Leon, D.F., et al., 2018. Exciton transport in strained monolayer WSe_2. Appl. Phys. Lett. 113 (25), 252101. https://doi.org/10.1063/1.5063263.

Cordovilla Leon, D.F., et al., 2019. Hot exciton transport in WSe_2 monolayers. Phys. Rev. B 100 (24), 241401. https://doi.org/10.1103/PhysRevB.100.241401.

Couto, O.D.D., et al., 2009. Photon anti-bunching in acoustically pumped quantum dots. Nat. Photonics 3 (11), 645–648. https://doi.org/10.1038/nphoton.2009.191.

Dai, Z., et al., 2019. Strain engineering of 2D materials: Issues and opportunities at the interface. Adv. Mater. 31 (45), 1805417. https://doi.org/10.1002/ADMA.201805417.

Datta, K., et al., 2021. Piezoelectric modulation of excitonic properties in monolayer WSe_2 under strong dielectric screening. ACS Nano 15 (7), 12334–12341. https://doi.org/10.1021/ACSNANO.1C04269.

Datta, K., et al., 2022. Spatiotemporally controlled room-temperature exciton transport under dynamic strain. Nat. Photonics 16 (3), 242–247. https://doi.org/10.1038/s41566-021-00951-3.

Dean, C.R., et al., 2010. Boron nitride substrates for high-quality graphene electronics. Nat. Nanotechnol. 5 (10), 722–726. https://doi.org/10.1038/nnano.2010.172.

Delsing, P., et al., 2019. The 2019 surface acoustic waves roadmap. J. Phys. D: Appl. Phys. 52 (35), 353001. https://doi.org/10.1088/1361-6463/ab1b04.

Desai, S.B., et al., 2014. Strain-induced indirect to direct bandgap transition in multilayer WSe_2. Nano Lett. 14 (8), 4592–4597. https://doi.org/10.1021/nl501638a.

Dorow, C.J., et al., 2018. High-mobility indirect excitons in wide single quantum well. Appl. Phys. Lett. 113 (21), 212102. https://doi.org/10.1063/1.5063844.

Ellis, J.K., Lucero, M.J., Scuseria, G.E., 2011. The indirect to direct band gap transition in multilayered MoS_2 as predicted by screened hybrid density functional theory. Appl. Phys. Lett. 99 (26), 261908. https://doi.org/10.1063/1.3672219.

Flatten, L.C., et al., 2018. Microcavity enhanced single photon emission from two-dimensional WSe2. Appl. Phys. Lett. 112 (19). https://doi.org/10.1063/1.5026779.

Fu, X., Jacopin, G., et al., 2014. Exciton drift in semiconductors under uniform strain gradients: application to bent ZnO microwires. ACS Nano 8 (4), 3412–3420. https://doi.org/10.1021/nn4062353.

Fu, X., Su, C., et al., 2014b. Tailoring exciton dynamics by elastic strain-gradient in semiconductors. Adv. Mater. 26, 2572–2579. https://doi.org/10.1002/adma.201305058.

Fuhrmann, D.A., et al., 2009. Surface acoustic wave mediated exciton dissociation in a ZnCdSe/$LiNbO_3$ hybrid. Appl. Phys. Lett. 94 (19), 193505. https://doi.org/10.1063/1.3136862.

Fuhrmann, D.A., et al., 2011. Dynamic modulation of photonic crystal nanocavities using gigahertz acoustic phonons. Nat. Photonics 5 (10), 605–609. https://doi.org/10.1038/nphoton.2011.208.

García-Cristóbal, A., et al., 2004. Spatiotemporal carrier dynamics in quantum wells under surface acoustic waves. Phys. Rev. B - Condens. Matter Mater. Phys. 69 (20), 1–13. https://doi.org/10.1103/PhysRevB.69.205301.

Gärtner, A., et al., 2006. Drift mobility of long-living excitons in coupled GaAs quantum wells. Appl. Phys. Lett. 89 (5), 52108. https://doi.org/10.1063/1.2267263.

Gell, J.R., et al., 2008. Modulation of single quantum dot energy levels by a surface-acoustic-wave. Appl. Phys. Lett. 93 (8). https://doi.org/10.1063/1.2976135/923500.

Ghorbani-Asl, M., et al., 2013. Strain-dependent modulation of conductivity in single-layer transition-metal dichalcogenides. Phys. Rev. B - Condens. Matter Mater. Phys. 87 (23), 235434. https://doi.org/10.1103/PHYSREVB.87.235434/FIGURES/6/MEDIUM.

Glazov, M.M., 2019. Phonon wind and drag of excitons in monolayer semiconductors. Phys. Rev. B 100 (4), 045426. https://doi.org/10.1103/PhysRevB.100.045426.

Golter, D.A., et al., 2016. Optomechanical quantum control of a nitrogen-vacancy center in diamond. Phys. Rev. Lett. 116 (14), 143602. https://doi.org/10.1103/PHYSREVLETT.116.143602/FIGURES/4/MEDIUM.

Golub, J.E., et al., 1990. Long-lived spatially indirect excitons in coupled GaAs/$Al_xGa_{1-x}As$ quantum wells. Phys. Rev. B 41 (12), 8564. https://doi.org/10.1103/PhysRevB.41.8564.

Goodman, A.J., et al., 2020. Substrate-dependent exciton diffusion and annihilation in chemically treated MoS_2 and WS_2. J. Phys. Chem. C. 124 (22), 12175–12184. https://doi.org/10.1021/acs.jpcc.0c04000.

Grosso, G., et al., 2009. Excitonic switches operating at around 100K. Nat. Photonics 3 (10), 577–580. https://doi.org/10.1038/nphoton.2009.166.

Gu, J., et al., 2019. A room-temperature polariton light-emitting diode based on monolayer WS2. Nat. Nanotechnol. 14 (11), 1024–1028. https://doi.org/10.1038/s41565-019-0543-6.

Harats, M.G., et al., 2020. Dynamics and efficient conversion of excitons to trions in non-uniformly strained monolayer WS_2. Nat. Photonics 14 (5), 324–329. https://doi.org/10.1038/s41566-019-0581-5.

Harats, M.G., Bolotin, K.I., 2021. Limits of funneling efficiency in non-uniformly strained 2D semiconductors. 2D Mater. 8 (1), 015010. https://doi.org/10.1088/2053-1583/abbabf.

Hashimoto, K., 2000. Surface Acoustic Wave Devices in Telecommunications. https://doi.org/10.1007/978-3-662-04223-6.

He, M., et al., 2020. Valley phonons and exciton complexes in a monolayer semiconductor. Nat. Commun. 11 (1), 1–7. https://doi.org/10.1038/s41467-020-14472-0.

Hermelin, S., et al., 2011. Electrons surfing on a sound wave as a platform for quantum optics with flying electrons. Nature 477 (7365), 435–438. https://doi.org/10.1038/nature10416.

Hernández-Mínguez, A., et al., 2012. Acoustically driven photon antibunching in nanowires. Nano Lett. 12 (1), 252–258. https://doi.org/10.1021/nl203461m.

Hernández-Minguez, A., Liou, Y.T., Santos, P.V., 2018. Interaction of surface acoustic waves with electronic excitations in graphene. J. Phys. D: Appl. Phys. 51, 383001. https://doi.org/10.1088/1361-6463/aad593.

High, A.A., et al., 2008. Control of exciton fluxes in an excitonic integrated circuit. Science 321 (5886), 229–231. https://doi.org/10.1126/science.1157845.
Hillmer, H., et al., 1989. Optical investigations on the mobility of two-dimensional excitons in GaAs/Ga_{1-x} Al_xAs quantum wells. Phys. Rev. B 39 (15), 10901.
Hoshi, Y., et al., 2017. Suppression of exciton-exciton annihilation in tungsten disulfide monolayers encapsulated by hexagonal boron nitrides. Phys. Rev. B 95 (24), 241403. https://doi.org/10.1103/PhysRevB.95.241403.
Hotta, T., et al., 2020. Exciton diffusion in a h BN-encapsulated monolayer $MoSe_2$. Phys. Rev. B 102 (11), 115424. https://doi.org/10.1103/PhysRevB.102.115424.
Huang, J., Hoang, T.B., Mikkelsen, M.H., 2016. Probing the origin of excitonic states in monolayer WSe_2. Sci. Rep. 6 (1), 1–7. https://doi.org/10.1038/srep22414.
Huang, T., et al., 2017. Theoretical study on dynamic acoustic modulation of free carriers, excitons, and trions in 2D MoS_2 flake. J. Phys. D: Appl. Phys. 50 (11), 114005. https://doi.org/10.1088/1361-6463/aa590d.
Hwang, E.H., Adam, S., Sarma, S.D., 2007. Carrier transport in two-dimensional graphene layers. Phys. Rev. Lett. 98 (18), 186806. https://doi.org/10.1103/PhysRevLett.98.186806.
Island, J.O., et al., 2016. Precise and reversible band gap tuning in single-layer $MoSe_2$ by uniaxial strain. Nanoscale 8 (5), 2589–2593. https://doi.org/10.1039/C5NR08219F.
Ivanov, A.L., Littlewood, P.B., 2001. Acoustically induced stark effect for excitons in intrinsic semiconductors. Phys. Rev. Lett. 87 (13), 136403. https://doi.org/10.1103/PhysRevLett.87.136403.
Jauregui, L.A., et al., 2019. Electrical control of interlayer exciton dynamics in atomically thin heterostructures. Science 366 (6467), 870–875. https://doi.org/10.1126/science.aaw4194.
Jia, L., et al., 2022. Fabrication technologies for the on-chip integration of 2D materials. Small Methods 6 (3), 2101435. https://doi.org/10.1002/SMTD.202101435.
Jiang, Y., et al., 2021. Interlayer exciton formation, relaxation, and transport in TMDCs van der Waals heterostructures. Light: Sci. Appl. 10 (1), 2047–7538. https://doi.org/10.1038/s41377-021-00500-1.
Jing, Y., et al., 2020. Tunable electronic structure of two-dimensional transition metal chalcogenides for optoelectronic applications. Nanophotonics 9 (7), 1675–1694. https://doi.org/10.1515/NANOPH-2019-0574.
Kalameitsev, A.V., Kovalev, V.M., Savenko, I.G., 2019. Valley acoustoelectric effect. Phys. Rev. Lett. 122 (25), 256801. https://doi.org/10.1103/PHYSREVLETT.122.256801/FIGURES/3/MEDIUM.
Kang, J.W., et al., 2019. Enhancing exciton diffusion in monolayer WS_2 with h-BN bottom layer. Phys. Rev. B 100 (20), 205304. https://doi.org/10.1103/PhysRevB.100.205304.
Kim, J., et al., 2017. Observation of ultralong valley lifetime in WSe_2/MoS_2 heterostructures. Sci. Adv. 3 (7). https://doi.org/10.1126/SCIADV.1700518/SUPPL_FILE/1700518_SM.PDF.
Kinzel, J.B., et al., 2011. Directional and dynamic modulation of the optical emission of an individual GaAs nanowire using surface acoustic waves. Nano Lett. 11 (4), 1512–1517. https://doi.org/10.1021/nl1042775.
Klein, J., et al., 2016. Stark effect spectroscopy of mono- and few-layer MoS_2. Nano Lett. 16 (3), 1554–1559. https://doi.org/10.1021/acs.nanolett.5b03954.
Komider, K., González, J.W., Fernández-Rossier, J., 2013. Large spin splitting in the conduction band of transition metal dichalcogenide monolayers. Phys. Rev. B - Condens. Matter Mater. Phys. 88 (24), 245436. https://doi.org/10.1103/PHYSREVB.88.245436/FIGURES/4/MEDIUM.
Komsa, H.P., Krasheninnikov, A.V., 2012. Effects of confinement and environment on the electronic structure and exciton binding energy of MoS_2 from first principles. Phys. Rev. B - Condens. Matter Mater. Phys. 86 (24), 241201. https://doi.org/10.1103/PhysRevB.86.241201.

Koo, Y., et al., 2021. Tip-induced nano-engineering of strain, bandgap, and exciton funneling in 2D semiconductors. Adv. Mater. 33 (17), 2008234. https://doi.org/10.1002/adma.202008234.

Kuc, A., Zibouche, N., Heine, T., 2011. Influence of quantum confinement on the electronic structure of the transition metal sulfide TS_2. Phys. Rev. B - Condens. Matter Mater. Phys. 83 (24), 245213. https://doi.org/10.1103/PHYSREVB.83.245213/FIGURES/6/MEDIUM.

Kulig, M., et al., 2018. Exciton diffusion and Halo effects in monolayer semiconductors. Phys. Rev. Lett. 120 (20), 207401. https://doi.org/10.1103/PhysRevLett.120.207401.

Kuznetsova, Y.Y., et al., 2010. All-optical excitonic transistor. Opt. Lett. 35 (10), 1587–1589. https://doi.org/10.1364/OL.35.001587.

Latini, S., et al., 2019. Cavity control of excitons in two-dimensional materials. Nano Lett. 19 (6), 3473–3479. https://doi.org/10.1021/ACS.NANOLETT.9B00183/ASSET/IMAGES/LARGE/NL-2019-001837_M011.JPEG.

Lazić, S., et al., 2014. Scalable interconnections for remote indirect exciton systems based on acoustic transport. Phys. Rev. B - Condens. Matter Mater. Phys. 89 (8), 1–8. https://doi.org/10.1103/PhysRevB.89.085313.

Lazić, S., et al., 2015. Dynamic control of the optical emission from GaN/InGaN nanowire quantum dots by surface acoustic waves. AIP Adv. 5 (9), 97217. https://doi.org/10.1063/1.4932147.

Lazić, S., et al., 2019. Dynamically tuned non-classical light emission from atomic defects in hexagonal boron nitride. Commun. Phys. 2 (1), 1–8. https://doi.org/10.1038/s42005-019-0217-6.

Lee, H., et al., 2022. Drift-dominant exciton funneling and trion conversion in 2D semiconductors on the nanogap. Sci. Adv. 8 (5), 5236. https://doi.org/10.1126/SCIADV.ABM5236/SUPPL_FILE/SCIADV.ABM5236_SM.PDF.

Leonard, J.R., et al., 2012. Transport of indirect excitons in a potential energy gradient. Appl. Phys. Lett. 100 (23), 1–5. https://doi.org/10.1063/1.4722938.

Li, Z., et al., 2021. Dielectric engineering for manipulating exciton transport in semiconductor monolayers. Nano Lett. 21, 8409–8417. https://doi.org/10.1021/ACS.NANOLETT.1C02990.

De Lima, M.M., Santos, P.V., 2005. Modulation of photonic structures by surface acoustic waves. Rep. Prog. Phys. 68 (7), 1639–1701. https://doi.org/10.1088/0034-4885/68/7/R02.

Lin, Y., et al., 2014. Dielectric screening of excitons and trions in single-layer MoS_2. Nano Lett. 14 (10), 5569–5576. https://doi.org/10.1021/nl501988y.

Liu, G.B., et al., 2015. Electronic structures and theoretical modelling of two-dimensional group-VIB transition metal dichalcogenides. Chem. Soc. Rev. 44 (9), 2643–2663. https://doi.org/10.1039/C4CS00301B.

Liu, X., Hersam, M.C., 2018. Interface characterization and control of 2D materials and heterostructures. Adv. Mater. 30 (39), 1801586. https://doi.org/10.1002/ADMA.201801586.

Liu, H., et al., 2019a. Neutral and defect-induced exciton annihilation in defective monolayer WS_2. Nanoscale 11 (16), 7913–7920. https://doi.org/10.1039/c9nr00967a.

Liu, Y., et al., 2019b. Recent progress in the fabrication, properties, and devices of heterostructures based on 2D materials. Nano-Micro Lett. 11 (1), 1–24. https://doi.org/10.1007/S40820-019-0245-5.

Liu, Y., et al., 2020. Electrically controllable router of interlayer excitons. Sci. Adv. 6 (41), eaba1830. https://doi.org/10.1126/sciadv.aba1830.

Luo, Y., et al., 2018. Deterministic coupling of site-controlled quantum emitters in monolayer WSe2 to plasmonic nanocavities. Nat. Nanotechnol. 13 (12), 1137–1142. https://doi.org/10.1038/s41565-018-0275-z.

Mak, K.F., et al., 2012. Control of valley polarization in monolayer MoS2by optical helicity. Nat. Nanotechnol. 7 (8), 494–498. https://doi.org/10.1038/nnano.2012.96.

Mandal, D., Banerjee, S., 2022. Surface acoustic wave (SAW) sensors: physics, materials, and applications. Sensors 22 (3), 820. https://doi.org/10.3390/S22030820.

Manenti, R., et al., 2016. Surface acoustic wave resonators in the quantum regime. Phys. Rev. B 93 (4), 041411. https://doi.org/10.1103/PHYSREVB.93.041411/FIGURES/4/MEDIUM.

Mangu, V.S., et al., 2017. Strain engineering, efficient excitonic photoluminescence, and exciton funnelling in unmodified MoS_2 nanosheets. Nanoscale 9 (43), 16602–16606. https://doi.org/10.1039/C7NR03537C.

McNeil, R.P.G., et al., 2011. On-demand single-electron transfer between distant quantum dots. Nature 477 (7365), 439–442. https://doi.org/10.1038/nature10444.

Metcalfe, M., et al., 2010. Resolved sideband emission of InAs/GaAs quantum dots strained by surface acoustic waves. Phys. Rev. Lett. 105 (3), 037401. https://doi.org/10.1103/PHYSREVLETT.105.037401/FIGURES/4/MEDIUM.

Mikhnenko, O.V., Blom, P.W.M., Nguyen, T.-Q., 2015. Exciton diffusion in organic semiconductors. Energy Environ. Sci. 8 (7), 1867–1888. https://doi.org/10.1039/C5EE00925A.

Miller, D.A.B., et al., 1985. Electric field dependence of optical absorption near the band gap of quantum-well structures. Phys. Rev. B 32 (2), 1043–1060. https://doi.org/10.1103/PhysRevB.32.1043.

Miseikis, V., et al., 2012. Acoustically induced current flow in graphene. Appl. Phys. Lett. 100 (13), 133105. https://doi.org/10.1063/1.3697403.

Moon, H., et al., 2020. Dynamic exciton funneling by local strain control in a monolayer semiconductor. Nano Lett. 20 (9), 6791–6797.

Moores, B.A., et al., 2018. Cavity quantum acoustic device in the multimode strong coupling regime. Phys. Rev. Lett. 120 (22), 227701. https://doi.org/10.1103/PHYSREVLETT.120.227701/FIGURES/4/MEDIUM.

Morgan, D., 2007. Surface Acoustic Wave Filters, Surface Acoustic Wave Filters. Elsevier Ltd, Cambridge, Massachusetts. https://doi.org/10.1016/B978-0-12-372537-0.X5000-6.

Morozov, S.V., et al., 2008. Giant intrinsic carrier mobilities in graphene and its bilayer. Phys. Rev. Lett. 100 (1), 016602. https://doi.org/10.1103/PhysRevLett.100.016602.

Mouri, S., et al., 2014. Nonlinear photoluminescence in atomically thin layered WSe_2 arising from diffusion-assisted exciton-exciton annihilation. Phys. Rev. B - Condens. Matter Mater. Phys. 90 (15), 1–5. https://doi.org/10.1103/PhysRevB.90.155449.

Nie, X., et al., 2023. Surface acoustic wave induced phenomena in two-dimensional materials. Nanoscale Horiz. 8 (2), 158–175. https://doi.org/10.1039/D2NH00458E.

Niehues, I., et al., 2018. Strain control of exciton-phonon coupling in atomically thin semiconductors. Nano Lett. 18 (3), 1751–1757. https://doi.org/10.1021/acs.nanolett.7b04868.

Novoselov, K.S., et al., 2016. 2D materials and van der Waals heterostructures. Science. https://doi.org/10.1126/science.aac9439.

Okada, A., et al., 2018. Cavity enhancement of anti-Stokes scattering via optomechanical coupling with surface acoustic waves. Phys. Rev. Appl. 10 (2), 024002. https://doi.org/10.1103/PHYSREVAPPLIED.10.024002/FIGURES/5/MEDIUM.

Palacios-Berraquero, C., et al., 2017. Large-scale quantum-emitter arrays in atomically thin semiconductors. Nat. Commun. 8 (1), 1–6. https://doi.org/10.1038/ncomms15093.

Patel, S.D., et al., 2023. Cavity optomechanics with WSe_2 single photon emitters. In: CLEO 2023, p. FW3J.2. Available at ⟨https://opg.optica.org/abstract.cfm?uri=CLEO_FS-2023-FW3J.2⟩.

Pei, J., et al., 2019. Many-body complexes in 2D semiconductors. Adv. Mater. 31 (2), 1706945. https://doi.org/10.1002/ADMA.201706945.

Peng, R., et al., 2022. Long-range transport of 2D excitons with acoustic waves. Nat. Commun. 13 (1), 1–7. https://doi.org/10.1038/s41467-022-29042-9.
Peng, Z., et al., 2020. Strain engineering of 2D semiconductors and graphene: from strain fields to band-structure tuning and photonic applications. Light: Sci. Appl. 9 (1), 1–25. https://doi.org/10.1038/s41377-020-00421-5.
Perea-Causín, R., et al., 2019. Exciton propagation and halo formation in two-dimensional materials. Nano Lett. 19 (10), 7317–7323. https://doi.org/10.1021/acs.nanolett.9b02948.
Pettenkofer, C., et al., 2001. Electronic band structure of single-crystal and single-layer WS_2: Influence of interlayer van der Waals interactions. Phys. Rev. B 64 (20), 205416. https://doi.org/10.1103/PhysRevB.64.205416.
Preciado, E., et al., 2015. Scalable fabrication of a hybrid field-effect and acousto-electric device by direct growth of monolayer $MoS_2/LiNbO_3$. Nat. Commun. 6 (1), 1–8. https://doi.org/10.1038/ncomms9593.
Premasiri, K., Gao, X.P.A., 2019. Tuning spin–orbit coupling in 2D materials for spintronics: A topical review. J. Phys.: Condens. Matter 31 (19), 193001. https://doi.org/10.1088/1361-648X/AB04C7.
Qi, P., et al., 2022. Molding 2D exciton flux toward room temperature excitonic devices. Adv. Mater. Technol. 2200032. https://doi.org/10.1002/ADMT.202200032.
Qiu, Z., et al., 2019. Giant gate-tunable bandgap renormalization and excitonic effects in a 2D semiconductor. Sci. Adv. 5 (7), eaaw2347. https://doi.org/10.1126/SCIADV.AAW2347/SUPPL_FILE/AAW2347_SM.PDF.
Raja, A., et al., 2017. Coulomb engineering of the bandgap and excitons in two-dimensional materials. Nat. Commun. 8 (1), 1–7. https://doi.org/10.1038/ncomms15251.
Rasmussen, F.A., Thygesen, K.S., 2015. Computational 2D materials database: Electronic structure of transition-metal dichalcogenides and oxides. J. Phys. Chem. C 119 (23), 13169–13183. https://doi.org/10.1021/ACS.JPCC.5B02950/SUPPL_FILE/JP5B02950_SI_001.PDF.
Rezk, A.R., et al., 2015. Acoustic-excitonic coupling for dynamic photoluminescence manipulation of quasi-2D MoS_2 nanoflakes. Adv. Optical Mater. 3 (7), 888–894. https://doi.org/10.1002/adom.201500034.
Rezk, A.R., et al., 2016. Acoustically-driven trion and exciton modulation in piezoelectric two-dimensional MoS_2. Nano Lett. 16 (2), 849–855. https://doi.org/10.1021/acs.nanolett.5b02826.
Rivera, P., et al., 2016. Valley-polarized exciton dynamics in a 2D semiconductor heterostructure. Science 351 (6274), 688–691. https://doi.org/10.1126/SCIENCE.AAC7820/SUPPL_FILE/RIVERA-SM.PDF.
Rivera, P., et al., 2021. Intrinsic donor-bound excitons in ultraclean monolayer semiconductors. Nat. Commun. 12 (1), 1–8. https://doi.org/10.1038/s41467-021-21158-8.
Roch, J.G., et al., 2018. Quantum-confined Stark effect in a MoS_2 monolayer van der Waals heterostructure. Nano Lett. 18 (2), 1070–1074. https://doi.org/10.1021/acs.nanolett.7b04553.
Rocke, C., et al., 1997. Acoustically driven storage of light in a quantum well. Phys. Rev. Lett. 78 (21), 4099–4102. https://doi.org/10.1103/PhysRevLett.78.4099.
Roldán, R., et al., 2014. Electronic properties of single-layer and multilayer transition metal dichalcogenides MX_2 (M=Mo, W and X=S, Se). Ann. Phys. 526 (9–10), 347–357. https://doi.org/10.1002/ANDP.201400128.
Rosenberger, M.R., et al., 2019. Quantum calligraphy: Writing single-photon emitters in a two-dimensional materials platform. ACS Nano 13 (1), 904–912. https://doi.org/10.1021/ACSNANO.8B08730/ASSET/IMAGES/LARGE/NN-2018-087303_0005.JPEG.

Rudolph, J., Hey, R., Santos, P.V., 2007a. Exciton transport by surface acoustic waves. Superlattices Microstructures 41 (5–6), 293–296. https://doi.org/10.1016/j.spmi.2007.03.008.

Rudolph, J., Hey, R., Santos, P.V., 2007b. Long-range exciton transport by dynamic strain fields in a GaAs quantum well. Phys. Rev. Lett. 99 (4), 1–4. https://doi.org/10.1103/PhysRevLett.99.047602.

Santos, P.V., et al., 2004. Band mixing and ambipolar transport by surface acoustic waves in GaAs quantum wells. Phys. Rev. B - Condens. Matter Mater. Phys. 69 (15), 155318. https://doi.org/10.1103/PhysRevB.69.155318.

Santos, P.V., Ramsteiner, M., Jungnickel, F., 1998. Spatially resolved photoluminescence in GaAs surface acoustic wave structures. Appl. Phys. Lett. 72 (17), 2099–2101. https://doi.org/10.1063/1.121288.

Scalise, E., et al., 2011. Strain-induced semiconductor to metal transition in the two-dimensional honeycomb structure of MoS_2. Nano Res. 5 (1), 43–48. https://doi.org/10.1007/S12274-011-0183-0.

Schaibley, J.R., et al., 2016. Valleytronics in 2D materials. Nat. Rev. Mater. 1 (11), 1–15. https://doi.org/10.1038/natrevmats.2016.55.

Schütz, M.J.A., 2017. Universal Quantum Transducers Based on Surface Acoustic Waves. pp. 143–196. https://doi.org/10.1007/978-3-319-48559-1_4.

Scolfaro, D., et al., 2021. Acoustically driven Stark effect in transition metal dichalcogenide monolayers. ACS Nano 15 (9), 15371–15380. https://doi.org/10.1021/ACSNANO.1C06854/ASSET/IMAGES/MEDIUM/NN1C06854_M002.svg.

Shi, H., et al., 2013. Quasiparticle band structures and optical properties of strained monolayer MoS_2 and WS_2. Phys. Rev. B - Condens. Matter Mater. Phys. 87 (15), 1–8. https://doi.org/10.1103/PhysRevB.87.155304.

Sim, S., et al., 2013. Exciton dynamics in atomically thin MoS_2: Interexcitonic interaction and broadening kinetics. Phys. Rev. B - Condens. Matter Mater. Phys. 88 (7), 075434. https://doi.org/10.1103/PhysRevB.88.075434.

Song, W., Yang, L., 2017. Quasiparticle band gaps and optical spectra of strained monolayer transition-metal dichalcogenides. Phys. Rev. B 96 (23), 235441. https://doi.org/10.1103/PHYSREVB.96.235441/FIGURES/4/MEDIUM.

Song, Y., Dery, H., 2013. Transport theory of monolayer transition-metal dichalcogenides through symmetry. Phys. Rev. Lett. 111 (2), 026601. https://doi.org/10.1103/PhysRevLett.111.026601.

Sonowal, K., et al., 2020. Acoustoelectric effect in two-dimensional Dirac materials exposed to Rayleigh surface acoustic waves. Phys. Rev. B 102 (23), 235405. https://doi.org/10.1103/PHYSREVB.102.235405/FIGURES/4/MEDIUM.

Sonowal, K., et al., 2022. Valley spin-acoustic resonance in MoS2 monolayers. Phys. Rev. B 106 (15), 155426. https://doi.org/10.1103/PHYSREVB.106.155426/FIGURES/5/MEDIUM.

Stier, A.V., et al., 2016. Probing the influence of dielectric environment on excitons in monolayer WSe_2: Insight from high magnetic fields. Nano Lett. 16 (11), 7054–7060. https://doi.org/10.1021/acs.nanolett.6b03276.

Tadesse, S.A., Li, M., 2015. Sub-optical wavelength acoustic wave modulation of integrated photonic resonators at microwave frequencies. Nat. Commun. 5 (May), 1–7. https://doi.org/10.1038/ncomms6402.

Unuchek, D., et al., 2018. Room-temperature electrical control of exciton flux in a van der Waals heterostructure. Nature 560 (7718), 340–344. https://doi.org/10.1038/s41586-018-0357-y.

Villa, B., et al., 2017. Surface acoustic wave modulation of a coherently driven quantum dot in a pillar microcavity. Appl. Phys. Lett. 111 (1), 011103. https://doi.org/10.1063/1.4990966/33684.

Violante, A., et al., 2014. Dynamics of indirect exciton transport by moving acoustic fields. New J. Phys. 16 (3), 033035. https://doi.org/10.1088/1367-2630/16/3/033035.
Völk, S., et al., 2010. Enhanced sequential carrier capture into individual quantum dots and quantum posts controlled by surface acoustic waves. Nano Lett. 10, 3399–3407. https://doi.org/10.1021/nl1013053.
Wang, G., et al., 2018. Colloquium: Excitons in atomically thin transition metal dichalcogenides. Rev. Mod. Phys. 90 (2), 021001. https://doi.org/10.1103/RevModPhys.90.021001.
Wang, K., et al., 2019. Exciton funneling in light-harvesting organic semiconductor microcrystals for wavelength-tunable lasers. Sci. Adv. 5 (6), 2953–2967. https://doi.org/10.1126/SCIADV.AAW2953/SUPPL_FILE/AAW2953_SM.PDF.
Wang, S., Liu, X., Zhou, P., 2022. The road for 2D semiconductors in the silicon age. Adv. Mater. 34 (48), 2106886. https://doi.org/10.1002/ADMA.202106886.
Wang, Y., et al., 2015. Strain-induced direct–indirect bandgap transition and phonon modulation in monolayer WS 2. Nano Res. 8 (8), 2562–2572. https://doi.org/10.1007/S12274-015-0762-6.
Weiß, M., Krenner, H.J., 2018. Interfacing quantum emitters with propagating surface acoustic waves. J. Phys. D: Appl. Phys. 51 (37), 373001. https://doi.org/10.1088/1361-6463/aace3c.
Wierzbowski, J., et al., 2017. Direct exciton emission from atomically thin transition metal dichalcogenide heterostructures near the lifetime limit. Sci. Rep. 7 (1), 1–6. https://doi.org/10.1038/s41598-017-09739-4.
Yan, T., et al., 2014. Photoluminescence properties and exciton dynamics in monolayer WSe_2. Appl. Phys. Lett. 105 (10), 101901. https://doi.org/10.1063/1.4895471.
Yang, S., Chen, Y., Jiang, C., 2021. Strain engineering of two-dimensional materials: Methods, properties, and applications. InfoMat 3 (4), 397–420. https://doi.org/10.1002/INF2.12177.
Yao, W., Xiao, D., Niu, Q., 2008. Valley-dependent optoelectronics from inversion symmetry breaking. Phys. Rev. B - Condens. Matter Mater. Phys. 77 (23), 235406. https://doi.org/10.1103/PHYSREVB.77.235406/FIGURES/3/MEDIUM.
You, Y., et al., 2015. Observation of biexcitons in monolayer WSe_2. Nat. Phys. 11 (6), 477–481. https://doi.org/10.1038/nphys3324.
Yu, M.W., et al., 2021. Quantifying photoinduced carriers transport in exciton–polariton coupling of MoS_2 monolayers. npj 2D Mater. Appl. 5 (1), 1–7. https://doi.org/10.1038/s41699-021-00227-y.
Yu, S., et al., 2017. 2D materials for optical modulation: Challenges and opportunities. Adv. Mater. 29 (14). https://doi.org/10.1002/ADMA.201606128.
Yu, W., et al., 2022. Flexible 2D materials beyond graphene: Synthesis, properties, and applications. Small 18 (14), 2105383. https://doi.org/10.1002/SMLL.202105383.
Yuan, L., et al., 2017. Exciton dynamics, transport, and annihilation in atomically thin two-dimensional semiconductors. J. Phys. Chem. Lett. 8, 3371–3379. https://doi.org/10.1021/acs.jpclett.7b00885.
Zhang, X.X., et al., 2015. Experimental evidence for Dark excitons in monolayer WSe_2. Phys. Rev. Lett. 115 (25), 257403. https://doi.org/10.1103/PhysRevLett.115.257403.
Zhao, W., Ribeiro, R.M., Eda, G., 2015. Electronic structure and optical signatures of semiconducting transition metal dichalcogenide nanosheets. Acc. Chem. Res. 48 (1), 91–99. https://doi.org/10.1021/AR500303M/ASSET/IMAGES/LARGE/AR-2014-00303M_0007.JPEG.
Zheng, S., et al., 2016. Acoustic charge transport induced by the surface acoustic wave in chemical doped graphene. Appl. Phys. Lett. 109 (18), 183110. https://doi.org/10.1063/1.4967192.

Zheng, S., et al., 2018. Acoustically enhanced photodetection by a black phosphorus-MoS_2 van der Waals heterojunction *p-n* diode. Nanoscale 10 (21), 10148–10153. https://doi.org/10.1039/c8nr02022a.

Zhu, Z.Y., Cheng, Y.C., Schwingenschlögl, U., 2011. Giant spin-orbit-induced spin splitting in two-dimensional transition-metal dichalcogenide semiconductors. Phys. Rev. B - Condens. Matter Mater. Phys. 84 (15), 153402. https://doi.org/10.1103/PHYSREVB.84.153402/FIGURES/4/MEDIUM.

Zrenner, A., et al., 1992. Indirect excitons in coupled quantum well structures. Surf. Sci. 263 (1–3), 496–501. https://doi.org/10.1016/0039-6028(92)90396-N.

CHAPTER FIVE

Excitonic devices in 2D heterostructures

L.V. Butov*
University of California at San Diego, La Jolla, CA, United States
*Corresponding author. e-mail address: lvbutov@physics.ucsd.edu

Contents

Abstract

Indirect excitons (IXs) can be controlled by voltage, can travel over large distances, and can cool down below the temperature of quantum degeneracy. These properties form the basis for the development of excitonic devices with IXs. In this contribution, we overview our studies of excitonic devices in GaAs heterostructures and in TMDCs heterostructures. We outline IX transport and excitonic devices, including excitonic traps, excitonic lattices, excitonic conveyers, excitonic ramps, excitonic split gates, and excitonic transistors. These devices can be used both for studying quantum systems of bosons in materials and for developing excitonic signal processing.

Semiconductors and Semimetals, Volume 112
ISSN 0080-8784, https://doi.org/10.1016/bs.semsem.2023.09.003

1. Introduction

Spatially indirect excitons (IXs), also known as interlayer excitons, dipolar excitons, or charge transfer excitons, have specific properties, which make them a platform both for studying quantum systems of bosons in materials and for developing excitonic devices. In this chapter, we briefly overview excitonic devices and relevant IX properties.

An IX is a bound pair of an electron and a hole confined in separated layers (Lozovik and Yudson, 1976) (Fig. 1). Due to the separation of the electron and hole layers (i) IXs have long lifetimes, typically orders of magnitude longer than lifetimes of spatially direct excitons (DXs), and (ii) IXs have out-of-plane electric dipole ed_z (e is the electron charge, d_z the separation between the layers). This leads to the following properties of IXs:

1. Due to their long lifetimes, IXs can thermalize below the temperature of quantum degeneracy that allows the realization of quantum excitonic systems (Lozovik and Yudson, 1976).
2. The long IX lifetimes allow long-range IX transport (Hagn et al., 1995).
3. The IX electric dipoles enable creating tailored potential landscapes for IXs by voltage $\delta E(x,y,t) \sim ed_zF_z(x,y,t)$ where F_z is out-of-plane electric field controlled in space and time by voltage $V(x,y,t)$ (Hagn et al., 1995).
4. The IX electric dipoles lead to repulsive dipolar interaction between IXs.

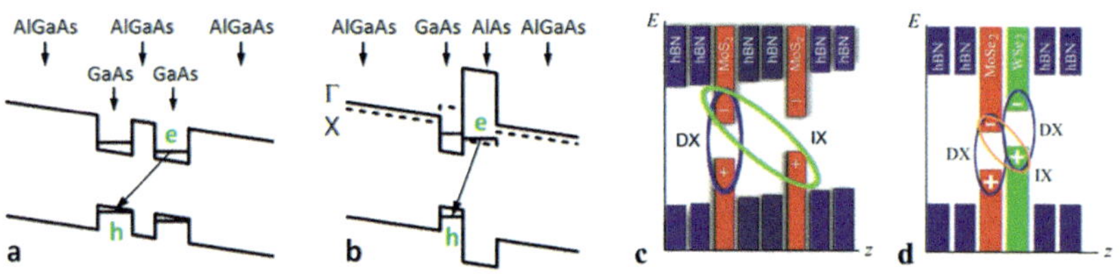

Fig. 1 Indirect excitons, IXs. (a,b) Energy band diagram of GaAs HS: GaAs/AlGaAs (a) and AlAs/GaAs (b). An IX is a bound pair of an electron, e (−), and a hole, h (+), confined in separated layers. IXs are shown by arrows. (c,d) Energy band diagram of TMDCs HS: MoS_2/hBN (c) and $MoSe_2$/WSe_2 (d). IXs and DXs are shown by ovals. IXs become the lowest energy exciton state under applied voltage in type I HS (a,c). IXs are the lowest energy exciton state even at no applied voltage in type II HS (b,d). The separated electron and hole layers can be realized in coupled quantum wells (CQW) (a,b) or in coupled atomic monolayers (c,d).

a. The repulsive interaction is essential for the realization of quantum excitonic systems (point 1) because the repulsive interaction stabilizes the exciton state against the formation of metallic electron–hole (e–h) liquid (Yoshioka and MacDonald, 1990; Butov et al., 1994; Zhu et al., 1995; Lozovik and Berman, 1996; Ben-Tabou de-Leon and Laikhtman, 2001), which otherwise can form the ground state in the e–h system (Keldysh, 1986).

b. The repulsive interaction is essential for facilitating IX transport (point 2) because the repulsive interaction screens the in-plane HS disorder and, as a result, enhances the IX mobility (Ivanov, 2002).

Various HS, both type I (Fig. 1a,c) and type II (Fig. 1b,d), composed from various materials, are used for creating IXs, in particular GaAs HS (Zrenner et al., 1992), ZnO HS (Morhain et al., 2005), GaN HS (Chiaruttini et al., 2019), and van der Waals HS based on single atomic layers of transition-metal dichalcogenides, TMDCs (Rivera et al., 2015). Typically, in all these HS, the widths of the electron and hole layers are small (from a monolayer to a few tens of monolayers), only a single 2D subband is occupied and, therefore, the IX systems are 2D. In this chapter, we overview IXs in GaAs HS (Fig. 1a,b, Sections 1–3) and in TMDCs HS (Fig. 1c,d, Section 4). GaAs HS form a low-disorder platform for IXs. TMDCs HS form a platform for the realization of IXs with high binding energies and also give an opportunity to explore IXs in periodic potentials due to moiré superlattices.

In the optically created IX systems considered in this chapter, the IX density n can be controlled by the optical excitation power P_{ex}, e.g. for a cw excitation $n = P_{ex}\tau$. The long IX lifetimes τ allow achieving high IX densities with small P_{ex} that facilitates lowering the excitation-induced heating of the IX system. At low e-h densities, $n << 1/a_B^2$ (a_B is the exciton Bohr radius), electrons and holes bind to hydrogen-like excitons, which can form a Bose–Einstein condensate (BEC) at low temperatures (Keldysh and Kozlov, 1968). In dense e–h systems, $n \gtrsim 1/a_B^2$, hydrogen-like excitons dissociate and an e–h plasma can be realized at high temperatures, and the Cooper-pair-like excitons and a BCS-like exciton condensate can be realized at low temperatures (Keldysh and Kopaev, 1965; Comte and Noziéres, 1982; Choksy et al., 2023). Typical a_B is ~20 nm for IXs in GaAs HS (Dignam and Sipe, 1991). The high-density regime, $n \gtrsim 1/a_B^2$, is not considered in this chapter.

The exciton temperature is determined by the ratio of the energy relaxation time and recombination time. The long IX lifetimes allow IXs to cool down essentially to the lattice temperature. For instance, in GaAs HS (Fig. 1a,b), IX lifetimes are controlled by voltage, typically in the range from tens of ns to tens of μs (Winbow et al., 2007), and are orders of magnitude longer than DX lifetimes, which are in the range of tens of ps in GaAs quantum wells (QWs) (Feldmann et al., 1987; Andreani et al., 1991; Deveaud et al., 1991). Due to their long lifetimes, IXs in GaAs HS can cool down to $\sim$ 0.1 K, below the temperature of quantum degeneracy $T_q = 2\pi\hbar^2 n/m$, which reaches $\sim$ 3 K for the exciton mass $m = 0.22m_0$ and density $n \sim 10^{10}$ cm^{-2} relevant to the experiments (Butov et al., 2001). This gives an opportunity to study quantum excitonic systems. The phenomena in quantum IX systems include in particular spontaneous coherence and BEC of IXs (Yang et al., 2006; High et al., 2012a, 2012b; Alloing et al., 2014; Gorbunov and Timofeev, 2016), the macroscopic spatial ordering in the IX condensate (Butov et al., 2002; Levitov et al., 2005), the commensurability effect of exciton density wave in the IX condensate (Yang et al., 2015), the spin textures in the IX condensate (High et al., 2013), the Pancharatnam–Berry phase in the IX condensate (Leonard et al., 2018), and the interference dislocations in the IX condensate (Leonard et al., 2021). The phenomena in quantum IX systems also include the stimulated exciton scattering (Butov et al., 2001), the enhanced exciton radiative recombination (Butov and Filin, 1998), the enhanced exciton tunneling recombination (Spielman et al., 2000; Eisenstein and MacDonald, 2004), the perfect Coulomb drag (Nandi et al., 2012), and other transport and optical phenomena consistent with IX condensation (Butov et al., 1994; Butov and Filin, 1998; Tutuc et al., 2004; Pellegrini et al., 2007; Tiemann et al., 2008; Liu et al., 2017; Li et al., 2017). Works (Butov et al., 1994, 2001; Yang et al., 2006; High et al., 2012a, 2012b; Alloing et al., 2014; Gorbunov and Timofeev, 2016; Butov et al., 2002; Levitov et al., 2005; Yang et al., 2015; High et al., 2013; Leonard et al., 2018, 2021; Butov and Filin, 1998) refer to IXs optically created in GaAs HS and works (Spielman et al., 2000; Eisenstein and MacDonald, 2004; Nandi et al., 2012; Tutuc et al., 2004; Pellegrini et al., 2007; Tiemann et al., 2008; Liu et al., 2017; Li et al., 2017) to IXs in e–e (or h–h) bilayers in GaAs (Spielman et al., 2000; Eisenstein and MacDonald, 2004; Nandi et al., 2012; Tutuc et al., 2004; Pellegrini et al., 2007; Tiemann et al., 2008) and graphene (Liu et al., 2017; Li et al., 2017) HS. Some of these phenomena were overviewed in Butov (2016, 2020) the recently studied interference dislocations (Leonard et al., 2021) are outlined in Section 2.4, the phenomena in e–e (or h–h) bilayers are not considered in this chapter.

The long IX lifetimes and the screening of in-plane disorder due to the repulsive IX interaction allow IXs to travel over long distances. The studies of IX transport in GaAs HS (Hagn et al., 1995; Chiaruttini et al., 2019; High et al., 2012a; Alloing et al., 2014; Gorbunov and Timofeev, 2016; Butov et al., 2002; Levitov et al., 2005; Yang et al., 2015; High et al., 2013; Leonard et al., 2018, 2021; Butov and Filin, 1998; Larionov et al., 2000; Vörös et al., 2005; Ivanov et al., 2006; Gärtner et al., 2006; Rudolph et al., 2007; Remeika et al., 2009; Vögele et al., 2009; Hammack et al., 2009; Lazić et al., 2010; Cohen et al., 2011; Winbow et al., 2011; Remeika et al., 2012; Leonard et al., 2012; Alloing et al., 2012; Lazić et al., 2014; Violante et al., 2014; Fedichkin et al., 2015; Kuznetsova et al., 2015a; Remeika et al., 2015; Dorow et al., 2016, 2018a, 2018b; Xu et al., 2019), led to the observation of various exciton transport phenomena, including, in particular, the inner ring in exciton emission patterns (Butov et al., 2002; Ivanov et al., 2006; Hammack et al., 2009; Alloing et al., 2012) (Section 2.2), the exciton delocalization (Butov et al., 2002; Ivanov et al., 2006; Remeika et al., 2009; Winbow et al., 2011; Remeika et al., 2012) (Section 2.3), and coherent exciton transport with suppressed scattering (High et al., 2012a, 2013; Leonard et al., 2018, 2021) (Section 2.4). The long IX lifetimes also allow studying spin transport in IXs (High et al., 2013; Leonard et al., 2018, 2009; Violante et al., 2015; Finkelstein et al., 2017), spin-related phenomena are not considered in this chapter.

The IX electric dipoles ed_z give an opportunity to explore exciton transport and other phenomena in tailored in-plane potential landscapes, which are created and controlled by voltage. In particular, IXs in GaAs HS are explored in electrostatically controlled traps (High et al., 2012b; Huber et al., 1998; Hammack et al., 2006a; Gorbunov and Timofeev, 2006; Chen et al., 2006; Rontani and Sham, 2009; High et al., 2009a, 2009b; Kuznetsova et al., 2010a; Schinner et al., 2011; Kowalik-Seidl et al., 2012; Gorbunov and Timofeev, 2012, 2013; Schinner et al., 2013a, 2013b; Shilo et al., 2013; Kuznetsova et al., 2015b; Cohen et al., 2016; Lobanov et al., 2016) (Section 3.1), lattices (Remeika et al., 2009, 2012; Zimmermann et al., 1997, 1998; Lagoin et al., 2022a, 2022b, 2023) (Section 3.2), conveyers (Winbow et al., 2011) (Section 3.3), ramps (Hagn et al., 1995; Gärtner et al., 2006; Leonard et al., 2012; Dorow et al., 2016) (Section 3.4), narrow channels (Vögele et al., 2009; Cohen et al., 2011; Grosso et al., 2009), and split-gates (Dorow et al., 2018a; Xu et al., 2019) (Section 3.5).

Besides forming a platform for studying basic physics of bosons in semiconductors, excitons can be explored for the development of signal processing based on computational state variables beyond magnetism and charge. Excitonic devices can be well suited to the development of an advanced signal processing due to the exciton properties (Butov, 2017):

I. Excitons are bosons and can form a condensate with vanishing resistance for exciton currents and low switching voltage for excitonic transistors. This allows creating energy-efficient computation with a small power consumption per switch.

II. Excitons can directly transform to photons. This allows efficient coupling of excitonic signal processing to optical communication.

III. The sizes of excitonic devices are limited by the exciton radius that can be much smaller than the photon wavelength. This allows creating nanoscale excitonic devices.

The excitonic devices require specially designed excitons to carry the signal. The IX properties make them suitable for excitonic signal processing devices (Butov, 2017):

a. The IX energy and fluxes can be controlled by voltage and light that allows the realization of transistors operating with excitons in place of electrons (High et al., 2007, 2008; Grosso et al., 2009; Kuznetsova et al., 2010b; Andreakou et al., 2014).

b. The IX emission rate can be tuned over orders of magnitude by voltage (Winbow et al., 2007, 2008). Turning the emission off allows the realization of multi-element excitonic circuits with suppressed losses, turning it on allows fast write and readout.

c. The low e–h overlaps in IXs allow the realization of coherent condensate (Yang et al., 2006; High et al., 2012a, 2012b; Alloing et al., 2014; Gorbunov and Timofeev, 2016) with suppressed thermal tails and dissipationless exciton currents, enabling the energy-efficient computation.

In this chapter, we overview our studies of excitonic devices. Section 2 concerns basic physics of IX transport. In Section 3, we overview excitonic traps, lattices, conveyers, ramps, split gates, and transistors in GaAs HS, and in Section 4—excitonic transistors in TMDCs HS.

2. Exciton transport

The operation of excitonic devices is based on the controlled exciton transport. In this Section, we overview basic transport phenomena in IXs. We outline both diffusive IX transport in IX systems and coherent ballistic IX transport in IX BEC.

2.1 Drift-diffusion model for IX transport

Classical diffusive IX transport is described by the drift-diffusion model (Ivanov, 2002; Ivanov et al., 2006; Hammack et al., 2009):

$$\frac{\partial n}{\partial t} = \nabla\left[D\nabla n + \mu n \nabla (u_0 n)\right] + \Lambda - \frac{n}{\tau} \tag{1}$$

n is the IX density. The first and second terms in square brackets in Eq. (1) describe IX diffusion and drift currents, respectively. The latter originates from the IX repulsive dipolar interaction and is approximated by the mean-field 'plate capacitor' formula for the IX energy shift with density $\delta E = nu_0$, $u_0 = 4\pi e^2 d_z/\varepsilon$, ε is the dielectric constant (Yoshioka and MacDonald, 1990). The diffusion coefficient $D = D^{(0)}\exp\left[-U^{(0)}/(k_B T + nu_0)\right]$ accounts for the temperature- and density-dependent screening of the long-range-correlated in-plane potential landscape by interacting IXs. $D^{(0)}$ is the diffusion coefficient in the absence of in-plane potential. $U^{(0)}/2$ is the amplitude of the in-plane potential. μ is the IX mobility. The IX generation rate Λ has a profile of the laser excitation spot. τ is the IX lifetime.

Fitting the IX cloud expansion by $R^2 \sim D^* \delta t$ probes the effective IX diffusion coefficient $D^* = D + \mu n u_0$, which includes both the diffusion and the drift due to the density gradient (Dorow et al., 2017). μ can be estimated using the Einstein relation $\mu = D/(k_B T)$. Therefore, the enhancement of IX transport due to the IX-interaction-induced drift from the origin is described by the enhanced $D^* = D\left[1 + nu_0/(k_B T)\right]$ (Dorow et al., 2017). Within the classical drift-diffusion model (Ivanov, 2002; Ivanov et al., 2006; Hammack et al., 2009), the IX transport should enhance with density due to the enhanced screening of in-plane (disorder) potential and enhanced drift from the origin.

The drift-diffusion Eq. (1) can be supplemented by the thermalization equation, which describes heating of excitons by photoexcitation and cooling via interaction with phonons, and by including the HS in-plane potential in the drift term (Ivanov et al., 2006; Hammack et al., 2009).

In particular, the exciton thermalization can lead to the appearance of the inner ring in exciton luminescence patterns (Section 2.2). The Einstein relation can be extended to the generalized Einstein relation $\mu = D(e^{T_q/T} - 1)/(k_B T_q)$, which gives $\mu = D/(k_B T)$ for the classical drift and diffusion (Ivanov, 2002; Ivanov et al., 2006; Hammack et al., 2009). The exciton correlations (Remeika et al., 2015) can be included to the IX interaction.

The exciton inner ring (Section 2.2), the exciton delocalization in in-plane potentials (Section 2.3), and the exciton transport in excitonic devices (Section 3) are adequately described by this drift-diffusion model (Ivanov, 2002; Ivanov et al., 2006; Hammack et al., 2009). Coherent ballistic IX transport in IX BEC is outlined in Section 2.4.

2.2 Exciton rings

The exciton rings, including the inner ring, external ring, and localized bright spot (LBS) rings, were observed in IX emission patterns (Fig. 2) (Butov et al., 2002). The origin of these rings is outlined below.

The external exciton ring forms on the interface between the electron-rich and hole-rich regions (Butov et al., 2004; Rapaport et al., 2004; Chen et al., 2005; Haque, 2006; Yang et al., 2010). For the experiments in Ref

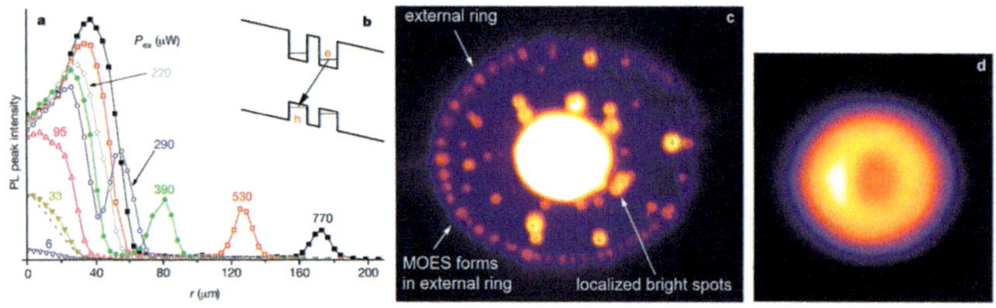

Fig. 2 Exciton rings and spatial ordering. (a) The peak intensity of IX luminescence vs. the distance from the center of focused laser excitation spot at different laser excitation powers P_{ex}. The excitation spot profile is shown by the dashed line. The inner and external exciton rings form around the laser excitation spot. The ring radius increases with P_{ex}. (b) The HS band diagram; e, electron; h, hole. (c) The IX luminescence pattern. The area of view is 410 × 330 μm. The focused laser excitation spot is in the center. The inner ring is not seen for the contrast of the image. The LBS rings form within the external ring. The macroscopically ordered exciton state (MOES) with periodic spatial modulation is observed in the external ring at low temperatures. (d) The inner ring. The focused laser excitation spot is in the center. The area of view is 60 × 60 μm. *From Butov, L.V., Gossard, A.C., Chemla, D.S., 2002. Macroscopically ordered state in an exciton system. Nature 418, 751 and Ivanov, A.L., Smallwood, L.E., Hammack, A.T., Yang, S., Butov, L.V., Gossard, A.C., 2006. Origin of the inner ring in photoluminescence patterns of quantum well excitons. Europhys. Lett. 73, 920.*

(Butov et al., 2002), the hole-rich (electron-rich) region is inside (outside) the external ring (the sign of carrier imbalance can depend on the HS and excitation). The laser excitation can generate an excess number of holes at the laser excitation spot (e.g. due to a more effective collection of holes generated in the HS barrier layers to the CQW). The electron current through the gated HS creates an electron gas in the CQW. The holes created at the excitation spot diffuse out in the hole layer and bind with electrons, forming excitons. This process depletes electrons, creating a hole-rich region around the laser excitation spot. A sharp interface between the hole-rich region and the outer electron-rich area can form, since a carrier crossing into a minority region binds rapidly with an opposite carrier to form an exciton. This interface creates the external ring (Fig. 2a,b). The external ring radius is controlled by the laser excitation power P_{ex} and by voltage across the HS V_g: The external ring expands with increasing P_{ex} (Fig. 2a) due to the increasing number of photogenerated holes and collapses with increasing V_g due to the increasing number of electrically injected electrons (Butov et al., 2004; Rapaport et al., 2004; Chen et al., 2005; Haque, 2006; Yang et al., 2010).

The LBS rings also form on the interface between the electron-rich and hole-rich regions (Butov et al., 2004; Yang et al., 2010). For the LBS rings, the hole-rich region forms around the laser excitation spot as for the external ring, however, the electron-rich regions form around filaments of the electron current through the gated HS in the regions where the filaments cross the CQW plane. For the LBS ring inside the external ring, this localized source of electrons is embedded in the hole-rich area and the electron-rich region is surrounded by the hole-rich region.

The external rings and LBS rings can occur in a classical exciton system and can be adequately described by the coupled diffusion equations (Butov et al., 2004; Rapaport et al., 2004; Chen et al., 2005; Haque, 2006; Yang et al., 2010). However, these rings are the regions where a cold IX system, separated from the heated excitation regions, is realized. Therefore, the rings can be used for the realization of quantum IX systems. Cooling IX systems in the ring regions below a few Kelvin led to the realization of IX spontaneous coherence and BEC (Yang et al., 2006; High et al., 2012a; Alloing et al., 2014) and the phenomena in the IX condensate, including the macroscopic spatial ordering (Butov et al., 2002; Levitov et al., 2005), the commensurability effect of exciton density wave (Yang et al., 2015), the spin textures (High et al., 2013), the Pancharatnam–Berry phase (Leonard et al., 2018), and the interference dislocations (Leonard et al., 2021). These phenomena disappear above the IX condensation temperature (Butov et al., 2002; Levitov et al., 2005; Yang et al., 2015; High et al., 2013; Leonard et al., 2018, 2021). The IX spontaneous coherence, BEC, macroscopic spatial ordering, commensurability effect, spin textures, and Pancharatnam–Berry phase are overviewed in Butov (2016, 2020) and the recently studied interference dislocations in Section 2.4.

The inner ring forms due to the IX transport from the laser excitation spot and cooling (Butov et al., 2002; Ivanov et al., 2006; Hammack et al., 2009; Alloing et al., 2012). The laser excitation increases the IX temperature in the excitation spot and, as a result, reduces the IX occupation of the radiative zone where IXs can transform to photons (in the radiative zone, the exciton momenta $k \lesssim E_g\sqrt{\varepsilon}/\hbar c$, E_g is the gap) (Feldmann et al., 1987; Andreani et al., 1991). As IXs travel away from the excitation spot, they cool down, the occupation of the radiative zone increases, and the IX luminescence intensity increases that produces the inner ring in the luminescence pattern (Fig. 2a,d). The inner ring can occur in a classical exciton system and can be adequately described by diffusive exciton transport (Ivanov et al., 2006; Hammack et al., 2009) (Section 2.1).

The laser-induced trap for cold excitons (Hammack et al., 2006b; Gorbunov et al., 2011; Alloing et al., 2013) is a geometrical variation of the inner ring. The optical trap is created by a ring-shaped laser excitation. The IX transport from the ring-shaped excitation spot and cooling create a cold IX system in the center of the optical trap. Both the center of the optical trap and the inner ring are the regions where a cold IX system, separated from the heated excitation region, is realized. Therefore, both the optical trap and the inner ring can be used for creating quantum IX systems (Ivanov et al., 2006; Hammack et al., 2006b).

2.3 Exciton delocalization

At low IX densities, IXs are localized by the HS in-plane disorder potential and the IX luminescence profile essentially follows the laser excitation profile (Fig. 2a). However, as outlined in Section 2.1, IXs screen the in-plane disorder potential due to the repulsive IX interaction. When the screening is strong enough, IXs can delocalize and travel over long distances. In particular, the high-P_{ex} data in the inner ring region in Fig. 2a show the IX transport overs tens of microns from the excitation spot. For instance, in GaAs HS with a low in-plane disorder and IX lifetime $\tau \sim 1$ μs, IX transport over ca. 100 μm from the excitation spot with D^* reaching $\sim$350 cm^2/s is observed (Dorow et al., 2018b).

The crossover from IX localization to IX delocalization was quantified using controlled in-plane potentials for IXs (Remeika et al., 2009). In these experiments, we studied IX transport in in-plane linear lattices created by interdigitated gates (Fig. 3). Base voltage V_0 realized the indirect regime and voltage modulation ΔV controlled the lattice amplitude U_l. The laser excitation controlled the IX density and, in turn, the IX interaction energy ΔE. The IX cloud extension was measured by IX luminescence imaging.

At low P_{ex}, IXs are localized in the laser excitation spot. At high lattice amplitudes, the IX delocalization in the x direction across the linear lattice is observed when the IX interaction energy ΔE is close to the lattice amplitude U_l (Fig. 3c). At vanishing lattice amplitude, the IX delocalization occurs when ΔE is close to the amplitude of the disorder potential Γ (Fig. 3c). The IX transport in the y direction along the linear lattice is only weakly affected by the lattice (Fig. 3c).

The observed IX localization–delocalization transition (LDT) was attributed (Remeika et al., 2009) to the interaction-induced percolation of IXs through the total external potential $E_{tot}(\mathbf{r})$, which is the sum of the electrostatic lattice potential and the HS disorder potential. If the local IX density is low, IXs are concentrated in the minima of the potential $E_{tot}(\mathbf{r})$. As a result, the IX transport from one potential minimum to the next through thermal activation or quantum tunneling is exponentially slow. As

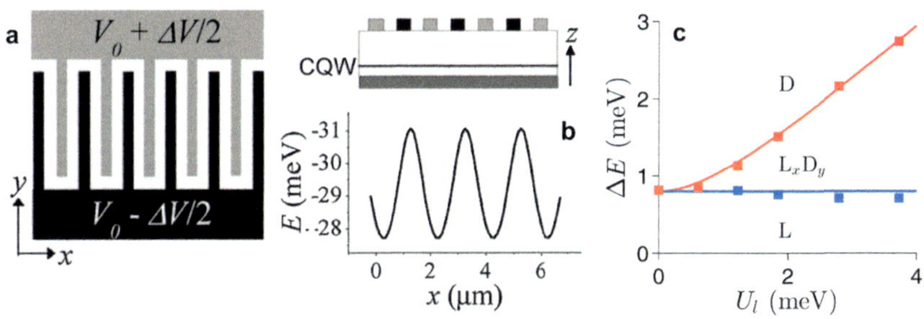

Fig. 3 Localization–delocalization crossovers in in-plane potentials. (a,b) Schematic electrode pattern (a) and calculated lattice potential for IXs at $\Delta V = 1$ V (b). The voltage ΔV controls the lattice amplitude U_l. The laser excitation controls the IX density and, in turn, the IX interaction energy ΔE. (c) Localization–delocalization transition. IXs are localized in region L, localized along x but delocalized along y in region L_xD_y, and delocalized in both directions in region D. *From Remeika, M., Graves, J.C., Hammack, A.T., Meyertholen, A.D., Fogler, M.M., Butov, L.V., et al., 2009. Localization-delocalization transition of indirect excitons in lateral electrostatic lattices. Phys. Rev. Lett. 102, 186803 and Remeika, M., Leonard, J.R., Dorow, C.J., Fogler, M.M., Butov, L.V., Hanson, M., et al., 2015. Measurement of exciton correlations using electrostatic lattices. Phys. Rev. B 92, 115311.*

the IX average density increases and reaches a certain threshold, 'percolation point', the potential is screened by interacting IXs that permits a faster IX transport, i.e., the IX delocalization. This scenario leads to $\Delta E_{\mathrm{LDT}} \sim E_{\mathrm{tot}}$, which reduces to $\Delta E_{\mathrm{LDT}} \sim U_l$ and $\Delta E_{\mathrm{LDT}} \sim \Gamma$ for large and small lattice amplitude, respectively (Remeika et al., 2009).

2.4 Interference dislocations

In this Section, we outline dislocation-like singularities in interference patterns in the condensate of IXs (Leonard et al., 2021). The results of Leonard et al. (2021): (1) The moiré effect can create dislocations in interference patterns, this mechanism is applicable for interference patterns in general for various physics systems; (2) The interference dislocations in IX condensate (Fig. 4) originate from the moiré effect; (3) The interference dislocations are formed by IXs ballistically propagating over dramatically long times without scattering that evidences IX superfluidity.

LBS rings (Section 2.2) are stable and well-defined sources of cold IXs (Yang et al., 2015). We use them in Ref (Leonard et al., 2021). for exploring phenomena in propagating condensate matter waves. Fig. 4a shows the interference pattern of IX emission measured by shift-interferometry. As a result of IX recombination, IXs transform to photons. These photons go through the Mach-Zehnder interferometer. Each of the two interferometer arms forms an image of the IX emission on CCD detector.

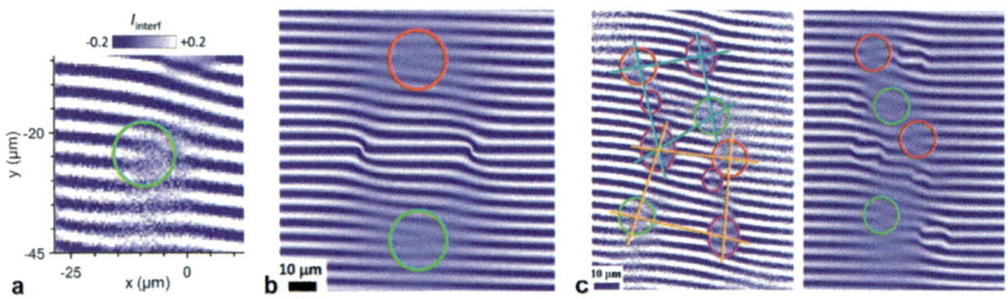

Fig. 4 Interference dislocations in IX condensate. (a) Measured interference pattern. The isolated interference dislocation is marked by a green circle. (b) Simulations: The combined interference patterns of radial IX condensate matter waves propagating from two sources produce interference dislocations due to the moiré effect. (c) Measured (left) and simulated (right) moiré patterns of interference dislocations. Right- and left-dislocations are marked by red and green circles. IX sources are embraced by magenta lines. Cyan and orange lines are drawn between the dislocations and stronger sources forming them. *From Leonard, J.R., Hu, L., High, A.A., Hammack, A.T., Wu, C., Butov, L.V., et al., 2021. Moiré pattern of interference dislocations in condensate of indirect excitons. Nat. Commun. 12, 1175.*

These two images are shifted relative to each other in x-direction so that the measured interference pattern is produced by the interference between the emission of IXs separated by the shift δx in CQW plane. Fig. 4a shows a dislocation-like phase singularity in the interference pattern.

Dislocations (forks) in interference patterns can be associated with vortices. Forks in interference patterns were reported for vortices in atom condensates (Inouye et al., 2001), polariton vortices (Lagoudakis et al., 2008, 2009; Roumpos et al., 2011), magnon vortices (Nowik-Boltyk et al., 2012), and optical vortices (Scheuer and Orenstein, 1999). However, a vortex can lead to the appearance of dislocation in an interference pattern only for certain interferometric experiments, in particular for the interference of a vortex field with a plane wave. For the shift-interferometry experiments, simulations show that a vortex should lead to the appearance of a pair of left- and right-oriented dislocations of interference fringes separated by a distance equal to the shift δx (Leonard et al., 2021). The interference pattern for a vortex is drastically different from the isolated interference dislocation, which is separated from other dislocations by distances much larger than the shift δx, in the experiment (Fig. 4a). This indicates that the observed interference dislocation is not associated with a vortex.

We also simulated the interference patterns for exciton polarization vortices, half-vortices, skyrmions, and half-skyrmions (Leonard et al., 2021). These interference patterns are also drastically different from the isolated interference dislocation in the experiment (Fig. 4a). This indicates that the observed interference dislocation is not associated with a polarization vortex, or half-vortex, or skyrmion, or half-skyrmion.

The origin of the interference dislocations is outlined below. IX condensate produced by a single source is described by a single macroscopic wavefunction approximated by a radial IX matter wave with $\psi(\mathbf{r}) \propto e^{ikR}$, where $R = |\mathbf{r} - \mathbf{r}_s|$ is the distance to the source, $\mathbf{r}_s$ the source location, and $\mathbf{k} = k(\mathbf{r} - \mathbf{r}_s)/R$ the IX momentum. For the shift $\delta\mathbf{r} = \delta\mathbf{x}$, the interference pattern is given by $I(x, y) \sim \cos\left(\mathbf{k}\delta\mathbf{x} + q_t y\right)$, where $q_t = 2\pi\alpha/\lambda$ sets the period of the interference fringes, α is a small tilt angle between the image planes of the interferometer arms, and λ is the emission wavelength.

Now, we consider two radial IX condensate matter waves produced by two IX sources. For simplicity, to outline the origin of the phase singularities, we consider the waves of equal strength and with equal k (Fig. 4b). For the two IX condensate matter waves, the interference pattern is given by

$$I(x, y) \sim \frac{1}{R_1}\cos\left(\mathbf{k}_1\delta\mathbf{x} + q_t y\right) + \frac{1}{R_2}\cos\left(\mathbf{k}_2\delta\mathbf{x} + q_t y\right), \tag{2}$$

where $R_i = |\mathbf{r} - \mathbf{r}_{s,i}|$ and $\mathbf{k}_i = k(\mathbf{r} - \mathbf{r}_{s,i})/R_i$ are the distance to the source and momentum for IXs propagating from source i with $i = 1, 2$. In this expression, each source generates a coherent IX condensate matter wave propagating ballistically from the source with no dissipation and $I(x,y)$ intensities produced by source 1 and source 2 are added. The addition of intensities corresponds to the lack of coherence between the IX matter waves propagating from source 1 and source 2 and is in accord with the experiment: the IX creation from optically generated holes and electrically generated electrons in LBS sources (Section 2.2) suggests the lack of coherence between different LBS sources.

The simulations of converging IX condensate matter waves (Eq. 2, Fig. 4b) reproduce the isolated dislocations observed in the experiment (Fig. 4a). The interference dislocations (Fig. 4b) originate from the moiré effect in the combined interference patterns of propagating condensate matter waves. The radial condensate matter waves propagating from source 1 and from source 2 create similar interference patterns shifted relative to each other by the separation between the sources $\mathbf{r}_{s,1} - \mathbf{r}_{s,2}$. Combining these two patterns is presented by Eq. (2). Factors $1/R_i$ in Eq. (2) correspond to 2D radial condensate matter waves propagating without dissipation from their sources. This combining of separated interference patterns forms a moiré pattern with the interference dislocations.

The locations of the interference dislocations originating from the moiré effect are given by the ballistic propagation of IXs. For two radial IX condensate matter waves produced by two sources separated along the shift δx, on the bisector line, Eq. (2) gives $I \sim \cos(q_t y)\cos(k\delta x \cos\gamma)$, where γ is the angle between $\mathbf{r} - \mathbf{r}_{s,1}$ and δx. In this equation, the first factor oscillates along y, and the second factor produces the beating pattern. When $k\delta x \cos \gamma = (n + 1/2)\pi$ with n an integer, the second factor changes sign and this phase slip creates the phase singularity in interference pattern. Numerical simulations (Fig. 4b) confirm that the dislocations in interference pattern are observed at the locations determined by this equation.

In the experimental system, there are several LBS sources of IXs (Fig. 4c). Three strong sources are clearly seen due to the pronounced phase domains at the source locations. Two weaker sources are also seen in Fig. 4c. Sharp phase shifts of interference fringes embracing a phase domain of interference fringes in a circular region around each LBS source of IXs are marked by magenta lines. These sharp phase shifts and, in turn, the phase domains are associated with the Pancharatnam–Berry phase and are described in Leonard et al. (2018). The right- and left-oriented interference dislocations are marked by red and green circles. All sources participate in the formation of these dislocations. A stronger contribution to the upper (lower) pair of right- and left-oriented dislocations is given by the two upper (lower) strong sources. This is illustrated by cyan (orange) lines between the dislocations and the sources giving stronger contribution. These lines form diamond shapes similar to the diamond shapes formed by the lines connecting interference dislocations and two sources producing them in Fig. 4b.

Simulated interference pattern for five radial IX condensate matter waves propagating from five sources with $I(x, y) \sim \sum_{i=1}^{5} P_i/R_i \cos\left(\mathbf{k}_i \delta \mathbf{x} + q_t y\right)$ is presented in Fig. 4c (R_i and k_i are the distance to the source and momentum for IXs propagating from source i). The values of k are estimated from the sharp phase shifts of interference fringes around the source as described in Leonard et al. (2018). The source powers P_i are estimated from the IX emission intensities in the source region. The simulations qualitatively reproduce the interference dislocations observed in the experiment (Fig. 4c).

The interference dislocations are formed by the IX condensate matter waves ballistically propagating from the IX sources to the locations of interference dislocations over distances ca. 30 μm. The long-range ballistic IX propagation over these distances shows that IXs propagate without scattering over dramatically long times, orders of magnitude longer than in classical exciton gas (Leonard et al., 2021), and, therefore, is the evidence for IX condensate superfluidity.

3. Excitonic devices in GaAs heterostructures

3.1 Excitonic traps

Potential traps made possible the realization of atom BEC (Cornell and Wieman, 2002; Ketterle, 2002). Traps also become an effective tool for studying excitons. Excitons were studied in a variety of static traps whose profile cannot be changed in situ, i.e., on a time scale shorter than the exciton lifetime. These traps include strain-induced traps (Trauernicht et al., 1983; Kash et al., 1988; Negoita et al., 1999; Naka and Nagasawa, 2005; Yoshioka et al., 2011; Morita et al., 2022), traps created by laser-induced interdiffusion (Brunner et al., 1992), and magnetic traps

(Christianen et al., 1998). Recently, excitons were also studied in traps whose shape and depth can be controlled in situ—electrostatic traps (High et al., 2012b; Huber et al., 1998; Hammack et al., 2006a; Gorbunov and Timofeev, 2006; Chen et al., 2006; Rontani and Sham, 2009; High et al., 2009a, 2009b; Kuznetsova et al., 2010a; Schinner et al., 2011; Kowalik-Seidl et al., 2012; Gorbunov and Timofeev, 2012, 2013; Schinner et al., 2013a, 2013b; Shilo et al., 2013; Kuznetsova et al., 2015b; Cohen et al., 2016; Lobanov et al., 2016) and laser-induced traps (Hammack et al., 2006b; Gorbunov et al., 2011; Alloing et al., 2013). In this Section, we overview our studies of various electrostatic traps for IXs: diamond-shaped traps (High et al., 2012b, 2009b; Lobanov et al., 2016), snowflake traps (Kuznetsova et al., 2010a, 2015b), and elevated traps (High et al., 2009a, 2009b).

Of particular interest are traps that can provide confining potentials with the IX energy gradually reducing toward the trap center. A confining potential could collect IXs from a large area, creating a dense IX gas at the trap center. This, in turn, can facilitate the creation of a quantum IX gas in the trap since the temperature of quantum degeneracy increases with density. Furthermore, a parabolic-like confining potential of the trap increases the IX condensation temperature (Lobanov et al., 2016).

Diamond-shaped traps create confining potentials for IXs (High et al., 2009b). The diamond-shaped trap is formed by a top electrode of a diamond shape (Fig. 5b). Because a thinner electrode produces a smaller electric field in the z direction, F_z, due to field divergence near the electrode edges, a voltage between the diamond-shaped top electrode and homogeneous bottom electrode creates a confining potential with the IX energy gradually reducing toward the trap center (Fig. 5c,d) (High et al., 2009b).

The diamond trap devices, as all excitonic devices outlined in Section 3, were fabricated using GaAs HS grown by molecular-beam epitaxy. An n^+-GaAs layer with $n_{Si} \sim 10^{18}$ cm^{-3} serves as a homogeneous bottom electrode. Two 8 nm GaAs QWs separated by a 4 nm $Al_{0.33}Ga_{0.67}As$ barrier are positioned 0.1 μm above the n^+-GaAs layer within an undoped 1 μm thick $Al_{0.33}Ga_{0.67}As$ layer. An in-plane electric field F_r near electrode edges can lead to IX dissociation. Therefore, the CQW layers are positioned closer to the homogeneous bottom electrode. This design suppresses F_r making the field-induced IX dissociation negligible (Hammack et al., 2006a).

The IX confinement in a diamond-shaped trap was demonstrated in High et al. (2009b). Top electrodes forming the diamond-shaped trap include a diamond-shaped electrode, a wire electrode surrounding the diamond, and an outer plane electrode (Fig. 5b). By varying the diamond

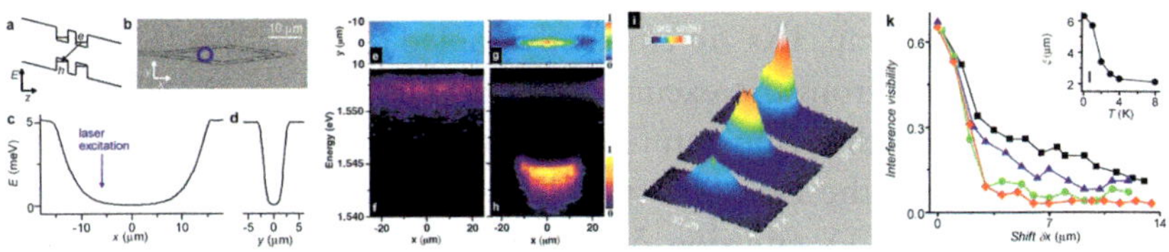

Fig. 5 Diamond-shaped excitonic traps. (a) HS energy-band diagram. e, electron; h, hole. (b) SEM image of electrodes forming the diamond trap: a diamond-shaped electrode is surrounded by a thin wire electrode followed by an outer plane electrode. (c,d) Simulation of IX energy profile through the trap center along *x* (c) and *y* (d) for $V_{\text{diamond}} = -2.5$ V, $V_{\text{wire}} = -2$ V, and $V_{\text{plane}} = -2$ V. The diamond-shaped trap realizes parabolic-like confining potential for IXs. The position of the laser excitation spot in experiment on exciton condensation (i–l) is indicated by the circle in (b) and by the arrow in (c). (e,g) *x–y* and (f,h) *x–E* emission images of IXs for (e,f) flat potential ($V_{\text{diamond}} = V_{\text{wire}} = V_{\text{plane}} = -2$ V) and for (g,h) diamond-shaped trap ($V_{\text{diamond}} = -3$ V, $V_{\text{wire}} = -2.5$ V, $V_{\text{plane}} = -2$ V). The laser excitation is defocused over the device area. (i–l) IX condensation in diamond-shaped trap shown in (c,d). (i) IX emission patterns for $T_{\text{bath}} = 7$, 4, and 0.05 K. (k) $A_{\text{interf}}(\delta x)$ for IXs in the trap for $T_{\text{bath}} = 50$ mK (black squares), 2 K (blue triangles), 4 K (green circles), and 8 K (red diamonds). (l) The IX coherence length ξ as a function of temperature. The IX condensation in the trap (i) with the entire IX cloud becoming coherent (l) is observed at low temperature. *From High, A.A., Leonard, J.R., Remeika, M., Butov, L.V., Hanson, M., Gossard, A.C., 2012b. Condensation of excitons in a trap. Nano Lett. 12, 2605 and High, A.A., Thomas, A.K., Grosso, G., Remeika, M., Hammack, A.T., Meyertholen, A.D., et al., 2009b. Trapping indirect excitons in a GaAs quantum-well structure with a diamond-shaped electrostatic trap. Phys. Rev. Lett. 103, 087403.*

electrode voltage V_{diamond} with respect to the wire voltage V_{wire} and outer plane voltage V_{plane}, we go from a flat IX energy (Fig. 5e,f) to a trap (Fig. 5g,h). The parabolic-like shape of the trap is evident in the *x–E* emission image (Fig. 5h). The collection of IXs to trap center is seen in both the *x–y* and *x–E* images (Fig. 5g,h).

The wire and outer plane electrodes are not necessary for creating a trap. However, voltage on the outer plane electrode creates the indirect regime, where long-life IXs are lower in energy than direct excitons, facilitating the IX collection to the trap from a large area. Due to the repulsive interaction, IXs effectively screen the in-plane disorder potential in the trap and collect to trap center that realizes a cold and dense exciton gas at the trap center (High et al., 2009b).

To realize IX BEC in the diamond-shaped trap, the excitation beam was focused to a spot ~5 μm in diameter on a side of the trap (Fig. 5b,c). This excitation scheme allows further cooling of the photogenerated IXs when they travel toward the trap center, thus facilitating the realization of a cold IX gas in the trap. IXs are efficiently cooled to the lattice temperature few microns away from the excitation spot (Hammack et al., 2009).

Fig. 5i presents the temperature dependence of IX emission pattern. At high temperatures, the IX cloud spreads over the trap resulting in a broad spatial profile of the IX emission. With lowering temperature, IX collect at the trap center. Studies of atoms in traps also show the collection of atoms at the trap center with lowering temperature due to the reduction of the thermal spreading of atoms over the trap and, eventually, condensation of atoms (Cornell and Wieman, 2002; Ketterle, 2002).

To verify IX condensation in the trap, we measured IX spontaneous coherence by shift-interferometry. The shift-interferometry probes the first-order coherence function $g_1(\delta x)$ as outlined in High et al. (2012b). $g_1(\delta x)$ is given by the amplitude of the interference fringes $A_{\mathrm{interf}}(\delta x)$ in 'the ideal experiment' with perfect spatial resolution. In practice, the measured $A_{\mathrm{interf}}(\delta x)$ is given by the convolution of $g_1(\delta x)$ with the point-spread function (PSF) of the optical system used in the experiment (Fogler et al., 2008). The PSF width corresponds to the spatial resolution of the optical system (~2 μm in High et al., 2012b).

Fig. 5k presents the amplitude of the interference fringes $A_{\mathrm{interf}}(\delta x)$ for different temperatures. The spatial extension of $A_{\mathrm{interf}}(\delta x)$ is characterized by a coherence length ξ. A strong enhancement of the IX coherence length is observed at low temperatures (Fig. 5l). While at high temperatures ξ is considerably smaller than the IX cloud width, at low temperatures the entire IX cloud becomes coherent (Fig. 5l).

Both for a classical gas and quantum gas $g_1(\delta x)$ is close to 1 at $\delta x = 0$ and drops with increasing δx within the coherence length ξ. For a classical gas, ξ_{cl} is close to the de Broglie wavelength λ_{dB}, which is well below the PSF width in the studied temperature range [for IX temperature $T > 0.1$ K, $\xi_{\mathrm{cl}} < 0.5$ μm; the PSF width is ~2 μm]. Therefore, for a classical gas at temperatures above few K, the measured ξ is given by the PSF width (Fig. 5l).

In contrast, for a condensate the coherence length becomes much larger than the thermal de Broglie wavelength and reaches the size of the exciton cloud in the trap so the entire exciton cloud becomes coherent. This signature of BEC is observed at the lowest temperatures in the experiment (Fig. 5l). The measured BEC temperature (High et al., 2012b) is in agreement with the theoretical estimates of IX BEC in the trap (Lobanov et al., 2016).

Below, we outline two other types of traps for IXs: the elevated traps (High et al., 2009a, 2009b) and snowflake traps (Kuznetsova et al., 2010a, 2015b). The wire and outer plane electrodes (Fig. 5b) are also used for creating an elevated trap in which the trap energy is at a higher energy than its surroundings (High et al., 2009b). An elevated trap was also realized using another electrode pattern (High et al., 2009a). The elevated traps allow evaporative cooling of the exciton gas, since higher energy excitons are more likely to escape the trap than lower energy excitons (High et al., 2009a, 2009b). The evaporative cooling can be used as additional to cooling of IXs via the phonon emission. However, cooling of IXs via the phonon emission is sufficient to cool IXs well below the temperature of quantum degeneracy (Butov et al., 2001) and realize IX BEC (Yang et al., 2006; High et al., 2012a, 2012b; Alloing et al., 2014; Gorbunov and Timofeev, 2016; Butov et al., 2002; Levitov et al., 2005; Yang et al., 2015; High et al., 2013; Leonard et al., 2018, 2021).

As outlined above, a trap providing a confining potential with the IX energy gradually reducing toward the trap center facilitates the creation of exciton BEC. Creating a trap with a confining in-plane potential on a large area is challenging. (i) A trap can be created by a circular opening in an electrode (Hammack et al., 2006a) or by a circular electrode (Chen et al., 2006; Shilo et al., 2013). In these traps, the in-plane potential varies within a small length from the electrode edge $\sim D_z$, where D_z is the distance between the top and bottom electrode (High et al., 2009b). Since D_z is typically small (1 μm in our samples) these traps produce a confining in-plane potential on a small area. Large traps of this type have a box-like shape (High et al., 2009b). (ii) A confining potential for IXs can be realized by electrodes at different voltages (Huber et al., 1998; Hammack et al., 2006a; High et al., 2009a; Schinner et al., 2011; Kowalik-Seidl et al., 2012; Schinner et al., 2013a, 2013b). The achievement of a large area of confining potentials in these traps requires a number of separately contacted electrodes. In-plane voltage gradients in these traps may produce in-plane electric currents and heat IXs. (iii) Confining potentials for IXs can be realized by using diamond-shaped traps as outlined above. The confining potential of diamond-shaped traps is sufficient for the realization of IX BEC (High et al., 2012b; Lobanov et al., 2016). However, the elongated shape of the diamond traps limits the area of IX collection and, as a result, limits the number of IXs in the trap.

Snowflake traps (Kuznetsova et al., 2015b) produce a 2D confining potential for IXs on a large area, an order of magnitude larger than the diamond traps, using a single electrode with no IX heating by in-plane electric currents. The principle of snowflake traps is based on the control of IX energy by electrode density (Kuznetsova et al., 2010a). The CQW is placed between a flat ground plane and a snowflake-shaped electrode, resulting in varying F_z in the CQW plane due to fringing electric fields from the shaped electrode. Decreasing electrode density toward the trap edges reduces F_z and, as a result, produces a confining potential for IXs (Kuznetsova et al., 2010a). Snowflake traps can be used for creating confined e–h systems with high densities.

3.2 Excitonic lattices

In-plane periodic potentials for IXs can be created in electrostatic lattices (Remeika et al., 2009, 2012; Zimmermann et al., 1997, 1998; Lagoin et al., 2022a, 2022b, 2023). As for cold atoms in optical lattices (Greiner et al., 2002; Chin et al., 2006; Billy et al., 2008), for IXs in electrostatic lattices, both the lattice and particle parameters can be controlled: the lattice amplitude by voltage and the IX density by the laser excitation power. IXs in electrostatic lattices give an opportunity to explore bosons in controlled periodic potentials in materials.

As outlined in Section 2.3, a linear lattice potential for IXs $E(x)$ can be created by interdigitated gates (Fig. 3). The linear lattices allowed exploring the IX localization–delocalization transition (Remeika et al., 2009) (Section 2.3) and allowed measuring IX correlations. It was found that the IX interaction deviates from the mean-field 'capacitor' formula due to correlations and the deviation is substantial at low densities (Remeika et al., 2009, 2015).

However, there are phenomena, in particular the Mott insulator—superfluid transition, which require in-plane energy modulation in both dimension (Greiner et al., 2002). A 2D lattice for IXs can be generated by a single electrode with a periodic array of holes (Hammack et al., 2006a). The lateral modulation of F_z, which determines the lattice amplitude, can be controlled by changing the voltage applied to the electrode. However, within this method, changing the lattice amplitude is accompanied by changing the average electric field F_z^{avg} and, in turn, lifetime and density of IXs. An independent control of F_z^{avg} and the lateral modulation of F_z can be realized using multiple electrodes separated by insulating layer(s) (Krauß et al., 2004). However, within this method, a considerable fraction of the applied voltage may drop in the deposited insulator and this fraction and, in turn, $F_z(x,y)$ may depend on the optical excitation.

A method for creating 2D lattices for IXs was proposed in Ref (Remeika et al., 2012). This method utilizes variable electrode density and allows creating various lattices, which can be controlled by voltage. Fig. 6

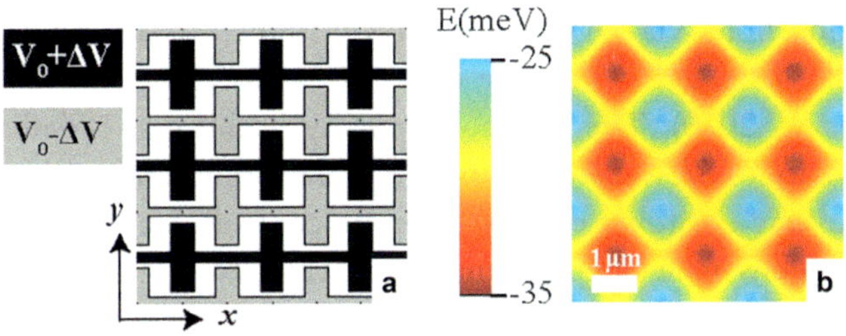

Fig. 6 2D excitonic lattices. Electrode schematics (a) and simulated IX energy at $\Delta V = 1$ V (b) for square lattice. *From Remeika, M., Fogler, M.M., Butov, L.V., Hanson, M., Gossard, A.C., 2012. Two-dimensional electrostatic lattices for indirect excitons. Appl. Phys. Lett. 100, 061103.*

shows the electrode pattern for creating square lattices. F_z^{avg} and spatial modulation of F_z can be independently controlled by voltages V_0 and ΔV. F_z^{avg} realizes the indirect regime and controls the IX lifetime. Modulation of F_z forms the lattice potential. The lattice amplitude can be controlled in situ by ΔV.

2D square lattice for IXs (Fig. 6) was created by this method in Remeika et al. (2012). Both IX intensity and energy were modulated with the period matching the lattice constant revealing the IX confinement in the 2D lattice. IX transport was also probed and IX localization–delocalization transition (similar to LDT in Section 2.3) was observed in the 2D lattice (Remeika et al., 2012).

As described in Ref (Remeika et al., 2012). The system of IXs in 2D lattices can be mapped to the Bose–Hubbard model characterized by two parameters, the inter-site hopping and the on-site interaction. Estimates in Remeika et al. (2012) indicate that for IXs in 2D lattices with the period ~200 nm for the case of one particle per site a finite-T crossover between incompressible Mott insulator state and normal state can be observed below ~50 mK, while superfluid—normal state transition can be observed at higher temperatures (Remeika et al., 2012).

3.3 Excitonic conveyers

In this section, we outline an electrostatic conveyer for excitons—a moving lattice created by a set of ac voltages (Winbow et al., 2011). The excitonic conveyer realizes controlled transport of excitons as charged coupled devices (CCD) realize controlled transport of electrons.

Moving potential lattices can be created by surface acoustic waves (SAW). Transport of excitons, exciton-polaritons, and laterally separated electrons and holes via SAW is intensively studied (Rudolph et al., 2007; Lazić et al., 2010, 2014; Violante et al., 2014; Rocke et al., 1997; Cerda-Méndez et al., 2010). The transport velocity in this case is defined by the sound velocity of SAW propagation ~3 μm/ns. In contrast, the velocity of the electrostatic excitonic conveyer can be controlled by the ac frequency from well below to well above the sound velocity.

The conveyer potential is created by a set of semitransparent 1 μm wide electrodes on the sample surface. The distance between the electrode centers is 2 μm, the conveyer periodicity is 7 electrodes, and the wavelength of the conveyer potential is $\lambda_{conv} = 14$ μm. A set of differentially phase-delayed ac sine waves at frequency f_{conv} creates a traveling potential lattice for IXs—the excitonic conveyer. The amplitude of the conveyer potential for IX A_{conv} is controlled by voltage. The conveyer velocity is controlled by the ac frequency $v_{conv} = \lambda_{conv} f_{conv}$ (Winbow et al., 2011).

Fig. 7a,b show *x–y* PL images for conveyer off and on. IX transport via conveyer is presented by the extension of the IX cloud along the direction of the moving potential. Fig. 7c presents IX transport via conveyer vs A_{conv}.

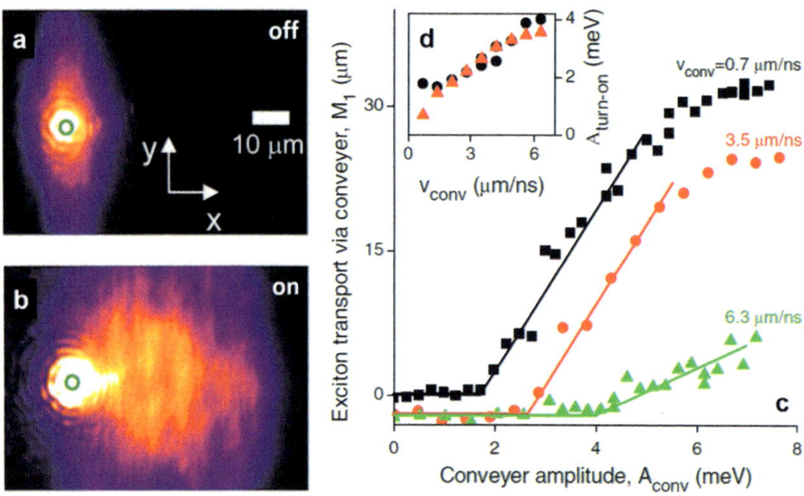

Fig. 7 Excitonic conveyers. (a,b) *x–y* IX PL images for conveyer off (a) and on (b). $A_{conv} = 7.5$ meV, $v_{conv} = 0.7$ μm/ns. (c) The average transport distance of IXs via conveyer M_1 vs the conveyer amplitude, A_{conv}. Lines are a guide to the eye, $A_{turn\text{-}on}$ is A_{conv} at the line intersection. (d) The measured (points) and calculated (triangles) $A_{turn\text{-}on}$ vs the conveyer velocity v_{conv}. *From Winbow, A.G., Leonard, J.R., Remeika, M., Kuznetsova, Y.Y., High, A.A., Hammack, A.T., et al., 2011. Electrostatic conveyer for excitons. Phys. Rev. Lett. 106, 196806.*

For a shallow conveyer, the IX cloud extension M_1 is not affected by the conveyer motion indicating that IXs do not follow the moving lattice, i.e., are dynamically delocalized in the lattice. In contrast, at higher A_{conv}, IXs are moved by the moving lattice, i.e., are dynamically localized in the lattice. At the dynamical localization–delocalization transition (DLDT), the IX cloud starts to follow the conveyer and M_1 changes from constant to increasing with A_{conv}. The DLDT is a dynamical counterpart of the LDT for IXs in static lattices (Remeika et al., 2009, 2012) (Sections 2.3 and 3.2).

Fig. 7c,d show that IX transport via conveyer is less efficient for higher v_{conv}. In particular, the conveyer amplitude at the DLDT, $A_{turn\text{-}on}$, increases with v_{conv}. The IX density dependence shows that IXs are hardly moved by the conveyer at low densities, efficient IX transport via conveyer is achieved at intermediate densities, and IX transport via conveyer becomes less efficient at high densities (Winbow et al., 2011).

Transport of IXs subject to the conveyer potential $U_{conv}(x)$ was also considered theoretically (Winbow et al., 2011). The model included IX drift due to the conveyer potential and IX interaction, screening of the disorder and conveyer potential by IXs, IX diffusion, and creation and decay of IXs. The theoretical simulations of IX transport via conveyer are in qualitative agreement with the experimental data exhibiting the dynamical IX delocalization in shallow conveyers and the DLDT with increasing conveyer amplitude. The simulated and measured $A_{turn\text{-}on}$ are in agreement (Fig. 7d). The simulated density dependence of IX transport via conveyer is also in qualitative agreement with the experiment (Winbow et al., 2011). The results are discussed below.

Conveyer amplitude dependence (Fig. 7c). When A_{conv} is smaller than the IX interaction energy or disorder amplitude, IXs are not localized in the minima of the moving conveyer potential and, therefore, are not moved by the conveyer. When A_{conv} becomes larger than both the IX interaction energy and disorder amplitude, IXs can localize in the minima of the moving conveyer potential. This results in efficient transport of IXs via conveyer.

Conveyer velocity dependence (Fig. 7c,d). IXs can efficiently follow the moving conveyer potential when the maximum IX drift velocity in the conveyer is higher than the conveyer velocity, $v_{drift} = \mu(\partial U_{conv}/\partial x)_{max} > v_{conv}$ (μ is the IX mobility). This leads to an estimate $A_{turn\text{-}on} \sim v_{conv}\lambda_{conv}/\mu$, showing that a higher conveyer amplitude is required for efficient IX transport via conveyer at a higher v_{conv}.

Density dependence. At low densities, IXs are localized in local minima of the disorder potential and hardly follow the moving conveyer. At the intermediate densities, IXs effectively screen the disorder and can be efficiently moved by the conveyer. IX transport via conveyer becomes less efficient at the high densities when IXs screen the conveyer potential. The requirement for efficient IX transport via conveyers $\mu(\partial U_{conv}/\partial x)_{max} > v_{conv}$ is relevant, where screening of disorder results in the enhancement of μ while screening of the conveyer potential results in the reduction of the conveyer amplitude.

3.4 Excitonic ramps

Excitonic ramps can be created by laterally modulated voltage (Hagn et al., 1995; Gärtner et al., 2006). An alternative method for creating potential landscapes, proposed in Ref (Kuznetsova et al., 2010a) is based on the lateral modulation of the electrode density. Divergence of the electric field around the periphery of an electrode reduces the normal component F_z. This enables one to control F_z, and thereby the potential energy landscape for IXs, by adjusting only the electrode density while keeping the entire device at same voltage. An advantage of this method is the absence of an energy-dissipating voltage gradient. In particular, this method was used to create snowflake traps (Kuznetsova et al., 2015b) (Section 3.1) and ramps (Leonard et al., 2012; Dorow et al., 2016) for IXs.

In this Section, we present an excitonic ramp based on the electrode density modulation in which a potential energy gradient is created by a perforated electrode at a constant voltage (Dorow et al., 2016). In contrast to the shaped electrode method (Leonard et al., 2012), the perforated electrode method (Dorow et al., 2016) allows the width of the channel to be varied independent of the energy gradient. As a result, the perforated electrode method allow supporting larger IX fluxes.

The modulation of the electrode density along x (Fig. 8a) is designed to create a ramp potential for IXs with a nearly constant energy gradient (Fig. 8b) (Dorow et al., 2016). Fig. 8c shows the profile of the IX PL intensity in a flat channel with no potential energy gradient. As there is no preferred direction of transport in the flat channel, two maxima that are nearly symmetric relative to the excitation spot position are observed. This is consistent with the inner ring effect (Section 2.2). Fig. 8d shows the profile of the IX PL intensity in the ramp. The IX transport in the ramp is directed toward lower IX energy. Such a directed transport of IXs in the ramp is analogous to directed transport of electrons in a diode.

The ramp device provides a proof of principle for controlling IX transport with electrode density gradients. The perforated electrode

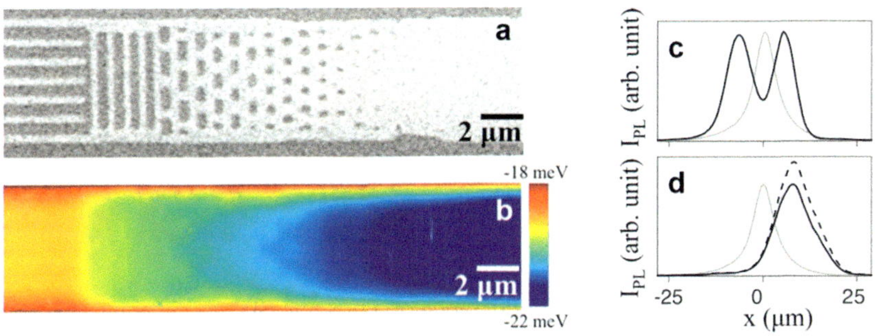

Fig. 8 Excitonic ramps. (a) SEM image of the ramp electrode. (b) Calculated IX energy shift $E(x,y) = -edF_z$ in the ramp. (c) IX PL profile (black) in a channel with constant potential energy. The channel is formed by a 5 μm wide electrode with no perforations. (d) IX PL profile (solid black) in the ramp. Estimation of IX PL profile (dashed black) with correction for the electrode transparency. Gray line shows the laser excitation profile (c,d). *From Dorow, C.J., Kuznetsova, Y.Y., Leonard, J.R., Chu, M.K., Butov, L.V., Wilkes, J., et al., 2016. Indirect excitons in a potential energy landscape created by a perforated electrode. Appl. Phys. Lett. 108, 073502.*

method gives an opportunity to create versatile potential landscapes for IXs and create channels for directing IX fluxes with the required geometry and energy profile.

3.5 Excitonic split gates

Split-gates are used for creating and controlling narrow channels (quantum point contacts, QPC) for electrons in mesoscopic electronic devices. Studies of electronic split-gate devices led to a number of findings including, in particular, electron focusing (Sharvin, 1965; Tsoi, 1974; van Houten et al., 1988), conductance quantization (van Wees et al., 1988; Wharam et al., 1988), and electron flow branching (Topinka et al., 2000, 2001).

Due to their properties (Section 1), IXs can be used to explore transport of cold bosons through split-gate devices, providing a counterpart to the transport studies of cold fermions through electronic split-gate devices. The predicted conductance quantization in QPC is qualitatively different for bosons and fermions: In contrast to conductance steps for fermions (van Wees et al., 1988; Wharam et al., 1988), the theory predicts conductance peaks for bosons (Fig. 9a) (Xu et al., 2019). The conductance peaks for bosons can be much higher than the conductance steps for fermions (Xu et al., 2019). The theory (Xu et al., 2019) predicts that quantum transport phenomena, for example, conductance quantization, single-slit diffraction, and two-slit interference are experimentally realizable in IX systems.

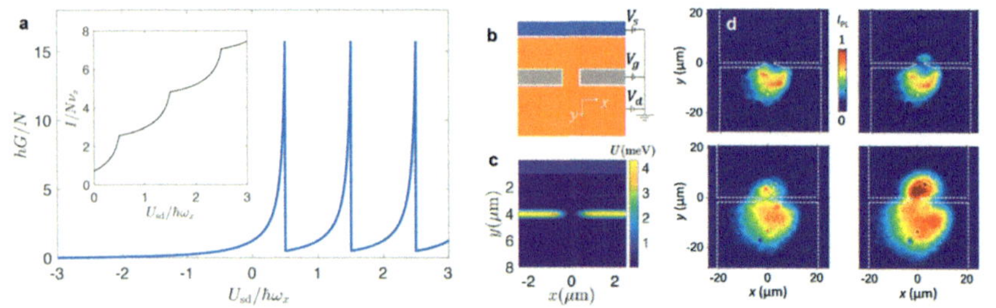

Fig. 9 Excitonic split gates. (a) Conductance of a bosonic QPC. (b) Schematic of an IX QPC device. The source (V_s), drain (V_d), and split-gate (V_g) voltages control the IX potential. (c) An example of IX potential in the split-gate device. (d) Proof of principle for IX split-gate devices. The laser spot is at (0,−4 μm). IX PL images for V_g = 4.5 and 5.5 V (left and right) and P_{ex} = 5 and 100 μW (top and bottom) show control of IX flux through split gate by V_g and P_{ex}. Dashed lines show the electrode contour. *From Dorow, C.J., Leonard, J.R., Fogler, M.M., Butov, L.V., West, K.W., Pfeiffer, L.N., 2018a. Split-gate device for indirect excitons. Appl. Phys. Lett. 112, 183501 and Xu, C., Leonard, J.R., Dorow, C.J., Butov, L.V., Fogler, M.M., Nikonov, D.E., et al., 2019. Exciton gas transport through nanoconstrictions. Nano Lett. 19, 5373.*

The proof-of-principle demonstration of IX split gate devices is outlined below. The electric field $F_z(x,y)$ creates the designed IX potential energy landscape $U(x,y) = -edF_z(x,y)$. For type II IX HS, such as $MoSe_2/WSe_2$ HS or AlAs/GaAs HS (Fig. 1b,d), where IXs form the lowest energy excitonic state at zero bias, the gate designs for IX split gate devices (Dorow et al., 2018a; Xu et al., 2019) and electronic split gate devices (van Wees et al., 1988; Wharam et al., 1988; Topinka et al., 2000, 2001) are similar. For type I HS, such as MoS_2/hBN HS or GaAs/AlGaAs HS (Fig. 1a,c), an additional electrode implementing the indirect regime in which IXs form the lowest energy excitonic state, is used (Dorow et al., 2018a). Gate voltage V_g on the split-gate electrode controls a narrow channel for IXs (Fig. 9b,c). IXs are photogenerated on one side of the split-gate, and their transport through the split-gate is detected by measuring the IX PL pattern. Fig. 9d shows that IX transport through split-gate is controlled both by the laser excitation power P_{ex} and split-gate voltage V_g (Dorow et al., 2018a). In the proof-of-principle demonstration of IX split-gate devices, the source and drain voltages were the same and IX transport through the split gate was caused by diffusion and interaction-induced drift (Eq. (1)) (Dorow et al., 2018a). The difference between the source and drain voltages provides an additional control for IX transport through the split gate (Xu et al., 2019).

IXs in split-gate devices were probed so far in the regime of classical IX transport at high temperatures. The experiments at low temperatures in the quantum regime yet need to be performed. The field of mesoscopic exciton systems is currently in its infancy. However, IXs have the potential to extend the phenomena studied in voltage-controllable electron systems to bosons.

3.6 Excitonic transistors

Potential advantages of excitonic signal processing are outlined in Section 1. Some excitonic devices, which can be used for excitonic signal processing, were noted in previous Sections: Excitonic conveyers (Section 3.3) realize controlled transport of excitons as CCD realize controlled transport of electrons, excitonic ramps (Section 3.4) realize directed transport of excitons that is analogous to directed transport of electrons in diodes, and excitonic storage (Winbow et al., 2007, 2008) may be used for storage (memory) applications in excitonic devices. In this Section, we overview proof of principle for excitonic transistors (High et al., 2007, 2008; Grosso et al., 2009; Kuznetsova et al., 2010b; Andreakou et al., 2014).

A transistor—a three-terminal device with the current through two terminals controlled by the third terminal, is the basic device of electronic logic. The operation principle of excitonic transistor is similar to that of electronic field-effect-transistor (FET). The excitonic transistor is a three-terminal device in which the exciton flux between the source and drain electrodes is controlled by voltage applied to the gate electrode located between source and drain (Fig. 10a). Excitons are generated at the source (input) and travel to the drain (output) due to the potential energy gradient $\propto edV_{sd}$ created by voltage V_{sd} between source and drain. The exciton flux from source to drain is controlled by voltage on the gate electrode V_g, which creates (removes) a barrier for the exciton flux in the region of the gate electrode when the excitonic transistor is off (on). The excitons at the input of excitonic transistor can be generated optically and the excitons can transform to photons at the output of excitonic transistor (Fig. 10a). Alternatively, the excitons at the input of excitonic transistor can arrive from the output of another excitonic device (e.g. another excitonic transistor) and the excitons from the output of excitonic transistor can go to the input of another excitonic device (e.g. another excitonic transistor), similarly to electrons in electronic circuits. This allows developing integrated excitonic circuits with an optical input and an optical output and

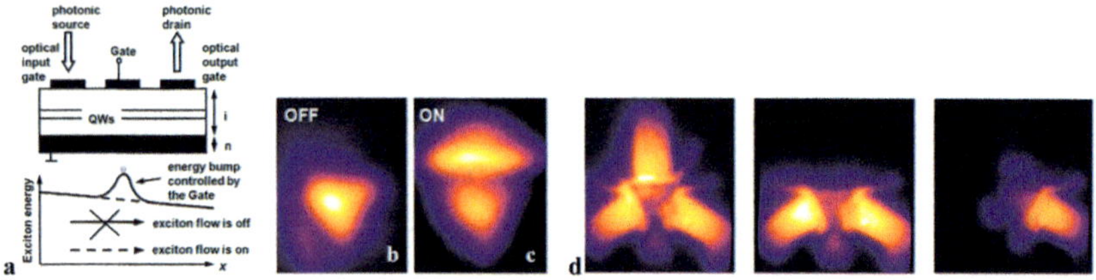

Fig. 10 Excitonic transistors. (a) Schematic of excitonic transistor. (b,c) Proof-of-principle demonstration of excitonic transistor: Emission images for IXs in off (b) and on (c) states of the excitonic transistor. The flux of IXs from source (at the bottom) to drain (at the top) is controlled by gate voltage. (d) Proof-of-principle demonstration of excitonic circuit with three IX transistors. Exciton flux from the center is controlled by voltage and directed to three, two, or one paths. *From High, A.A., Hammack, A.T., Butov, L.V., Hanson, M., Gossard, A.C., 2007. Exciton optoelectronic transistor. Opt. Lett. 32, 2466 and High, A.A., Novitskaya, E.E., Butov, L.V., Hanson, M., Gossard, A.C., 2008. Control of exciton fluxes in an excitonic integrated circuit. Science 321, 229.*

multiple excitonic devices between the input and output performing excitonic signal processing.

Optoelectronic excitonic transistors were demonstrated in GaAs HS (High et al., 2007, 2008; Grosso et al., 2009) (Fig. 10b,c). All-optical excitonic transistors, where the exciton flux from source to drain is controlled optically, were also demonstrated in GaAs HS (Kuznetsova et al., 2010b; Andreakou et al., 2014). Proof-of-principle excitonic circuits demonstrated a directional-switch, star-switch, and flux-merger operations with excitonic fluxes (High et al., 2008). The basic requirements for excitonic devices include the ability of excitons to travel over long distances exceeding the device dimensions and the ability to control the exciton transport. IXs meet both these requirements (Section 1) and are used as the operation medium for excitonic transistors and excitonic circuits (High et al., 2007, 2008; Grosso et al., 2009; Kuznetsova et al., 2010b; Andreakou et al., 2014).

Another type of excitonic devices is based on exciton-polaritons in microcavities (Amo et al., 2010; Gao et al., 2012; Ballarini et al., 2013; Nguyen et al., 2013; Sturm et al., 2014). The polaritons with a strong exciton–photon coupling typically lack e–h separation, which is used in control of IXs by voltage as outlined above. However, polariton fluxes are optically controlled (Amo et al., 2010; Gao et al., 2012; Ballarini et al., 2013; Nguyen et al., 2013; Sturm et al., 2014). A review of polaritonic devices is beyond the scope of this chapter.

4. Excitonic devices in TMDCs heterostructures

4.1 Toward high-temperature excitonic devices

Excitons exist at temperatures roughly below E_X/k_B (E_X is the exciton binding energy, k_B the Boltzmann constant) (Chemla et al., 1984). IXs in GaAs HS exist at low temperatures due to their low binding energies: E_X is typically ~4 meV in GaAs/AlGaAs HS (Sivalertporn et al., 2012) and ~10 meV in GaAs/AlAs HS (Zrenner et al., 1992). Furthermore, the temperature of quantum degeneracy, which can be achieved with increasing density before excitons dissociate to e–h plasma, scales proportionally to E_X (Fogler et al., 2014). Therefore, material systems where IXs have high-binding energies can provide the medium for the realization of both high-temperature quantum phenomena and high-temperature excitonic devices. IXs are explored in III–V and II–VI semiconductor HS based on GaAs (Sections 2, 3), GaN (Chiaruttini et al., 2019; Fedichkin et al., 2015; Lefebvre et al., 2004; Fedichkin et al., 2016), and ZnO (Morhain et al., 2005; Kuznetsova et al., 2015a). Among these materials, the highest IX binding energy ~30 meV is in ZnO HS (Morhain et al., 2005).

Excitons with high binding energies can be realized in van der Waals HS composed of atomically thin layers of TMDCs (Geim and Grigorieva, 2013; Ye et al., 2014; Chernikov et al., 2014; Goryca et al., 2019). The IX binding energies in TMDCs HS reach hundreds of meV (Fogler et al., 2014; Deilmann and Thygesen, 2018), making IXs stable at room temperature (Calman et al., 2018) and making TMDCs HS a promising platform both for exploring high-T IX quantum phenomena (Fogler et al., 2014; Berman and Kezerashvili, 2017) and developing excitonic devices.

Since excitons should be able to travel over long distances exceeding the device dimensions, long-range exciton transport is required for excitonic devices as outlined in Sections 2 and 3.6. Transport of both DXs in TMDCs monolayers (MLs) (Cadiz et al., 2018; Cordovilla Leon et al., 2018, 2019; Datta et al., 2022; Hao et al., 2020; Kulig et al., 2018; Kumar et al., 2014) and IXs in TMDCs HS (Rivera et al., 2016; Jauregui et al., 2019; Unuchek et al., 2019a, 2019b; Liu et al., 2020; Choi et al., 2020; Huang et al., 2020; Yuan et al., 2020; Li et al., 2021; Wang et al., 2021; Shanks et al., 2022; Sun et al., 2022; Peng et al., 2022; Tagarelli et al., 2023) is intensively studied. Diffusive IX transport with 1/e decay distances $d_{1/e}$ up to a few μm was realized in these studies. Disordered potentials can be responsible for limiting the IX transport distances in the TMDCs HS.

TMDCs HS offer the opportunity to explore IXs in moiré superlattice potentials with the period $b \approx a/(\delta\theta^2 + \delta^2)^{1/2}$ typically in the ~10 nm range (a is the lattice constant, δ the lattice mismatch, $\delta\theta$ the deviation of the twist angle between the layers from $i\pi/3$, i is an integer) (Wu et al., 2018; Yu et al., 2018; Wu et al., 2017; Yu et al., 2017; Zhang et al., 2017; Rivera et al., 2018; Seyler et al., 2019; Tran et al., 2019; Jin et al., 2019; Alexeev et al., 2019; Shimazaki et al., 2020; Wilson et al., 2021; Gu et al., 2022). The moiré potentials can be affected by atomic reconstruction (Alden et al., 2013; Woods et al., 2014; Yoo et al., 2019; Weston et al., 2020; Rosenberger et al., 2020; Zhao et al., 2023) and by disorder. Disordered moiré potentials should have a significant effect on exciton transport and, in particular, can suppress exciton transport in TMDCs HS. The realization of fairly periodic moiré potentials requires that the disorder is weak.

Excitonic devices including, in particular, excitonic transistors (Unuchek et al., 2019a, 2019b; Liu et al., 2020), excitonic conveyers (Datta et al., 2022; Peng et al., 2022), and excitonic ramps (Shanks et al., 2022) are explored in TMDCs HS. The counterparts of these devices in GaAs HS are outlined in Sections 3.3, 3.4, and 3.6.

4.2 Excitonic transistor in TMDCs heterostructure

In this Section, we outline an excitonic transistor in TMDCs HS (Fowler-Gerace et al., 2021) in which the control of IX transport is governed by mechanisms, beyond the mechanism for controlling IX transport by an energy barrier (or a trap) created by the gate electrode that was used in excitonic transistors both in GaAs HS (High et al., 2007, 2008; Grosso et al., 2009) and in TMDCs HS (Unuchek et al., 2019a, 2019b; Liu et al., 2020) and described in Section 3.6. We discuss the origin of the IX transport control by voltage in the TMDCs HS, in particular, the electric-field control of the moiré potential (Fowler-Gerace et al., 2021).

The $MoSe_2/WSe_2$ HS is assembled by stacking mechanically exfoliated 2D crystals on a graphite substrate. The $MoSe_2$ and WSe_2 MLs are encapsulated by hexagonal boron nitride (hBN). IXs are formed from electrons and holes confined in adjacent $MoSe_2$ and WSe_2 MLs (Fig. 1d). The bias across the HS is created by the gate voltage V_g applied between the narrow semitransparent multilayer graphene top gate (Fig. 11) and the global graphite back gate.

The HS region to the left of the graphene gate ($x = -7$ to $2\,\mu$m) is referred to as the source and the HS region to the right of the gate ($x = 6$ to

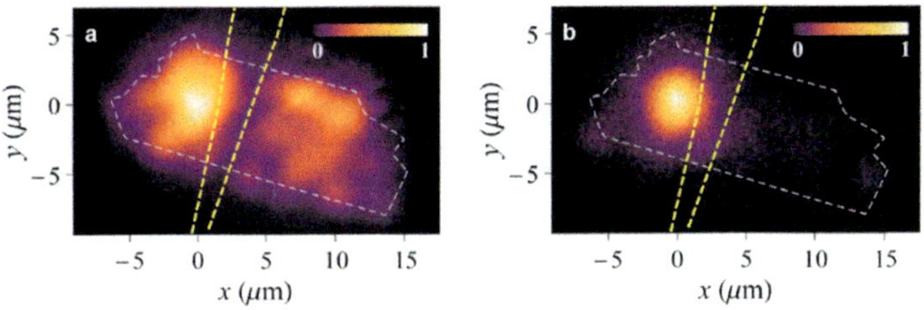

Fig. 11 Excitonic transistor in TMDCs HS. (a,b) *x–y* images of IX luminescence in on (a) and off (b) state of the excitonic transistor. The white and yellow dashed lines show the boundary of the $MoSe_2/WSe_2$ HS and graphene gate, respectively. The gate voltage V_g controls the IX propagation from the laser excitation spot at $x = 0$ to the other side of the graphene gate. $V_g = 10$ V (a), 0 (b). *From Fowler-Gerace, L.H., Choksy, D.J., Butov, L.V., 2021. Voltage-controlled long-range propagation of indirect excitons in a van der Waals heterostructure, Phys. Rev. B 104, 165302.*

13 μm) is referred to as the drain. In Fig. 11, IXs are optically generated by laser excitation focused in the source region.

When the device is in the off state, IX propagation from the source to the drain is suppressed and the IX luminescence profile follows the laser excitation profile (Fig. 11b). When the device is switched on, IXs spread out away from the laser excitation spot and propagate to the drain region (Fig. 11a). The total IX luminescence intensity in the drain region increases by 40 times as the gate voltage switches from off to on.

The properties of voltage-controlled IX propagation in $MoSe_2/WSe_2$ HS outlined above, are qualitatively different from those in GaAs HS (High et al., 2008; Grosso et al., 2009). GaAs/AlAs HS have a staggered band alignment (Zrenner et al., 1992) similar to $MoSe_2/WSe_2$ HS (Fig. 1b,d). This makes GaAs/AlAs HS a closer system for the comparison with $MoSe_2/WSe_2$ HS. For the excitation spot positioned in the source region, similar to the experimental geometry in Fig. 11, in the GaAs/AlAs HS IX propagation is long-range at $V_g = 0$ while both a positive and negative voltage on a gate creating a barrier and a trap for IXs, respectively, suppress the IX propagation (Grosso et al., 2009). This behavior is opposite to the voltage-controlled IX propagation in $MoSe_2/WSe_2$ HS where applied voltage strongly enhances the IX propagation (Fig. 11).

The opposite behavior is also observed for the excitation spot positioned on the gate electrode (Fig. 12b). For the GaAs/AlAs HS, trapping (antitrapping) IX potentials created by gate voltage cause the IX cloud confinement in (spreading away from) the gate region (Grosso et al., 2009). For the $MoSe_2/WSe_2$ HS, the IX energy reduces with increasing V_g that

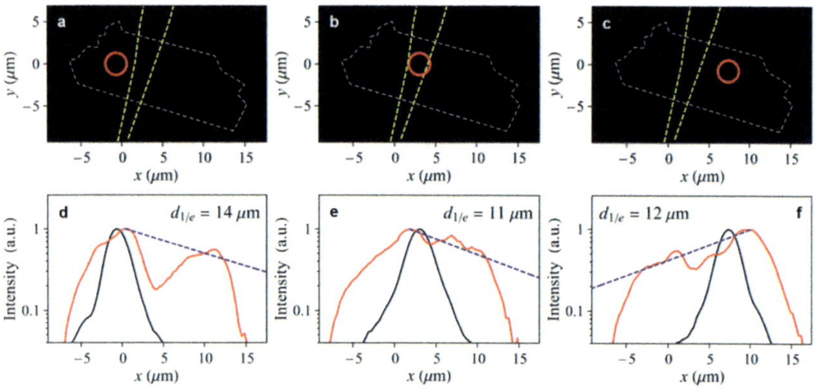

Fig. 12 Voltage-controlled IX transport in TMDCs HS. (a–c) The laser excitation spot [red circle in (a–c)] is positioned at the source (a,d), gate (b,e), and drain (c,f) region of the device. The gate voltage V_g controls the IX transport from the laser excitation spot. (d–f) Normalized IX luminescence profiles along $y = 0$ for $V_g = 10$ V (red lines) and 0 (black lines). Dashed lines show exponential signal reduction with 1/e decay distance $d_{1/e} = 14$, 11, and 12 μm. A lower IX luminescence intensity is seen in the region covered by the graphene gate, which is centered at $x = 4$ μm. *From Fowler-Gerace, L.H., Choksy, D.J., Butov, L.V., 2021. Voltage-controlled long-range propagation of indirect excitons in a van der Waals heterostructure, Phys. Rev. B 104, 165302.*

facilitates trapping IXs in the gate region. However, similar to the case of excitation in the source region (Fig. 11), an enhancement of the IX propagation away from the excitation spot with increasing V_g is observed in the case of excitation in the gate region (Fig. 12e).

In GaAs HS, the IX transport has been controlled by an energy barrier to IX propagation (or a trap for IXs) created by the gate electrode (High et al., 2007, 2008; Grosso et al., 2009). The control of IX transport realized in TMDCs HS was referred to the same mechanism (Unuchek et al., 2019a, 2019b; Liu et al., 2020). The qualitative differences outlined above show that the IX transport in the $MoSe_2/WSe_2$ HS is controlled by a mechanism, different from the control by an energy barrier or a trap for IXs created by the gate electrode.

This mechanism enables the IX transport away from the excitation spot not only through the gate but everywhere in the plane of the HS: Fig. 11 shows that increasing V_g enables the IX propagation away from the excitation spot both through the gate (toward positive x) and in the opposite direction (toward negative x); Fig. 12 shows that increasing V_g enables the IX propagation away from the excitation spot for any excitation spot position on the HS. A possible origin of the IX transport control by voltage in the $MoSe_2/WSe_2$ HS is discussed below.

IXs are the lowest energy exciton state in $MoSe_2/WSe_2$ HS even at no applied voltage (Fig. 1d). The absence of IX propagation at $V_g = 0$ indicates IX localization. The moiré superlattice potentials are predicted to cause modulations of IX energy reaching tens of meV (Wu et al., 2018; Yu et al., 2018; Wu et al., 2017; Yu et al., 2017) and these strong energy modulations can lead to the IX localization. The theory (Yu et al., 2017) predicts that the moiré potential is tuned by voltage in both R- and H-stacked TMDCs HS. In particular, increasing electric field can reduce the moiré potential amplitude and can cause percolation of IX states through the HS (Yu et al., 2017). The calculated electric field corresponding to the IX percolation is in the range of a few tenths of V/nm (Yu et al., 2017). This is of the same order of magnitude with the estimated electric field for the onset of IX propagation in the device. We note however that a comparison with the theory can be complicated by various factors including, for instance, atomic reconstruction, which can change the potential landscape and can create a network of propagation channels (Alden et al., 2013; Woods et al., 2014; Yoo et al., 2019; Weston et al., 2020; Rosenberger et al., 2020; Zhao et al., 2023). Understanding IX transport in electric-field-controlled (reconstructed) moiré potentials in TMDCs HS is the subject for future works.

5. Summary

Indirect excitons (IXs) can be controlled by voltage, can travel over large distances, and can cool down below the temperature of quantum degeneracy. These properties form the basis for the development of excitonic devices with IXs. In this contribution, we overviewed our studies of excitonic devices in GaAs HS and in TMDCs HS. We outlined IX transport and excitonic devices, including excitonic traps, excitonic lattices, excitonic conveyers, excitonic ramps, excitonic split gates, and excitonic transistors. These devices can be used both for studying quantum systems of bosons in materials and for developing excitonic signal processing.

Acknowledgements

Our studies of indirect excitons presented in this contribution were performed in collaboration with G. Abstreiter, P. Andreakou, G.O. Andreev, K.W. Baldwin, G. Böhm, T. Bretagnon, E.V. Calman, K.L. Campman, J.-M. Chauveau, D.S. Chemla, D.J. Choksy,

M.K. Chu, V.T. Dolgopolov, C.J. Dorow, F. Fedichkin, A.I. Filin, M.M. Fogler, L.H. Fowler-Gerace, A.K. Geim, N.A. Gippius, A.C. Gossard, J.C. Graves, M. Griswold, G. Grosso, A.A. Guenther, T. Guillet, A.T. Hammack, M. Hanson, M.W. Hasling, A.A. High, Lunhui Hu, Sheng Hu, A. Imamoglu, A.L. Ivanov, A.V. Kavokin, Y.Y. Kuznetsova, P. Lefebvre, J.R. Leonard, L.S. Levitov, T.C.H. Liew, P.B. Littlewood, S.V. Lobanov, Yu.E. Lozovik, A.D. Meyertholen, A.V. Mintsev, A. Mishchenko, C. Morhain, L. Mouchliadis, D.E. Nikonov, E.E. Novitskaya, K.S. Novoselov, T. Ostatnický, L.N. Pfeiffer, S.V. Poltavtsev, M. Remeika, A.A. Shashkin, B.D. Simons, L.E. Smallwood, E.A. Szwed, M. Vladimirova, G. Weimann, J. Wilkes, A.G. Winbow, Congjun Wu, Chao Xu, Sen Yang, I.A. Young, Zhiwen Zhou, and A. Zrenner and it is a pleasure to thank my collaborators. Our studies of indirect excitons were supported by DOE Office of Basic Energy Sciences under award DE-FG02-07ER46449 and by NSF.

References

Alden, J.S., Tsen, A.W., Huang, P.Y., Hovden, R., Brown, L., Park, J., et al., 2013. Strain solitons and topological defects in bilayer graphene. Proc. Natl. Acad. Sci. U. S. A. 110, 11256.

Alexeev, E.M., Ruiz-Tijerina, D.A., Danovich, M., Hamer, M.J., Terry, D.J., Nayak, P.K., et al., 2019. Resonantly hybridized excitons in moiré superlattices in van der Waals heterostructures. Nature 567, 81.

Alloing, M., Beian, M., Lewenstein, M., Fuster, D., González, Y., González, L., et al., 2014. Evidence for a Bose-Einstein condensate of excitons. Europhys. Lett. 107, 10012.

Alloing, M., Lemaître, A., Galopin, E., Dubin, F., 2012. Nonlinear dynamics and inner-ring photoluminescence pattern of indirect excitons. Phys. Rev. B 85, 245106.

Alloing, M., Lemaître, A., Galopin, E., Dubin, F., 2013. On-demand confinement of semiconductor excitons by all-optical control, Optically programmable excitonic traps. Sci. Rep. 3, 1578.

Amo, A., Liew, T.C.H., Adrados, C., Houdré, R., Giacobino, E., Kavokin, A.V., et al., 2010. Exciton-polariton spin switches. Nat. Photonics 4, 361.

Andreakou, P., Poltavtsev, S.V., Leonard, J.R., Calman, E.V., Remeika, M., Kuznetsova, Y.Y., et al., 2014. Optically controlled excitonic transistor. Appl. Phys. Lett. 104, 091101.

Andreani, L.C., Tassone, F., Bassani, F., 1991. Radiative lifetime of free excitons in quantum wells. Solid State Commun. 77, 641.

Ballarini, D., De Giorgi, M., Cancellieri, E., Houdré, R., Giacobino, E., Cingolani, R., et al., 2013. All-optical polariton transistor. Nat. Commun. 4, 1778.

Ben-Tabou de-Leon, S., Laikhtman, B., 2001. Exciton-exciton interactions in quantum wells: optical properties and energy and spin relaxation. Phys. Rev. B 63, 125306.

Berman, O.L., Kezerashvili, R.Y., 2017. Superfluidity of dipolar excitons in a transition metal dichalcogenide double layer. Phys. Rev. B 96, 094502.

Billy, J., Josse, V., Zuo, Z., Bernard, A., Hambrecht, B., Lugan, P., et al., 2008. Direct observation of Anderson localization of matter waves in a controlled disorder. Nature 453, 891.

Brunner, K., Bockelmann, U., Abstreiter, G., Walther, M., Böhm, G., Tränkle, G., et al., 1992. Photoluminescence from a single GaAs/AlGaAs quantum dot. Phys. Rev. Lett. 69, 3216.

Butov, L.V., 2016. Collective phenomena in cold indirect excitons. J. Exp. Theor. Phys. 122, 434.

Butov, L.V., 2017. Excitonic devices. Superlattices Microstructures 108, 2.

Butov, L.V., 2020. Condensation of indirect excitons. MRS Bull. 45, 380.

Butov, L.V., Filin, A.I., 1998. Anomalous transport and luminescence of indirect excitons in AlAs/GaAs coupled quantum wells as evidence for exciton condensation. Phys. Rev. B 58, 1980.

Butov, L.V., Gossard, A.C., Chemla, D.S., 2002. Macroscopically ordered state in an exciton system. Nature 418, 751.

Butov, L.V., Ivanov, A.L., Imamoglu, A., Littlewood, P.B., Shashkin, A.A., Dolgopolov, V.T., et al., 2001. Stimulated scattering of indirect excitons in coupled quantum wells: signature of a degenerate Bose-gas of excitons. Phys. Rev. Lett. 86, 5608.

Butov, L.V., Levitov, L.S., Mintsev, A.V., Simons, B.D., Gossard, A.C., Chemla, D.S., 2004. Formation mechanism and low-temperature instability of exciton rings. Phys. Rev. Lett. 92, 117404.

Butov, L.V., Zrenner, A., Abstreiter, G., Böhm, G., Weimann, G., 1994. Condensation of indirect excitons in coupled AlAs/GaAs quantum wells. Phys. Rev. Lett. 73, 304.

Cadiz, F., Robert, C., Courtade, E., Manca, M., Martinelli, L., Taniguchi, T., et al., 2018. Exciton diffusion in WSe_2 monolayers embedded in a van der Waals heterostructure. Appl. Phys. Lett. 112, 152106.

Calman, E.V., Fogler, M.M., Butov, L.V., Hu, S., Mishchenko, A., Geim, A.K., 2018. Indirect excitons in van der Waals heterostructures at room temperature. Nat. Commun. 9, 1895.

Cerda-Méndez, E.A., Krizhanovskii, D.N., Wouters, M., Bradley, R., Biermann, K., Guda, K., et al., 2010. Polariton condensation in dynamic acoustic lattices. Phys. Rev. Lett. 105, 116402.

Chemla, D.S., Miller, D.A.B., Smith, P.W., Gossard, A.C., Wiegmann, W., 1984. Room temperature excitonic nonlinear absorption and refraction in GaAs/AlGaAs multiple quantum well structures. IEEE J. Quantum Electron. 20, 265.

Chen, G., Rapaport, R., Pffeifer, L.N., West, K., Platzman, P.M., Simon, S., et al., 2006. Artificial trapping of a stable high-density dipolar exciton cloud. Phys. Rev. B 74, 045309.

Chen, G., Rapaport, R., Simon, S.H., Pfeiffer, L., West, K., 2005. Dynamics of the in-plane charge separation front in a two-dimensional electron-hole gas. Phys. Rev. B 71, 041301R.

Chernikov, A., Berkelbach, T.C., Hill, H.M., Rigosi, A., Li, Y., Aslan, O.B., et al., 2014. Exciton binding energy and nonhydrogenic Rydberg series in monolayer WS_2. Phys. Rev. Lett. 113, 076802.

Chiaruttini, F., Guillet, T., Brimont, C., Jouault, B., Lefebvre, P., Vives, J., et al., 2019. Trapping dipolar exciton fluids in GaN/(AlGa)N nanostructures. Nano Lett. 19, 4911.

Chin, J.K., Miller, D.E., Liu, Y., Stan, C., Setiawan, W., Sanner, C., et al., 2006. Evidence for superfluidity of ultracold fermions in an optical lattice. Nature 443, 961.

Choi, J., Hsu, W.T., Lu, L.S., Sun, L., Cheng, H.Y., Lee, M.H., et al., 2020. Moiré potential impedes interlayer exciton diffusion in van der Waals heterostructures. Sci. Adv. 6, eaba8866.

Choksy, D.J., Szwed, E.A., Butov, L.V., Baldwin, K.W., Pfeiffer, L.N., 2023. Fermi edge singularity in neutral electron-hole system. Nat. Phys. 19, 1275.

Christianen, P.C.M., Piazza, F., Lok, J.G.S., Maan, J.C., van der Vleuten, W., 1998. Magnetic trap for excitons. Physica B 249–251, 624.

Cohen, K., Rapaport, R., Santos, P.V., 2011. Remote dipolar interactions for objective density calibration and flow control of excitonic fluids. Phys. Rev. Lett. 106 126402.

Cohen, K., Shilo, Y., West, K., Pfeiffer, L., Rapaport, R., 2016. Dark high density dipolar liquid of excitons. Nano Lett. 16, 3726.

Comte, C., Noziéres, P., 1982. Exciton Bose condensation: the ground state of an electron–hole gas – I. Mean field description of a simplified model. J. Physique 43, 1069.

Cordovilla Leon, D.F., Li, Z., Jang, S.W., Cheng, C.H., Deotare, P.B., 2018. Exciton transport in strained monolayer WSe_2. Appl. Phys. Lett. 113, 252101.
Cordovilla Leon, D.F., Li, Z., Jang, S.W., Deotare, P.B., 2019. Hot exciton transport in WSe_2 monolayers. Phys. Rev. B 100, 241401R.
Cornell, E.A., Wieman, C.E., 2002. Nobel lecture: Bose-Einstein condensation in a dilute gas, the first 70 years and some recent experiments. Rev. Mod. Phys. 74, 875.
Datta, K., Lyu, Z., Li, Z., Taniguchi, T., Watanabe, K., Deotare, P.B., 2022. Spatiotemporally controlled room-temperature exciton transport under dynamic strain. Nat. Photonics 16, 242.
Deilmann, T., Thygesen, K.S., 2018. Interlayer trions in the MoS_2/WS_2 van der Waals heterostructure. Nano Lett. 18, 1460.
Deveaud, B., Clerot, F., Roy, N., Satzke, K., Sermage, B., Katzer, D.S., 1991. Enhanced radiative recombination of free excitons in GaAs quantum wells. Phys. Rev. Lett. 67, 2355.
Dignam, M.M., Sipe, J.E., 1991. Exciton states in coupled double quantum wells in a static electric field. Phys. Rev. B 43, 4084.
Dorow, C.J., Hasling, M.W., Calman, E.V., Butov, L.V., Wilkes, J., Campman, K.L., et al., 2017. Spatially resolved and time resolved imaging of transport of indirect excitons in high magnetic fields. Phys. Rev. B 95, 235308.
Dorow, C.J., Hasling, M.W., Choksy, D.J., Leonard, J.R., Butov, L.V., West, K.W., et al., 2018b. High-mobility indirect excitons in wide single quantum well. Appl. Phys. Lett. 113, 212102.
Dorow, C.J., Kuznetsova, Y.Y., Leonard, J.R., Chu, M.K., Butov, L.V., Wilkes, J., et al., 2016. Indirect excitons in a potential energy landscape created by a perforated electrode. Appl. Phys. Lett. 108, 073502.
Dorow, C.J., Leonard, J.R., Fogler, M.M., Butov, L.V., West, K.W., Pfeiffer, L.N., 2018a. Split-gate device for indirect excitons. Appl. Phys. Lett. 112, 183501.
Eisenstein, J.P., MacDonald, A.H., 2004. Bose-Einstein condensation of excitons in bilayer electron systems. Nature 432, 691.
Fedichkin, F., Andreakou, P., Jouault, B., Vladimirova, M., Guillet, T., Brimont, C., et al., 2015. Transport of dipolar excitons in (Al,Ga)N/GaN quantum wells. Phys. Rev. B 91, 205424.
Fedichkin, F., Guillet, T., Valvin, P., Jouault, B., Brimont, C., Bretagnon, T., et al., 2016. Room-temperature transport of indirect excitons in (Al,Ga)N/GaN quantum wells. Phys. Rev. Appl. 6, 014011.
Feldmann, J., Peter, G., Göbel, E.O., Dawson, P., Moore, K., Foxon, C., et al., 1987. Linewidth dependence of radiative exciton lifetimes in quantum wells. Phys. Rev. Lett. 59, 2337.
Finkelstein, R., Cohen, K., Jouault, B., West, K., Pfeiffer, L.N., Vladimirova, M., et al., 2017. Transition from spin-orbit to hyperfine interaction dominated spin relaxation in a cold fluid of dipolar excitons. Phys. Rev. B 96, 085404.
Fogler, M.M., Butov, L.V., Novoselov, K.S., 2014. High-temperature superfluidity with indirect excitons in van der Waals heterostructures. Nat. Commun. 5, 4555.
Fogler, M.M., Yang, Sen, Hammack, A.T., Butov, L.V., Gossard, A.C., 2008. Effect of spatial resolution on the estimates of the coherence length of excitons in quantum wells. Phys. Rev. B. 78, 035411.
Fowler-Gerace, L.H., Choksy, D.J., Butov, L.V., 2021. Voltage-controlled long-range propagation of indirect excitons in a van der Waals heterostructure. Phys. Rev. B 104, 165302.
Gao, T., Eldridge, P.S., Liew, T.C.H., Tsintzos, S.I., Stavrinidis, G., Deligeorgis, G., et al., 2012. Polariton condensate transistor switch. Phys. Rev. B 85, 235102.

Gärtner, A., Holleitner, A.W., Kotthaus, J.P., Schuh, D., 2006. Drift mobility of long-living excitons in coupled GaAs quantum wells. Appl. Phys. Lett. 89, 052108.

Geim, A.K., Grigorieva, I.V., 2013. Van der Waals heterostructures. Nature 499, 419.

Gorbunov, A.V., Timofeev, V.B., 2006. Large-scale coherence of the Bose condensate of spatially indirect excitons. JETP Lett. 84, 329.

Gorbunov, A.V., Timofeev, V.B., 2012. Phase diagram of the bose condensation of dipolar excitons in GaAs/AlGaAs quantum-well heterostructures. JETP Lett. 96, 138.

Gorbunov, A.V., Timofeev, V.B., 2013. Compensation of dipolar-exciton spin splitting in magnetic field. Solid. State Commun. 157, 6.

Gorbunov, A.V., Timofeev, V.B., 2016. Coherence of Bose-Einstein condensates of dipolar excitons in GaAs/AlGaAs heterostructures. Low. Temp. Phys. 42, 340.

Gorbunov, A.V., Timofeev, V.B., Demin, D.A., 2011. Electro-optical trap for dipolar excitons in a GaAs/AlAs Schottky diode with a single quantum well. JETP Lett. 94, 800.

Goryca, M., Li, J., Stier, A.V., Taniguchi, T., Watanabe, K., Courtade, E., et al., 2019. Revealing exciton masses and dielectric properties of monolayer semiconductors with high magnetic fields. Nat. Commun. 10, 4172.

Greiner, M., Mandel, O., Esslinger, T., Hänsch, T.W., Bloch, I., 2002. Quantum phase transition from a superfluid to a Mott insulator in a gas of ultracold atoms. Nature 415, 39.

Grosso, G., Graves, J., Hammack, A.T., High, A.A., Butov, L.V., Hanson, M., et al., 2009. Excitonic switches operating at around 100 K. Nat. Photonics 3, 577.

Gu, J., Ma, L., Liu, S., Watanabe, K., Taniguchi, T., Hone, J.C., et al., 2022. Dipolar excitonic insulator in a moiré lattice. Nat. Phys. 18, 395.

Hagn, M., Zrenner, A., Böhm, G., Weimann, G., 1995. Electric-field-induced exciton transport in coupled quantum well structures. Appl. Phys. Lett. 67, 232.

Hammack, A.T., Butov, L.V., Wilkes, J., Mouchliadis, L., Muljarov, E.A., Ivanov, A.L., et al., 2009. Kinetics of the inner ring in the exciton emission pattern in coupled GaAs quantum wells. Phys. Rev. B 80, 155331.

Hammack, A.T., Gippius, N.A., Yang, Sen, Andreev, G.O., Butov, L.V., Hanson, M., et al., 2006a. Excitons in Electrostatic Traps. J. Appl. Phys. 99 066104.

Hammack, A.T., Griswold, M., Butov, L.V., Smallwood, L.E., Ivanov, A.L., Gossard, A.C., 2006b. Trapping of cold excitons in quantum well structures with laser light. Phys. Rev. Lett. 96, 227402.

Hao, S., Bellus, M.Z., He, D., Wang, Y., Zhao, H., 2020. Controlling exciton transport in monolayer $MoSe_2$ by dielectric screening. Nanoscale Horiz. 5, 139.

Haque, M., 2006. Ring-shaped luminescence pattern in biased quantum wells studied as a steady-state reaction front. Phys. Rev. E 73, 066207.

High, A.A., Hammack, A.T., Butov, L.V., Hanson, M., Gossard, A.C., 2007. Exciton optoelectronic transistor. Opt. Lett. 32, 2466.

High, A.A., Hammack, A.T., Butov, L.V., Mouchliadis, L., Ivanov, A.L., Hanson, M., et al., 2009a. Indirect excitons in elevated traps. Nano Lett. 9, 2094.

High, A.A., Hammack, A.T., Leonard, J.R., Yang, S., Butov, L.V., Ostatnický, T., et al., 2013. Spin currents in a coherent exciton gas. Phys. Rev. Lett. 110, 246403.

High, A.A., Leonard, J.R., Hammack, A.T., Fogler, M.M., Butov, L.V., Kavokin, A.V., et al., 2012a. Spontaneous coherence in a cold exciton gas. Nature 483, 584.

High, A.A., Leonard, J.R., Remeika, M., Butov, L.V., Hanson, M., Gossard, A.C., 2012b. Condensation of excitons in a trap. Nano Lett. 12, 2605.

High, A.A., Novitskaya, E.E., Butov, L.V., Hanson, M., Gossard, A.C., 2008. Control of exciton fluxes in an excitonic integrated circuit. Science 321, 229.

High, A.A., Thomas, A.K., Grosso, G., Remeika, M., Hammack, A.T., Meyertholen, A.D., et al., 2009b. Trapping indirect excitons in a GaAs quantum-well structure with a diamond-shaped electrostatic trap. Phys. Rev. Lett. 103, 087403.
Huang, Z., Liu, Y., Dini, K., Tan, Q., Liu, Z., Fang, H., et al., 2020. Robust room temperature valley Hall effect of interlayer excitons. Nano Lett. 20, 1345.
Huber, T., Zrenner, A., Wegscheider, W., Bichler, M., 1998. Electrostatic exciton traps. Phys. Status Solidi A 166, R5.
Inouye, S., Gupta, S., Rosenband, T., Chikkatur, A.P., Görlitz, A., Gustavson, T.L., et al., 2001. Observation of vortex phase singularities in Bose-Einstein condensates. Phys. Rev. Lett. 87, 080402.
Ivanov, A.L., 2002. Quantum diffusion of dipole-oriented indirect excitons in coupled quantum wells. Europhys. Lett. 59, 586.
Ivanov, A.L., Smallwood, L.E., Hammack, A.T., Yang, S., Butov, L.V., Gossard, A.C., 2006. Origin of the inner ring in photoluminescence patterns of quantum well excitons. Europhys. Lett. 73, 920.
Jauregui, L.A., Joe, A.Y., Pistunova, K., Wild, D.S., High, A.A., Zhou, Y., et al., 2019. Electrical control of interlayer exciton dynamics in atomically thin heterostructures. Science 366, 870.
Jin, C., Regan, E.C., Yan, A., Utama, M., Iqbal Bakti, W., Zhao, D., et al., 2019. Observation of moiré excitons in WSe_2/WS_2 heterostructure superlattices. Nature 567, 76.
Kash, K., Worlock, J.M., Sturge, M.D., Grabbe, P., Harbison, J.P., Scherer, A., et al., 1988. Strain-induced lateral confinement of excitons in GaAs-AlGaAs quantum well microstructures. Appl. Phys. Lett. 53, 782.
Keldysh, L.V., 1986. The electron-hole liquid in semiconductors. Contemp. Phys. 27, 395.
Keldysh, L.V., Kopaev, Y.V., 1965. Possible instability of the semimetallic state toward Coulomb interaction. Sov. Phys. Solid. State 6, 2219.
Keldysh, L.V., Kozlov, A.N., 1968. Collective properties of excitons in semiconductors. Sov. Phys. JETP 27, 521.
Ketterle, W., 2002. Nobel lecture: when atoms behave as waves: Bose-Einstein condensation and the atom laser. Rev. Mod. Phys. 74, 1131.
Kowalik-Seidl, K., Vögele, X.P., Rimpfl, B.N., Schinner, G.J., Schuh, D., Wegscheider, W., et al., 2012. Tunable photoemission from an excitonic antitrap. Nano Lett. 12, 326.
Krauß, J., Kotthaus, J.P., Wixforth, A., Hanson, M., Driscoll, D.C., Gossard, A.C., et al., 2004. Capture and release of photonic images in a quantum well. Appl. Phys. Lett. 85, 5830.
Kulig, M., Zipfel, J., Nagler, P., Blanter, S., Schüller, C., Korn, T., et al., 2018. Exciton diffusion and Halo effects in monolayer semiconductors. Phys. Rev. Lett. 120, 207401.
Kumar, N., Cui, Q., Ceballos, F., He, D., Wang, Y., Zhao, H., 2014. Exciton diffusion in monolayer and bulk $MoSe_2$. Nanoscale 6, 4915.
Kuznetsova, Y.Y., Fedichkin, F., Andreakou, P., Calman, E.V., Butov, L.V., Lefebvre, P., et al., 2015a. Transport of indirect excitons in ZnO quantum wells. Opt. Lett. 40, 3667.
Kuznetsova, Y.Y., Andreakou, P., Hasling, M.W., Leonard, J.R., Calman, E.V., Butov, L.V., et al., 2015b. Two-dimensional snowflake trap for indirect excitons. Opt. Lett. 40, 589.
Kuznetsova, Y.Y., High, A.A., Butov, L.V., 2010a. Control of excitons by the laterally modulated electrode density. Appl. Phys. Lett. 97, 201106.
Kuznetsova, Y.Y., Remeika, M., High, A.A., Hammack, A.T., Butov, L.V., Hanson, M., et al., 2010b. All-optical excitonic transistor. Opt. Lett. 35, 1587.
Lagoin, C., Bhattacharya, U., Grass, T., Chhajlany, R.W., Salamon, T., Baldwin, K., et al., 2022a. Extended Bose-Hubbard model with dipolar excitons. Nature 609, 485.
Lagoin, C., Suffit, S., Baldwin, K., Pfeiffer, L., Dubin, F., 2022b. Mott insulator of strongly interacting two-dimensional semiconductor excitons. Nat. Phys. 18, 149.

Lagoin, C., Suffit, S., Baldwin, K., Pfeiffer, L., Dubin, F., 2023. Dual-density waves with neutral and charged dipolar excitons of GaAs bilayers. Nat. Mater. 22, 170.
Lagoudakis, K.G., Ostatnický, T., Kavokin, A.V., Rubo, Y.G., André, R., Deveaud-Plédran, B., 2009. Observation of half quantum vortices in an exciton-polariton condensate. Science 326, 974.
Lagoudakis, K.G., Wouters, M., Richard, M., Baas, A., Carusotto, I., André, R., et al., 2008. Quantized vortices in an exciton-polariton condensate. Nat. Phys. 4, 706.
Larionov, A.V., Timofeev, V.B., Hvam, J., Soerensen, K., 2000. Interwell excitons in GaAs/AlGaAs double quantum wells and their collective properties. J. Exp. Theor. Phys. 90, 1093.
Lazić, S., Santos, P.V., Hey, R., 2010. Exciton transport by moving strain dots in GaAs quantum wells. Phys. E 42, 2640.
Lazić, S., Violante, A., Cohen, K., Hey, R., Rapaport, R., Santos, P.V., 2014. Scalable interconnections for remote indirect exciton systems based on acoustic transport. Phys. Rev. B 89, 085313.
Lefebvre, P., Kalliakos, S., Bretagnon, T., Valvin, P., Taliercio, T., Gil, B., et al., 2004. Observation and modeling of the time-dependent descreening of internal electric field in a wurtzite $GaN/Al_{0.15}Ga_{0.85}N$ quantum well after high photoexcitation. Phys. Rev. B 69, 035307.
Leonard, J.R., High, A.A., Hammack, A.T., Fogler, M.M., Butov, L.V., Campman, K.L., et al., 2018. Pancharatnam-Berry phase in condensate of indirect excitons. Nat. Commun. 9, 2158.
Leonard, J.R., Hu, L., High, A.A., Hammack, A.T., Wu, C., Butov, L.V., et al., 2021. Moiré pattern of interference dislocations in condensate of indirect excitons. Nat. Commun. 12, 1175.
Leonard, J.R., Kuznetsova, Y.Y., Yang, Sen, Butov, L.V., Ostatnický, T., Kavokin, A., et al., 2009. Spin transport of excitons. Nano Lett. 9, 4204.
Leonard, J.R., Remeika, M., Chu, M.K., Kuznetsova, Y.Y., High, A.A., Butov, L.V., et al., 2012. Transport of indirect excitons in a potential energy gradient. Appl. Phys. Lett. 100, 231106.
Levitov, L.S., Simons, B.D., Butov, L.V., 2005. Pattern formation as a signature of quantum degeneracy in a cold exciton system. Phys. Rev. Lett. 94, 176404.
Li, J.I.A., Taniguchi, T., Watanabe, K., Hone, J., Dean, C.R., 2017. Excitonic superfluid phase in double bilayer graphene. Nat. Phys. 13, 751.
Li, Z., Lu, X., Cordovilla Leon, D.F., Lyu, Z., Xie, H., Hou, J., et al., 2021. Interlayer exciton transport in $MoSe_2/WSe_2$ heterostructures. ACS Nano 15, 1539.
Liu, X., Watanabe, K., Taniguchi, T., Halperin, B.I., Kim, P., 2017. Quantum Hall drag of exciton condensate in graphene. Nat. Phys. 13, 746.
Liu, Y., Dini, K., Tan, Q., Liew, T., Novoselov, K.S., Gao, W., 2020. Electrically controllable router of interlayer excitons. Sci. Adv. 6, eaba1830.
Lobanov, S.V., Gippius, N.A., Butov, L.V., 2016. Theory of condensation of indirect excitons in a trap. Phys. Rev. B 94, 245401.
Lozovik, Y.E., Yudson, V.I., 1976. A new mechanism for superconductivity: pairing between spatially separated electrons and holes. Sov. Phys. JETP 44, 389.
Lozovik, Yu.E., Berman, O.L., 1996. Phase transitions in a system of two coupled quantum wells. JETP Lett. 64, 573.
Morhain, C., Bretagnon, T., Lefebvre, P., Tang, X., Valvin, P., Guillet, T., et al., 2005. Internal electric field in wurtzite $ZnO/Zn_{0.78}Mg_{0.22}O$ quantum wells. Phys. Rev. B 72, 241305R.
Morita, Y., Yoshioka, K., Kuwata-Gonokami, M., 2022. Observation of Bose-Einstein condensates of excitons in a bulk semiconductor. Nat. Commun. 13, 5388.
Naka, N., Nagasawa, N., 2005. Bosonic stimulation of cold excitons in a harmonic potential trap in Cu_2O. J. Lumin. 112, 11.

Nandi, D., Finck, A.D.K., Eisenstein, J.P., Pfeiffer, L.N., West, K.W., 2012. Exciton condensation and perfect Coulomb drag. Nature 488, 481.

Negoita, V., Snoke, D.W., Eberl, K., 1999. Stretching quantum wells: a method for trapping free carriers in GaAs heterostructures. Appl. Phys. Lett. 75, 2059.

Nguyen, H.S., Vishnevsky, D., Sturm, C., Tanese, D., Solnyshkov, D., Galopin, E., et al., 2013. Realization of a double-barrier resonant tunneling diode for cavity polaritons. Phys. Rev. Lett. 110, 236601.

Nowik-Boltyk, P., Dzyapko, O., Demidov, V.E., Berloff, N.G., Demokritov, S.O., 2012. Spatially non-uniform ground state and quantized vortices in a two-component Bose-Einstein condensate of magnons. Sci. Rep. 2, 482.

Pellegrini, V., Luin, S., Karmakar, B., Pinczuk, A., Dennis, B.S., Pfeiffer, L.N., et al., 2007. Seeing emergent phases in quantum Hall double layers. J. Appl. Phys. 101, 081718.

Peng, R., Ripin, A., Ye, Y., Zhu, J., Wu, C., Lee, S., et al., 2022. Long-range transport of 2D excitons with acoustic waves. Nat. Commun. 13, 1334.

Rapaport, R., Chen, G., Snoke, D., Simon, S.H., Pfeiffer, L., West, K., et al., 2004. Charge separation of a dense two-dimensional electron-hole gases: mechanism for exciton ring pattern formation. Phys. Rev. Lett. 92, 117405.

Remeika, M., Fogler, M.M., Butov, L.V., Hanson, M., Gossard, A.C., 2012. Two-dimensional electrostatic lattices for indirect excitons. Appl. Phys. Lett. 100, 061103.

Remeika, M., Graves, J.C., Hammack, A.T., Meyertholen, A.D., Fogler, M.M., Butov, L.V., et al., 2009. Localization-delocalization transition of indirect excitons in lateral electrostatic lattices. Phys. Rev. Lett. 102, 186803.

Remeika, M., Leonard, J.R., Dorow, C.J., Fogler, M.M., Butov, L.V., Hanson, M., et al., 2015. Measurement of exciton correlations using electrostatic lattices. Phys. Rev. B 92, 115311.

Rivera, P., Schaibley, J.R., Jones, A.M., Ross, J.S., Wu, S., Aivazian, G., et al., 2015. Observation of long-lived interlayer excitons in monolayer $MoSe_2$-WSe_2 heterostructures. Nat. Commun. 6, 6242.

Rivera, P., Seyler, K.L., Yu, H., Schaibley, J.R., Yan, J., Mandrus, D.G., et al., 2016. Valley-polarized exciton dynamics in a 2D semiconductor heterostructure. Science 351, 688.

Rivera, P., Yu, H., Seyler, K.L., Wilson, N.P., Yao, W., Xu, X., 2018. Interlayer valley excitons in heterobilayers of transition metal dichalcogenides. Nat. Nanotechnol. 13, 1004.

Rocke, C., Zimmermann, S., Wixforth, A., Kotthaus, J.P., Böhm, G., Weimann, G., 1997. Acoustically driven storage of light in a quantum well. Phys. Rev. Lett. 78, 4099.

Rontani, M., Sham, L.J., 2009. Josephson oscillations between exciton condensates in electrostatic traps. Phys. Rev. B 80, 075309.

Rosenberger, M.R., Chuang, H.J., Phillips, M., Oleshko, V.P., McCreary, K.M., Sivaram, S.V., et al., 2020. Twist angle-dependent atomic reconstruction and Moiré patterns in transition metal dichalcogenide heterostructures. ACS Nano 14, 4550.

Roumpos, G., Fraser, M.D., Löffler, A., Höfling, S., Forchel, A., Yamamoto, Y., 2011. Single vortex-antivortex pair in an exciton-polariton condensate. Nat. Phys. 7, 129.

Rudolph, J., Hey, R., Santos, P.V., 2007. Long-range exciton transport by dynamic strain fields in a GaAs quantum well. Phys. Rev. Lett. 99, 047602.

Scheuer, J., Orenstein, M., 1999. Optical vortices crystals: spontaneous generation in nonlinear semiconductor microcavities. Science 285, 230.

Schinner, G.J., Repp, J., Schubert, E., Rai, A.K., Reuter, D., Wieck, A.D., et al., 2013a. Confinement and interaction of single indirect excitons in a voltage-controlled trap formed inside double InGaAs quantum wells. Phys. Rev. Lett. 110, 127403.

Schinner, G.J., Repp, J., Schubert, E., Rai, A.K., Reuter, D., Wieck, A.D., et al., 2013b. Many-body correlations of electrostatically trapped dipolar excitons. Phys. Rev. B 87, 205302.

Schinner, G.J., Schubert, E., Stallhofer, M.P., Kotthaus, J.P., Schuh, D., Rai, A.K., et al., 2011. Electrostatically trapping indirect excitons in coupled InGaAs quantum wells. Phys. Rev. B 83, 165308.

Seyler, K.L., Rivera, P., Yu, H., Wilson, N.P., Ray, E.L., Mandrus, D.G., et al., 2019. Signatures of moiré-trapped valley excitons in $MoSe_2/WSe_2$ heterobilayers. Nature 567, 66.

Shanks, D.N., Mahdikhanysarvejahany, F., Stanfill, T.G., Koehler, M.R., Mandrus, D.G., Taniguchi, T., et al., 2022. Interlayer exciton diode and transistor. Nano Lett. 22, 6599.

Sharvin, Y.V., 1965. A possible method for studying Fermi surfaces. Sov. Phys. JETP 21, 655.

Shilo, Y., Cohen, K., Laikhtman, B., Rapaport, R., West, K., Pfeiffer, L., 2013. Particle correlations and evidence for dark state condensation in a cold dipolar exciton fluid. Nat. Commun. 4, 2335.

Shimazaki, Y., Schwartz, I., Watanabe, K., Taniguchi, T., Kroner, M., Imamoğlu, A., 2020. Strongly correlated electrons and hybrid excitons in a moiré heterostructure. Nature 580, 472.

Sivalertporn, K., Mouchliadis, L., Ivanov, A.L., Philp, R., Muljarov, E.A., 2012. Direct and indirect excitons in semiconductor coupled quantum wells in an applied electric field. Phys. Rev. B 85, 045207.

Spielman, I.B., Eisenstein, J.P., Pfeiffer, L.N., West, K.W., 2000. Resonantly enhanced tunneling in a double layer quantum Hall ferromagnet. Phys. Rev. Lett. 84, 5808.

Sturm, C., Tanese, D., Nguyen, H.S., Flayac, H., Galopin, E., Lemaître, A., et al., 2014. All-optical phase modulation in a cavity-polariton Mach-Zehnder interferometer. Nat. Commun. 5, 3278.

Sun, Z., Ciarrocchi, A., Tagarelli, F., Marin, J.F.G., Watanabe, K., Taniguchi, et al., 2022. Excitonic transport driven by repulsive dipolar interaction in a van der Waals heterostructure. Nat. Photonics 16, 79.

Tagarelli, F., Lopriore, E., Erkensten, D., Perea-Causín, R., Brem, S., Hagel, J., et al., 2023. Electrical control of hybrid exciton transport in a van der Waals heterostructure. Nat. Photonics 17, 615.

Tiemann, L., Lok, J.G.S., Dietsche, W., von Klitzing, K., Muraki, K., Schuh, D., et al., 2008. Exciton condensate at a total filling factor of one in Corbino two-dimensional electron bilayers. Phys. Rev. B 77, 033306.

Topinka, M.A., LeRoy, B.J., Shaw, S.E.J., Heller, E.J., Westervelt, R.M., Maranowski, K.D., et al., 2000. Imaging coherent electron flow from a quantum point contact. Science 289, 2323.

Topinka, M.A., LeRoy, B.J., Westervelt, R.M., Shaw, S.E.J., Fleischmann, R., Heller, E.J., et al., 2001. Coherent branched flow in a two-dimensional electron gas. Nature 410, 183.

Tran, K., Moody, G., Wu, F., Lu, X., Choi, J., Kim, K., et al., 2019. Evidence for moiré excitons in van der Waals heterostructures. Nature 567, 71.

Trauernicht, D.P., Mysyrowicz, A., Wolfe, J.P., 1983. Strain confinement and thermodynamics of free excitons in a direct-gap semiconductor. Phys. Rev. B 28, 3590.

Tsoi, V.S., 1974. Focusing of electrons in a metal by a transverse magnetic field. JETP Lett. 19, 70.

Tutuc, E., Shayegan, M., Huse, D.A., 2004. Counterflow measurements in strongly correlated GaAs hole bilayers: evidence for electron-hole pairing. Phys. Rev. Lett. 93, 036802.

Unuchek, D., Ciarrocchi, A., Avsar, A., Watanabe, K., Taniguchi, T., Kis, A., 2019a. Room-temperature electrical control of exciton flux in a van der Waals heterostructure. Nature 560, 340.

Unuchek, D., Ciarrocchi, A., Avsar, A., Sun, Z., Watanabe, K., Taniguchi, T., et al., 2019b. Valley-polarized exciton currents in a van der Waals heterostructure. Nat. Nanotechnol. 14, 1104.

van Houten, H., van Wees, B.J., Mooij, J.E., Beenakker, C.W.J., Williamson, J.G., Foxon, C.T., 1988. Coherent electron focussing in a two dimensional electron gas. Europhys. Lett. 5, 721.
van Wees, B.J., van Houten, H., Beenakker, C.W.J., Williamson, J.G., Kouwenhoven, L.P., van der Marel, D., et al., 1988. Quantized conductance of point contacts in a two-dimensional electron gas. Phys. Rev. Lett. 60, 848.
Violante, A., Cohen, K., Lazić, S., Hey, R., Rapaport, R., Santos, P.V., 2014. Dynamics of indirect exciton transport by moving acoustic fields. New J. Phys. 16, 033035.
Violante, A., Hey, R., Santos, P.V., 2015. Coherent transport and manipulation of spins in indirect-exciton nanostructures. Phys. Rev. B 91, 125302.
Vögele, X.P., Schuh, D., Wegscheider, W., Kotthaus, J.P., Holleitner, A.W., 2009. Density enhanced diffusion of dipolar excitons within a one-dimensional channel. Phys. Rev. Lett. 103, 126402.
Vörös, Z., Balili, R., Snoke, D.W., Pfeiffer, L., West, K., 2005. Long-distance diffusion of excitons in double quantum well structures. Phys. Rev. Lett. 94, 226401.
Wang, J., Shi, Q., Shih, E.M., Zhou, L., Wu, W., Bai, Y., et al., 2021. Diffusivity reveals three distinct phases of interlayer excitons in $MoSe_2/WSe_2$ heterobilayers. Phys. Rev. Lett. 126, 106804.
Weston, A., Zou, Y., Enaldiev, V., Summerfield, A., Clark, N., Zólyomi, V., et al., 2020. Atomic reconstruction in twisted bilayers of transition metal dichalcogenides. Nat. Nanotechnol. 15, 592.
Wharam, D.A., Thornton, T.J., Newbury, R., Pepper, M., Ahmed, H., Frost, J.E.F., et al., 1988. One-dimensional transport and the quantisation of the ballistic resistance. J. Phys. C: Solid State Phys. 21, L209.
Wilson, N.P., Yao, W., Shan, J., Xu, X., 2021. Excitons and emergent quantum phenomena in stacked 2D semiconductors. Nature 599, 383.
Winbow, A.G., Butov, L.V., Gossard, A.C., 2008. Photon storage with subnanosecond readout rise time in coupled quantum wells. J. Appl. Phys. 104, 063515.
Winbow, A.G., Hammack, A.T., Butov, L.V., Gossard, A.C., 2007. Photon storage with nanosecond switching in coupled quantum well nanostructures. Nano Lett. 7, 1349.
Winbow, A.G., Leonard, J.R., Remeika, M., Kuznetsova, Y.Y., High, A.A., Hammack, A.T., et al., 2011. Electrostatic conveyer for excitons. Phys. Rev. Lett. 106, 196806.
Woods, C.R., Britnell, L., Eckmann, A., Ma, R.S., Lu, J.C., Guo, H.M., et al., 2014. Commensurate-incommensurate transition in graphene on hexagonal boron nitride. Nat. Phys. 10, 451.
Wu, F., Lovorn, T., MacDonald, A.H., 2017. Topological exciton bands in Moiré heterojunctions. Phys. Rev. Lett. 118, 147401.
Wu, F., Lovorn, T., MacDonald, A.H., 2018. Theory of optical absorption by interlayer excitons in transition metal dichalcogenide heterobilayers. Phys. Rev. B 97, 035306.
Xu, C., Leonard, J.R., Dorow, C.J., Butov, L.V., Fogler, M.M., Nikonov, D.E., et al., 2019. Exciton gas transport through nanoconstrictions. Nano Lett. 19, 5373.
Yang, S., Hammack, A.T., Fogler, M.M., Butov, L.V., Gossard, A.C., 2006. Coherence length of cold exciton gases in coupled quantum wells. Phys. Rev. Lett. 97, 187402.
Yang, S., Butov, L.V., Levitov, L.S., Simons, B.D., Gossard, A.C., 2010. Exciton front propagation in photoexcited GaAs quantum wells. Phys. Rev. B 81, 115320.
Yang, S., Butov, L.V., Simons, B.D., Campman, K.L., Gossard, A.C., 2015. Fluctuation and commensurability effect of exciton density wave. Phys. Rev. B 91, 245302.
Ye, Z., Cao, T., O'Brien, K., Zhu, H., Yin, X., Wang, Y., et al., 2014. Probing excitonic dark states in single-layer tungsten disulphide. Nature 513, 214.
Yoo, H., Engelke, R., Carr, S., Fang, S., Zhang, K., Cazeaux, P., et al., 2019. Atomic and electronic reconstruction at the van der Waals interface in twisted bilayer graphene. Nat. Mater. 18, 448.

Yoshioka, D., MacDonald, A.H., 1990. Double quantum well electron-hole systems in strong magnetic fields. J. Phys. Soc. Jpn. 59, 4211.
Yoshioka, K., Chae, E., Kuwata-Gonokami, M., 2011. Transition to a Bose-Einstein condensate and relaxation explosion of excitons at sub-Kelvin temperatures. Nat. Commun. 2, 328.
Yu, H., Liu, G.B., Tang, J., Xu, X., Yao, W., 2017. Moiré excitons: from programmable quantum emitter arrays to spin-orbit–coupled artificial lattices. Sci. Adv. 3, e1701696.
Yu, H., Liu, G.B., Yao, W., 2018. Brightened spin-triplet interlayer excitons and optical selection rules in van der Waals heterobilayers. 2D Mater. 5, 035021.
Yuan, L., Zheng, B., Kunstmann, J., Brumme, T., Kuc, A.B., Ma, C., et al., 2020. Twist-angle-dependent interlayer exciton diffusion in WS_2-WSe_2 heterobilayers. Nat. Mater. 19, 617.
Zhang, C., Chuu, C.P., Ren, X., Li, M.Y., Li, L.J., Jin, C., et al., 2017. Interlayer couplings, Moiré patterns, and 2D electronic superlattices in MoS_2/WSe_2 hetero-bilayers. Sci. Adv. 3, e1601459.
Zhao, S., Li, Z., Huang, X., Rupp, A., Göser, J., Vovk, I.A., et al., 2023. Excitons in mesoscopically reconstructed moiré heterostructures. Nat. Nanotechnol. 18, 572.
Zhu, X., Littlewood, P.B., Hybertsen, M.S., Rice, T.M., 1995. Exciton condensate in semiconductor quantum well structures. Phys. Rev. Lett. 74, 1633.
Zimmermann, S., Schedelbeck, G., Govorov, A.O., Wixforth, A., Kotthaus, J.P., Bichler, M., et al., 1998. Spatially resolved exciton trapping in a voltage-controlled lateral superlattice. App. Phys. Lett. 73, 154.
Zimmermann, S., Govorov, A.O., Hansen, W., Kotthaus, J.P., Bichler, M., Wegscheider, W., 1997. Lateral superlattices as voltage-controlled traps for excitons. Phys. Rev. B 56, 13414.
Zrenner, A., Leeb, P., Schäfler, J., Böhm, G., Weimann, G., Worlock, J.M., et al., 1992. Indirect excitons in coupled quantum well structures. Surf. Sci. 263, 496.

CHAPTER SIX

III-Nitride optoexcitonics: Physics, epitaxy, and emerging device applications

Yuanpeng Wu[a,1], Woncheol Lee[a,1], Emmanouil Kioupakis[b,*], and Zetian Mi[a,*]
[a]Department of Electrical Engineering and Computer Science, University of Michigan, Ann Arbor, MI, United States
[b]Department of Materials Science and Engineering, University of Michigan, Ann Arbor, MI, United States
*Corresponding authors. e-mail address: kioup@umich.edu; ztmi@umich.edu

Contents

Abstract

Within this chapter, we thoroughly explore the cutting-edge advancements in III-nitride optoexcitonics and the exciting possibilities in applications they bring forth. We unveil the intrinsic physics that underlies the distinctive properties of monolayer III-nitrides. Subsequently, we delve into the intricate processes of epitaxy and characterization, both for embedded and free-standing monolayer III-nitride heterostructures. Expanding our exploration, we shed light on the crucial role excitonic effects play in revolutionizing optoelectronic devices, encompassing an array of innovations like UV LEDs, visible micro-LEDs, and next-generation lasers.

[1] Y. W. and W. L. contributed equally to this work.

Semiconductors and Semimetals, Volume 112
ISSN 0080-8784, https://doi.org/10.1016/bs.semsem.2023.09.002

1. Introduction

Over the past decades, III-nitride semiconductors, consisting of InN, GaN, AlN, BN, and their ternary and quaternary alloys, have been extensively studied for a broad range of electronic and optoelectronic device applications, including LEDs (Wong et al., 2020; Liu et al., 2020a; Zhang et al., 2019a; Zhao et al., 2016a), laser diodes (Ra et al., 2020; Zhang et al., 2019b), and photodetectors operating from the UV and visible to the near-infrared (Aiello et al., 2020; Yang et al., 2014), as well as for high-frequency and high-power transistors (Wang et al., 2023a). While III-nitride semiconductors in bulk format have exciton binding energies ranging from 5 meV (InN) to 60 meV (AlN) (Wu et al., 2021, 2022a), recent theoretical studies have shown that exciton binding energy exceeding 1 eV can be achieved in atomically thin III-nitrides (Wu et al., 2023). Compared with conventional monolayer transition metal dichalcogenides (TMDCs), which have been extensively explored in excitonic devices (Chiu et al., 2015; Hu et al., 2018; Wang et al., 2018a; Alexeev et al., 2019; Jauregui et al., 2019; Calman et al., 2020), III-nitride materials offer advantages including scalable synthesis (Liu et al., 2021; Wen et al., 2020; Biswas et al., 2020; Sun et al., 2019), lack of moire patterns (that traps excitons) (Baek et al., 2020), the intrinsic polarization fields that separate electrons from holes (indirect excitons) without an external applied field (Calman et al., 2018), as well as the heterointegration with functional substrates such as Si, metal and insulation substrates (Wang et al., 2019a, 2019a, 2018b, 2019b). Compared with conventional 2D TMDCs, the direct bandgap nature for 2D III nitrides, except hBN, can be well maintained in any thickness (from a single atomic layer to a few monolayers or nanometers), while that for most of the conventional 2D materials can only be maintained in their monolayer structures. This makes ultrathin III-nitrides particularly suitable for optoelectronic devices, especially light-emitting devices with tunable emission energies, etc.

III-nitride nanostructures also exhibit new excitonic and quantum properties for applications in emerging quantum optoelectronics (Simon et al., 2010; Zheng et al., 2021; Stachurski et al., 2022). Recent studies suggest that the epitaxy of monolayer GaN in dislocation-free AlN nanocrystals can overcome the significant surface/interface roughness during conventional Al(Ga)N epitaxy and lead to unusually large exciton binding energy. Experimentally, it has been measured that monolayer GaN can exhibit exciton binding energy exceeding 200 meV, which is

~10 times larger than that in bulk GaN and in good agreement with theory (Wu et al., 2020a). The emission is TE polarized and can be controllably tuned in the UV-B and UV-C bands by varying the thicknesses and compositions of Al(Ga)N barrier layers. Moreover, hBN, albeit an indirect bandgap, can luminesce as strongly as a direct-gap material because of the unusually strong phonon-assisted optical matrix element (~600 times stronger than Si) and large exciton binding energy (~0.5 eV) (Laleyan et al., 2023). With the use of ultrahigh temperature molecular beam epitaxy, it has been shown that defects can be significantly reduced in hBN to achieve high-efficiency deep-UV emission.

In this article, we provide an overview of the recent developments of III-nitride optoexcitonics and their emerging applications in UV and visible light-emitting devices. In Section 2, the physics of monolayer III-nitrides is presented. The epitaxy and characterization of monolayer III-nitride heterostructures and nanostructures are described in Section 3. In Section 4, we discuss the emerging applications of monolayer III-nitrides with a particular focus on UV LEDs, visible micro-LEDs and lasers. Finally, conclusions and prospects are presented at the end of Section 4.

2. Physics of monolayer III-nitrides

Monolayer III-nitrides are heterostructures that consist of just one atomic layer of III-nitride materials. This can be either an embedded monolayer-thick quantum well (e.g., a monolayer GaN quantum well embedded in an AlN barrier) or a single atomic III-nitride layer deposited on top of a substrate (e.g., monolayer hBN deposited on a graphite substrate). Due to their extreme thickness, monolayer III-nitrides exhibit unique properties compared to their bulk counterparts.

First, the extreme degree of geometrical quantum confinement leads to a significant increase of the band-gap energy. In a quantum-confined system, in which the electron and hole wave functions are localized at the atomic scale, the energy levels of electrons and holes along the confined direction become discrete. As the dimensions of the confined system are reduced, the energy separation between the discrete electron states increases, resulting in a larger band gap. Monolayer III-nitrides correspond to the case where this quantum confinement effect is most extreme. For example, as depicted in Fig. 1, monolayer GaN quantum well embedded within AlN or $Al_{0.67}Ga_{0.33}N$ quantum barrier exhibits an electronic gap in

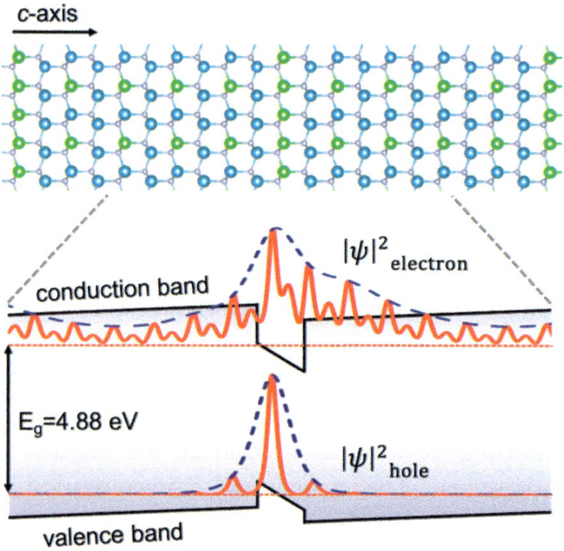

Fig. 1 Schematics and band diagram of monolayer GaN quantum well embedded within $Al_{0.67}Ga_{0.33}N$ quantum barrier. Wavefunctions of an electron and a hole ($|\psi|^2$, red solid lines) are strongly confined within the atomically thin GaN layer. Extreme quantum confinement effect increases the bandgap of GaN into the deep UV range. *Figure reproduced with permission from Aiello, A., Wu, Y., Mi, Z., Bhattacharya, P., 2020. Deep ultraviolet monolayer GaN/AlN disk-in-nanowire array photodiode on silicon. Appl. Phys. Lett. 116 (6), 061104.*

the deep UV range (5.44 eV and 4.88 eV, respectively), which is much wider than the band gap of bulk GaN (3.4 eV) (Bayerl et al., 2016; Bayerl and Kioupakis, 2019). In addition, freestanding monolayer hBN is also predicted to show an electronic gap of 7.95 eV, which is much higher than the electronic gap of bulk hBN (6.22 eV) (Laleyan et al., 2023).

In addition, in monolayer III-nitrides, the geometrical confinement of the electrons and holes can improve radiative efficiency. In the case of bulk III-nitride quantum wells, the spontaneous polarization field along the polar *c*-axis separates the electrons from holes to opposite ends of the quantum well (Leroux et al., 1998; Grandjean et al., 1999). The reduced overlap between the electron and hole wave functions leads to the suppression of the radiative efficiency, which has been shown to be one of the major constraints for III-nitride-based optoelectronic devices (Speck and Chichibu, 2011). However, the atomic dimensions of monolayer III-nitride quantum wells leave no room for the spatial separation of electrons from holes, ensuring a high degree of overlap between electron and hole wavefunctions (Bayerl et al., 2016; Bayerl and Kioupakis, 2019; Aiello et al., 2019). This characteristic improves the quantum efficiency of III-nitride-based optoelectronic devices.

Moreover, the enhanced Coulomb interaction in the atomically thin III-nitride structures results in a substantial increase of the exciton binding energy. Monolayer III-nitrides are a quasi-2D material, where the motion of electrons and holes is only allowed in two dimensions and is confined along the third. Compared to the case of bulk 3D systems, where the presence of additional dimension decreases the strength of the Coulomb interaction between electrons and holes, monolayer III-nitrides exhibit reduced dielectric screening and enhanced Coulomb interaction between electrons and holes. This results in a significant increase in the exciton binding energy (Bayerl et al., 2016; Bayerl and Kioupakis, 2019; Aiello et al., 2019). For ideal 2D systems, it has been known that the exciton binding energy becomes a factor of four larger than that in the 3D bulk system, assuming the same dielectric constant between the 2D material and the surrounding dielectric (Bastard et al., 1982; Yang et al., 1991). For 2D materials surrounded by vacuum or materials with lower dielectric constants, the enhancement factor can be even larger. Therefore, monolayer III-nitrides can exhibit exciton binding energies much larger than those of their 3D counterparts, although the exact enhancement factor varies depending on the degree of confinement and the surrounding dielectric material. This increase of the exciton binding energy will lead to the formation of room-temperature stable excitons, which is beneficial for the realization of optoelectronic devices with high radiative efficiency even at room temperature.

Furthermore, we can manipulate the physical properties of monolayer III-nitrides by engineering the surrounding environment. First, the band alignment at the interface between monolayer III-nitrides and other materials provides engineering opportunities. For monolayer GaN quantum wells, $Al_xGa_{(1-x)}N$ alloys can be utilized for the quantum barrier. By varying the Al alloy fraction of the $Al_xGa_{(1-x)}N$ barrier, we can control the level of energy barrier at the interface and the degree of quantum confinement effect (Bayerl et al., 2016; Aiello et al., 2019). As a result, embedded monolayer GaN wells can emit photons of various wavelengths, depending on the materials employed for the quantum barriers. Dielectric screening is also an important factor that affects the electronic and optical properties of semiconductor materials. In the case of monolayer III-nitrides, the extrinsic dielectric screening from the surrounding materials greatly affects the dielectric screening of electrons and holes, in addition to the intrinsic dielectric screening present within the monolayer itself. For example, the electronic gap of freestanding monolayer hBN is predicted to be 7.95 eV (Wang et al., 2022a). However, the electronic gap abruptly drops to 6.95 eV for monolayer hBN grown on the graphite substrate, demonstrating that the substrate screening induces a giant renormalization of the band gap, as shown in Fig. 2 (Wang et al., 2022a). In addition, it is predicted that monolayer hBN grown on the graphite substrate will result in a significant decrease in the exciton binding energy, decreasing from 2.3 eV to 0.7 eV.

3. Epitaxy and characterization of monolayer III-nitride heterostructures and nanostructures

The synthesis of monolayer III-nitride heterostructures has been performed by various synthesis methods including molecular beam epitaxy (Islam et al., 2017a; Toropov et al., 2020; Mazumder et al., 2012), metal–organic chemical vapor deposition (MOCVD) (Liu et al., 2011; Oto et al., 2010), and chemical synthesis (Ben et al., 2021; Chen et al., 2018). Ideally monolayer III-nitride can be synthesized in either embedded format or free-standing format, have a sharp interface with barrier layers, and can be

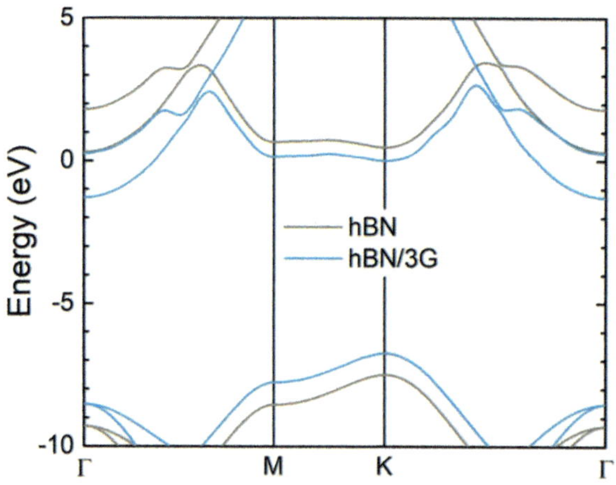

Fig. 2 Quasiparticle band structures of freestanding monolayer h-BN (gray curves) and monolayer h-BN on top of three graphene layers (blue curves). *Figure reproduced with permission from Wang, P., et al., 2022a. Scalable synthesis of monolayer hexagonal boron nitride on graphene with giant bandgap renormalization. Adv. Mater. 34 (21), e2201387.*

synthesized at large scale on various functional substrates. However, it remains challenging to achieve these goals. Severe interdiffusion has been widely observed at the heterostructure interface due to the difference in ionicity (Bogusławski et al., 2000; Gogneau et al., 2004; Gačević et al., 2013; Gao et al., 2019). During the epitaxy of free-standing monolayer III-nitrides, the nonuniformity in height and atomic edges of the epitaxial substrates prevents the achievement of wafer scale migration (Wang et al., 2022a). Various characterization techniques have been adopted to measure the structural properties of monolayer III-nitrides including X-ray diffraction (XRD), atom topological (ATP), and transmission electron microscopy (TEM).

3.1 Monolayer GaN

Utilizing growth interruption and migration-enhanced epitaxy, we successfully demonstrated the controlled creation of monolayer GaN within an AlN matrix with distinct excitonic emission properties (Aiello et al., 2020; Wu et al., 2020a; Aiello et al., 2019). We conducted the growth of GaN/AlN quantum heterostructures, where the GaN quantum disks have either monolayer or bilayer thickness, and the AlN barrier thickness measures approximately 10 nm. Following each growth cycle of the AlN barrier, we closed the Al shutter for a duration of 10 min, while keeping only the nitrogen shutter open. This step facilitates the consumption of excess Al adatoms at the growth front, resulting in an Al-free surface that allows for the subsequent formation of monolayer GaN within the AlN matrix (Wu et al., 2020a; Islam et al., 2017b). The thickness of the GaN quantum disks was precisely controlled by adjusting the duration of the Ga shutter opening. To verify the presence of embedded monolayer GaN in AlN matrix, we performed HAADF-STEM imaging as depicted in Fig. 3a. Fig. 3b presents the room-temperature photoluminescence (PL) spectra, which clearly demonstrates distinct emission peaks at 238 nm and 270 nm for GaN/AlN heterostructures with monolayer and bilayer GaN, respectively. These findings align well with theoretical calculations (Bayerl et al., 2016). Fig. 3c displays excitation-power-dependent PL measurements, where the integrated PL intensity (I_{PL}) originating from monolayer GaN exhibits a linear relationship with the excitation intensity (I_0), indicating strong suppression of Auger-Meitner recombination, as expected for systems dominated by excitons. Additionally, we compare the experimental data with previous theoretical predictions based on first-principles calculations (Bayerl et al., 2016). Fig. 3d shows the variation in both the

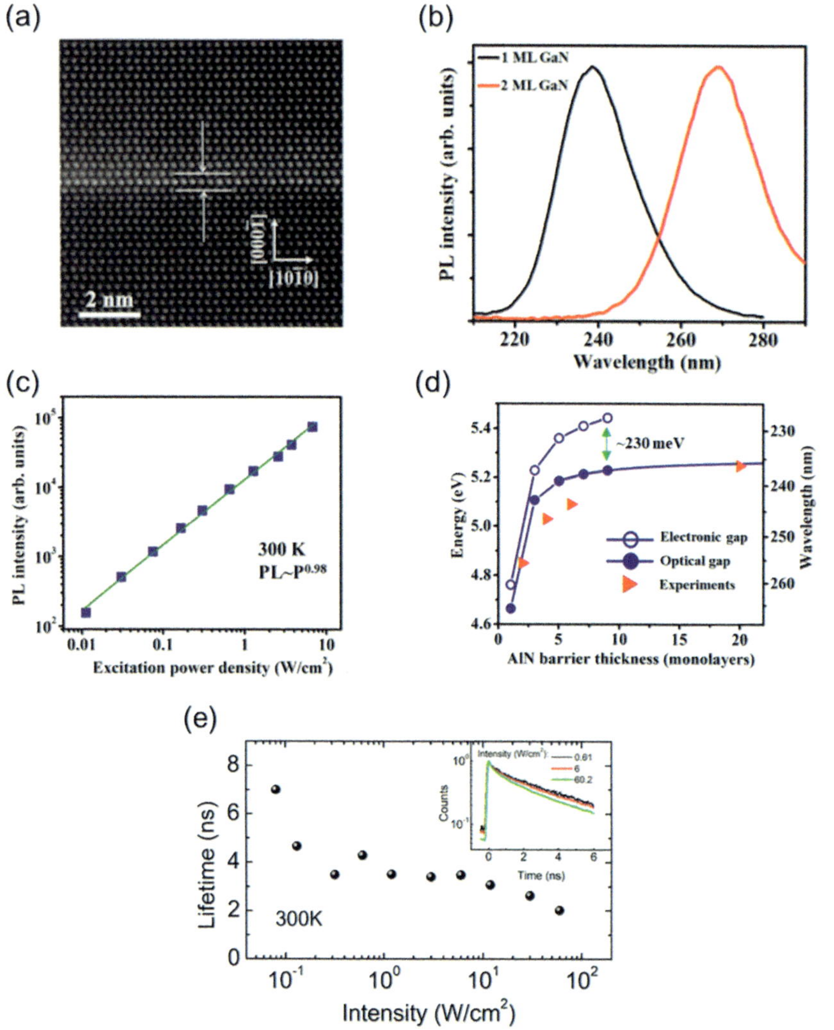

Fig. 3 (a) HAADF-STEM image of monolayer GaN embedded in AlN. (b) Room-temperature PL spectra from GaN/AlN heterostructure with monolayer GaN (black curve) and 2 monolayers of GaN (red curve). (c) Integrated PL intensity versus excitation power. (d) Electronic and optical gaps of monolayer GaN as a function of the AlN barrier thickness. Experimentally measured PL peaks at 300 K are indicated with red triangles. Lines connected to and extrapolated from the theory data points are guides to the eye. (e) Lifetime of PL emission from monolayer GaN/AlN heterostructure. *Figures reproduced with permission from Aiello, A., Wu, Y., Mi, Z., Bhattacharya, P., 2020. Deep ultraviolet monolayer GaN/AlN disk-in-nanowire array photodiode on silicon. Appl. Phys. Lett. 116 (6), 061104 and Wu, Y., et al., 2020a. Monolayer GaN excitonic deep ultraviolet light emitting diodes. Appl. Phys. Lett. 116 (1), 013101.*

electronic and optical gaps as the AlN barrier thickness increases. To explore this, we experimentally grew a series of GaN/AlN heterostructures with specific GaN and AlN barrier thicknesses. The measured PL emission peaks align favorably with the calculated optical gaps, providing evidence that the observed deep UV emission from monolayer GaN is excitonic in nature. Furthermore, by comparing it with the calculated electronic bandgap, we determined an exciton binding energy of approximately 200 meV, which is nearly ten times higher than that of bulk GaN (Wu et al., 2021, 2023).

Fig. 3e illustrates the relationship between the measured recombination lifetime (δ) and excitation intensity at a temperature of 300 K. The transient data is analyzed using a stretched exponential function, $I_{PL} \propto e^{-(t/\tau)^\wedge\gamma}$, where t represents time and γ ranges between 0.47 and 0.64 (Aiello et al., 2019). The inset in the figure presents typical transients obtained from time-resolved photoluminescence (TRPL) measurements. Notably, there is an initial decrease in δ as excitation intensity increases, which is likely attributed to the quantum-confined Stark effect (QCSE) (Park and Chuang, 2000; Muller-Caspary et al., 2019). While a significant polarization field is expected between GaN and AlN, the QCSE may be limited due to the strong confinement of the system. As excitation exceeds 0.3 W/cm^2, the recombination lifetime remains relatively constant. The unchanging δ value, coupled with the linear relationship between integrated PL and excitation, strongly indicates a transition dominated by tightly confined excitons.

3.2 Monolayer hBN

Monolayer hexagonal boron nitride (hBN) has been widely regarded as a fundamental component for 2D heterostructures and devices (Ben et al., 2021; Dean et al., 2010; Maity et al., 2021; Mengle and Kioupakis, 2019). However, achieving controlled and scalable synthesis of hBN and its 2D heterostructures has posed a significant challenge. Two-dimensional (2D) hBN, known for its potential in far ultraviolet (UV)-C light emission, offers greater effectiveness and safety in containing microbial diseases compared to traditional UV light (Laleyan et al., 2023). Recent efforts have focused on growing hBN epitaxially on metals (Schué et al., 2019; Chen et al., 2020), sapphire (Liu et al., 2020b), and graphene substrates (Cho et al., 2016) using various methods such as sputtering, CVD (Lu et al., 2017), MOCVD (Lee et al., 2019), and MBE (Laleyan et al., 2018). Previous studies on hBN have primarily relied on mechanical exfoliation or chemical vapor deposition, which suffers from poor interface properties, high impurity incorporation, and dense defects. While luminescence studies of hBN have mostly centered on defect-related emission in the visible or near-UV spectrum, there have been reports on near band-edge transitions, including cathodoluminescence (Watanabe et al., 2004), photoluminescence (Cassabois et al., 2016), and electroluminescence (Song et al., 2021). However, a fundamental challenge remains unresolved regarding the transformation of an indirect bandgap semiconductor into an efficient light emitter.

To address these challenges, we conducted first-principles density functional theory (DFT) calculations to examine the heterointerface between monolayer hBN and various substrates, such as sapphire, AlN, SiO_x, Ni, and Cu as shown in Fig. 4a (Laleyan et al., 2023). Our results indicate that Ni demonstrates optimal interfacial interaction with hBN,

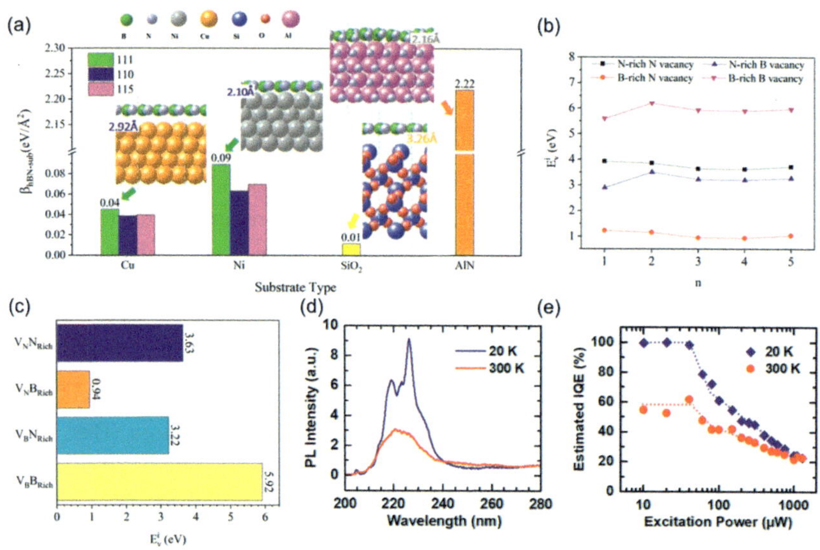

Fig. 4 (a) Interfacial energies for hBN on different substrates. (b) Boron and nitrogen vacancy formation energies under boron-rich and nitrogen-rich conditions. (c) The formation energies of B and N vacancies in hBN supported on the Ni (Wu et al., 2018b) substrate under N-rich and B-rich conditions, computed using different supercell size n. In all vacancy calculations, only one vacancy was created in the supercell. (d) Photoluminescence (PL) spectra of hBN/Ni (Wu et al., 2018b) at 20 K and 300 K. (e) Estimated internal quantum efficiency (IQE) of hBN/Ni at various excitation powers, plotted in logarithmic scale. The dashed lines serve as a guide to the reader. *Figures reproduced with permission from Laleyan, D.A., et al., 2023. Epitaxial hexagonal boron nitride with high quantum efficiency. APL Mater. 11 (5), 051103.*

characterized by minimal lattice mismatch (~0.5%) and favorable adhesion energy ($\beta_{hBN\text{-}sub}$). Different facets of Ni consistently exhibit good interfacial interaction without inducing significant distortion in the hBN structure. In contrast, SiO_2 shows weak interaction, while sapphire and AlN strongly interact with hBN, leading to notable distortion in the hBN lattice. Consequently, hBN is expected to have difficulty nucleating and growing on SiO_2, and hBN grown on sapphire is likely to be highly defective. Additionally, we investigated the impact of chemical environments on hBN growth, particularly vacancy formation, which directly affects hBN quality. As shown in Fig. 4b, under B-rich conditions, N vacancy formation energy is low, resulting in extensive N vacancies. Conversely, under N-rich conditions, both B and N vacancy formation energies are high. Thus, high-quality hBN is preferably grown under N-rich conditions. We also examined the influence of supercell size on single vacancy formation energy, finding that B vacancy has the lowest energy at the smallest supercell size ($n = 1$), while N vacancy energy decreases with increasing supercell size until $n = 3$ and stabilizes for $n > 3$.

In our experimental investigation, we focused on the growth of hBN on different substrates using N-rich epitaxy conditions. Notably, when hBN was directly grown on sapphire and SiO_2, we observed predominant defect-related emission spanning the UV bands. However, the epitaxy of hBN on Ni substrate displayed distinct characteristics. At room temperature, the emission was primarily dominated by deep UV excitonic emission, with suppressed defect-related emission as predicted by theoretical models. To gain further insight into the optical properties of epitaxial hBN on Ni, we conducted additional characterization using variable excitation power and temperature-dependent PL spectroscopy. The representative PL spectra are depicted in Fig. 4d. At cryogenic temperatures, we resolved distinct peaks at approximately 220 nm, consistent with previous reports highlighting high-quality hBN in the introduction.

To assess the quantum efficiency of the deep UV emission of hBN, we utilized temperature and power-dependent photoluminescence spectroscopy. Drawing from previous studies on GaN-based LED structures, we assumed that the highest IQE achievable at cryogenic temperatures (e.g., 20 K) is unity, owing to the suppression of nonradiative recombination (Park et al., 2015). By comparing the luminescence efficiency, defined as the integrated luminescence intensity versus excitation power, with the highest value obtained at cryogenic temperature, we derived the IQE at various temperatures and excitation powers. It is worth noting that although the optical transition in hBN is phonon-dependent due to its indirect nature, which might imply inefficiency at low temperatures, this transition actually benefits from phonon emission that remains unaffected by temperature. Remarkably, as depicted in Fig. 4e, the room-temperature IQE of epitaxial hBN reached 60%, surpassing the efficiency of other indirect bandgap materials by several orders of magnitude. It also stood on par with or outperformed commonly used direct bandgap semiconductors in commercial LEDs and laser diodes. Additionally, we observed that the efficiency remained relatively stable under low to moderate excitation conditions, consistent with emission dominated by excitons. However, with increasing excitation power, the IQE exhibited a droop due to augmented Auger-Meitner recombination or other higher-order carrier loss phenomena under high excitation conditions. This droop effect is similar to what is commonly observed in GaN-based LEDs (Kioupakis et al., 2011, 2012; Meyaard et al., 2013).

There is a growing recognition of the necessity to achieve control over the nucleation and crystallinity of 2D nitrides for functional device applications. The challenge lies in the lack of controllability arising from a complex nucleation environment. This environment consists of epitaxial substrate surfaces with varying atomic steps, interfacial energies along crystalline edges, and polycrystalline domains with diverse orientations. These factors hinder precise synthesis control of 2D nitrides. In our recent work, we have successfully developed a scalable and controlled synthesis method for 2D hBN on graphene by manipulating the initial nucleation process (Wang et al., 2022a). As depicted schematically in Fig. 5a and b, the graphene substrate surface comprises two distinct interfaces: zigzag (ZZ) atomic edges and armchair (AC) edges. In uncontrolled epitaxy, hBN nucleates on both interfaces, resulting in random interfaces and the absence of large-scale unidirectional hBN with single crystallinity. However, through our controlled synthesis approach, nucleation at the ZZ atomic edges of graphene is significantly suppressed. This suppression is achieved by performing epitaxy at a high temperature (1600 °C), which allows only nucleation at energetically stable interfaces. Consequently, hBN exclusively grows at AC atomic edges, enabling the achievement of uniform, flawless, and wafer-scale epitaxy of 2D hBN. This controlled synthesis of 2D hBN facilitates a significant band structure renormalization, leading to a substantial exciton binding energy.

In Fig. 5c, the collected spectra from the hBN/HOPG sample, measured at 12 K, are displayed. The peak at 5.86 eV corresponds to the

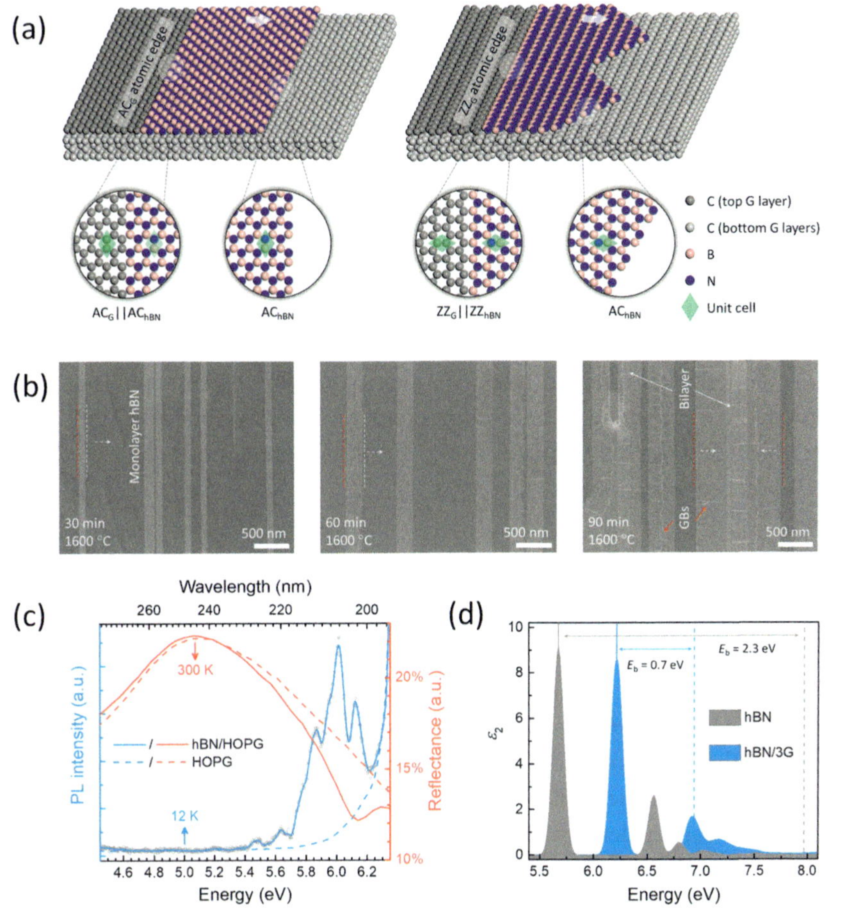

Fig. 5 (a) Schematics of monolayer hBN grown along the armchair (ACG) (left) and zigzag (ZZG) (right) graphene atomic edges, forming straight and jagged nanoribbons, respectively. All growth fronts of monolayer hBN are terminated with AC hBN edges. The magnifications show the atomic configurations of the ACG||AC hBN and ZZG||ZZ hBN in-plane interfaces, and the AC hBN growth fronts. The white arrows point to the growth direction. (b) Evolution of straight monolayer hBN nanoribbons is followed after 30, 60, and 90 min growth time at 1600 °C. (c) Measured, time-integrated PL spectra (blue curves, 12 K) and reflectance spectra (red curves, 300 K) of monolayer hBN/HOPG heterostructure (solid curves) and HOPG substrate (dashed curves). The gray circles are the PL data for monolayer hBN/HOPG heterostructures, while the blue solid curve is the corresponding smoothed curve. (d) Calculated absorption spectrum of a freestanding monolayer hBN (gray area) and monolayer hBN on three graphene layers (blue area). The vertical dashed lines indicate the quasiparticle bandgap, and the vertical solid lines show the 1s-exciton state position. *Figures reproduced with permission from Wang, P., et al., 2022a. Scalable synthesis of monolayer hexagonal boron nitride on graphene with giant bandgap renormalization. Adv. Mater. 34 (21), e2201387.*

multilayer hBN, while the 6.12 eV peak aligns well with the dip observed in the reflection spectrum at 300 K, indicating monolayer emission from hBN. According to DFT calculations, monolayer hBN exhibits a predicted bandgap of 8 eV. It has been discovered that the presence of the adjacent graphene layer causes a significant band structure renormalization of up to 1 eV, attributed to the metallic properties of graphene. This metallic nature results in strong screening of the Coulomb interaction among charge carriers within hBN. For freestanding monolayer hBN, the calculated exciton binding energy can reach 2.3 eV, as illustrated in Fig. 5d. However, in the presence of three monolayers of graphene, the exciton binding energy of hBN is reduced to 0.7 eV due to the intense metallic screening provided by graphene. The confinement of the wavefunction within the ultrathin hBN further validates the strong-light interactions associated with the observed emission.

4. Emerging device applications

While the formation of exciton can increase the electron–hole wavefunction overlap and the radiative recombination efficiency, it naturally suggests adopting heterostructures in III-nitrides optoelectronics to enhance the exciton binding energy for improved device performance. In optoelectronic devices, it is expected that the excitonic effect dominates under low carrier density regime before the Mott transition occurs. In this section, we discuss the excitonic effect in emerging light-emitting device applications including deep UV LEDs, visible micro-LEDs and UV lasers.

4.1 Deep UV LEDs

Fig. 6a presents the schematic of monolayer GaN deep UV LEDs, created through standard optical lithography and contact metallization techniques. To serve as the n-metal contact, Ti (80 nm)/Au (20 nm) was deposited on the backside of the Si substrate. For the p-metal contact, Ni (10 nm)/Au (10 nm) was deposited using a tilted angle deposition technique, which also defined the device area. Additionally, Ti (20 nm)/Al (300 nm)/Au (20 nm) metal grid contact was placed on the device top surface to enable current spreading and injection. The measured current–voltage characteristics at room temperature exhibited a turn-on voltage of approximately 5 V. Under a reverse bias of −10 V, the device leakage current remained negligibly small. This excellent electrical performance can be attributed to the

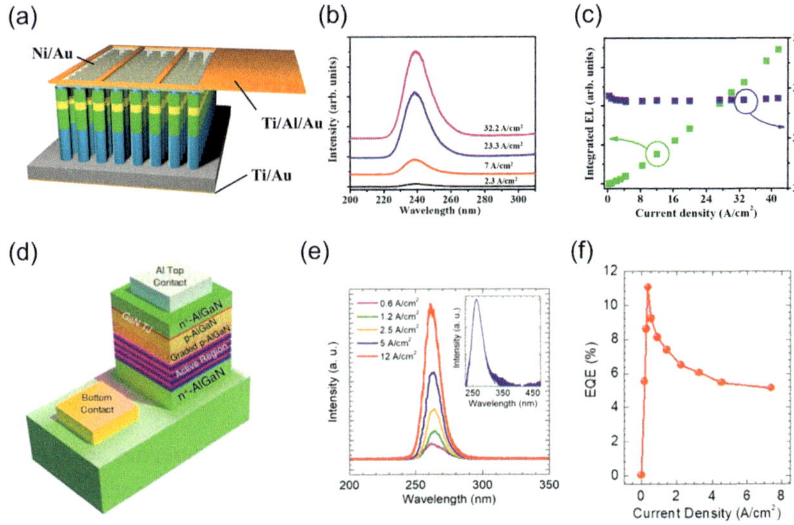

Fig. 6 (a) Schematic of deep UV nanowire LED structure with monolayer GaN/AlN incorporated as the active region. (b) Current-dependent EL spectra of the as-fabricated LED structure. (c) Variations of the integrated EL intensity (green) and peak position (blue) vs injection current. The measured device size is 300 μm × 300 μm. (d) Schematic illustration of the tunnel junction LED structures. (e) EL spectra measured at different injection currents for a representative tunnel junction LED. Inset shows an EL spectrum measured at 25 A/cm^2 current density with the intensity in log scale. (f) Variations of EQE with injected current density for an LED from the tunnel junction LED (Pandey et al., 2020a). *Reproduced with permission from Wu, Y., et al., 2020a. Monolayer GaN excitonic deep ultraviolet light emitting diodes. Appl. Phys. Lett. 116 (1), 013101 and Pandey, A., Shin, W.J., Gim, J., Hovden, R., Mi, Z., 2020a. High-efficiency AlGaN/GaN/AlGaN tunnel junction ultraviolet light-emitting diodes. Photonics Res. 8 (3), 331–337.*

efficient p-type conduction of Mg-doped AlN nanocrystals. Studies have indicated that the formation energy of Al-substitutional Mg acceptors is significantly reduced in AlN nanostructures compared to epilayers, thanks to effective surface strain relaxation. Consequently, there is enhanced incorporation of Mg dopants, along with reduced N-vacancy formation under N-rich epitaxy conditions, enabling the formation of a Mg-impurity band in AlN. This, in turn, facilitates hole-hopping conduction (Wu et al., 2020b).

Fig. 6b displays the electroluminescence (EL) spectra of the LEDs incorporating single and double monolayer GaN active regions, exhibiting peak emission at 238 nm, consistent with the PL measurements. The peak positions remain nearly constant as the current varies, as shown in Fig. 6c, indicating a negligible quantum-confined Stark effect. The atomically thin

GaN active region possesses extreme quantum confinement for electrons and holes, preventing the separation of their wavefunctions despite the significant polarization field between GaN and AlN (Wu et al., 2022b). Moreover, the integrated EL intensity demonstrates a near-linear increase with current. In-depth examinations of the relative external quantum efficiency versus injection current reveal minimal efficiency droop. This stands in contrast to the severe efficiency droop observed in conventional AlGaN LEDs. In monolayer GaN LEDs, the strong exciton binding energy enhances radiative decay while suppressing scattering, leading to a reduced radiative carrier lifetime and effective carrier density. Consequently, Auger-Meitner recombination and electron overflow-induced droop in deep UV devices can be effectively suppressed in strongly excitonic systems.

In conventional planar deep UV LEDs, the device properties are largely limited by issues including efficient p-type doping, high contact resistivity, etc. wherein the effects of excitonic interactions on the carrier dynamics and device performance are hindered. To achieve efficient hole injection in deep UV AlGaN LEDs, several challenges need to be addressed, including low hole concentration (Simon et al., 2010; Taniyasu et al., 2006; Carnevale et al., 2012), low hole mobility (Tran et al., 2017; Li et al., 2018; Liang and Towe, 2018; Wu et al., 2019, 2018a), and high p-type work function (Khan et al., 2008; Hirayama et al., 2009; Lee et al., 2010a; Wu et al., 2015). While heavily doped p-GaN can reduce contact resistance, the absorption of emitted deep UV light by GaN leads to significant degradation of light extraction efficiency (LEE) (Carnevale et al., 2012; Nam et al., 2003; Nakarmi et al., 2006). An alternative approach to hole injection is the integration of a tunnel junction, enabling non-equilibrium hole injection through tunneling. Numerous studies have investigated the injection of nonequilibrium holes via tunneling. The presence of a thin InGaN interlayer reduces the interband tunneling barrier due to the strong polarization field at the InGaN/AlGaN interfaces (Sadaf et al., 2016). Planar AlGaN tunnel junction LEDs have achieved emission wavelengths as low as 255 nm, and tunnel junctions have been successfully incorporated in nanocrystals (Pandey et al., 2020b).

A diagram of the LED structure is depicted in Fig. 6d (Pandey et al., 2020a). The growth process commenced with a 50 nm thick AlN layer, succeeded by the subsequent AlGaN growth. To form the bottom n-contact, the initial $Al_{0.65}Ga_{0.35}N$ layer, approximately 500 nm thick, was doped with Si. Preceding the active region, the Al composition of the AlGaN was graded up from 65% to 85% within an approximately 20 nm thickness.

The active region comprised four AlGaN quantum wells with compositions of around 60%, specifically designed for peak emission at approximately 265 nm. The AlGaN barriers, which possessed higher Al compositions, were grown with decreasing thicknesses, ranging from about 5 nm to approximately 3 nm closer to the Mg-doped AlGaN layer. Fig. 6e exhibits the EL spectra of a tunnel junction LED, measured at various current densities.

It is intriguing to observe that all the devices exhibited a pronounced droop even at a relatively low current injection of around ~0.5–1 A/cm^2, indicating that the cause of the droop is unrelated to the tunnel junction designs. This efficiency droop, frequently measured for InGaN-based blue and green LEDs at current densities of approximately ~5–10 A/cm^2, has been extensively investigated (Lee et al., 2010b). The fundamental factors contributing to the efficiency droop, such as carrier delocalization, Shockley–Read–Hall recombination, Auger-Meitner recombination, and device heating, have been the subject of intensive research (Kioupakis et al., 2011; Kim et al., 2007). At low current densities of roughly ~1 A/cm^2, the influence of device heating and Auger-Meitner recombination is expected to be negligible. Previous studies have shown that GaN-based LEDs can deviate from low-level injection conditions, even under relatively low current densities, due to the asymmetric transport of charge carriers. In Al-rich AlGaN, where the activation energy for Mg dopant is large, p-type conduction primarily occurs through hole hopping in the Mg impurity band at room temperature, resulting in limited mobility (Wu et al., 2020b). As a result, even at a modest current density of 1 A/cm^2, the device operates in a regime that significantly deviates from the low-carrier injection condition. The resulting electric field in the p-AlGaN layer, even at a seemingly small current density, disproportionately affects the transport of holes compared to electrons due to the substantial discrepancy in their mobility values. This disparity leads to a noteworthy increase in charge carrier recombination outside the active region of the device, specifically, electron overflow to the p-AlGaN layer, at low injection currents. Moreover, it is important to acknowledge that since the epitaxial growth of the entire LED structure was conducted under slightly Ga-rich conditions, the distribution of Ga in the epilayers may not be uniform. Ga-rich regions have been demonstrated to serve as highly effective radiative recombination sites due to their capability of locally confining excitons (Zhao et al., 2016b). Nevertheless, as the injected current into the device increases, carrier delocalization ensues, allowing carriers to recombine at nonradiative recombination centers, resulting in a decline in device efficiency.

4.2 Excitonic micro-LEDs

The efficiency of an LED is fundamentally constrained by Shockley–Read–Hall recombination (Bai et al., 2020; Hwang et al., 2017; Pasayat et al., 2020), Auger-Meitner recombination (Pynn et al., 2016; Olivier et al., 2017; Liu et al., 2015), and electron overflow/leakage (Bai et al., 2020; Hwang et al., 2017; Chen et al., 2021). Auger-Meitner recombination and electron overflow often lead to efficiency droop at high current densities, while Shockley–Read–Hall recombination severely limits quantum efficiency at low injection current. Consequently, the increased surface recombination of micro and nanoscale LEDs imposes a fundamental restriction on the external quantum efficiency of conventional quantum well structures. Notably, the radiative recombination efficiency can be significantly enhanced by excitons. This enhancement presents an opportunity to make micro and nanoscale LEDs relatively unaffected by defects/traps.

However, in conventional *c*-plane InGaN quantum well structures, exciton binding energy is considerably diminished, due to the substantial spontaneous and strain-induced piezoelectric polarization fields, along with the resultant quantum-confined Stark effect (QCSE) (Carnevale et al., 2012; Muziol et al., 2019; Sinito et al., 2019). This limitation prevents the utilization of the excitonic effect to overcome the efficiency bottleneck of micro-LEDs. Recent investigations have demonstrated that GaN-based heterostructures can significantly enhance the exciton binding energy through quantum and nanoscale engineering. For instance, in small-sized InGaN nanowires, the exciton oscillator strength can be boosted by nearly two orders of magnitude compared to conventional quantum wells, thanks to efficient strain relaxation (Simon et al., 2010; Pieniak et al., 2021). Theoretically predicted and experimentally demonstrated exciton binding energies as large as 200 meV to 1 eV have been achieved in the extreme monolayer limit in III-nitrides (Wu et al., 2020a; Wang et al., 2022a, 2022b). Moreover, studies on structures grown along semipolar and non-polar planes have revealed a greater exciton binding energy due to reduced polarization fields and, consequently, enhanced electron-hole wavefunction overlap (Wu et al., 2022b; Holder et al., 2014; Li et al., 2019). Considering these advancements, we anticipate that nanowire excitonic LEDs, empowered by strain and quantum engineering, could provide a viable solution to surpass the efficiency bottleneck of micro and nanoscale LEDs.

In an excitonic LED, the emission of EL arises from the binding of an electron and hole, rather than the recombination of free charge carriers. The detrimental impact of Shockley–Read–Hall recombination on the performance of an excitonic LED is substantially diminished due to the strong interaction between electrons and holes. Consequently, the efficiency of an excitonic LED exhibits a rapid increase in both EQE and device output power with injection current, surpassing what was previously achievable with conventional InGaN quantum well micro-LEDs.

To elucidate the effect of excitons on the micro-LED performance, we first focused on two nanowire arrays, named Array A and Array B, which differ in their dimensions. Array A comprises nanowires with diameters of approximately 125 nm and a lattice constant of 245 nm, while Array B consists of nanowires with diameters of approximately 165 nm and the same lattice constant. Following the fabrication of micro-LED devices, we designated the devices fabricated on Array A as Device A and those on Array B as Device B. The injection windows of both devices occupied an area of approximately 750 nm × 750 nm, with a representative injection window depicted in the inset of Fig. 7a. The J–V characteristics of the two devices are presented in Fig. 7a, demonstrating rectifying behavior with negligible reverse bias leakage, signifying the formation of well-defined p–n junctions. Device A, incorporating nanowires with smaller diameters, exhibits a pronounced excitonic effect, resulting in an exceptionally high peak EQE of 25.2% (Fig. 7a). This distinct phenomenon has not been previously observed in *c*-plane quantum well LEDs due to the weakened excitonic effect caused by the strong QCSE and the high plasma damage involved in mesa etching, which obscures this effect at low injection currents. At high injection currents, the indium-rich nanoclusters become saturated, and the majority of the emission arises from the diffuse *c*-plane InGaN through free electron–hole recombination, rather than excitonic emission. Consequently, the transition from excitonic emission to band-to-band recombination results in a similar performance for nanowires with different dimensions, as shown in Fig. 7b and c at relatively high current densities.

Examining the EL spectra in Fig. 7d, it is evident that at low injection currents corresponding to the peak EQE, the emission is predominantly governed by the lower energy peak located at approximately 515 nm, associated with excitonic emission from the indium-rich semi-polar quantum disks. With increasing injection currents, the emission from these nanoclusters becomes saturated, leading to the dominance of band-to-band recombination from the central region of the nanowire (*c*-plane quantum disks). Consequently, as the injection current rises, the higher energy emission at around 475–490 nm gradually becomes more prominent, ultimately becoming the dominant emission peak. In the case of Device B,

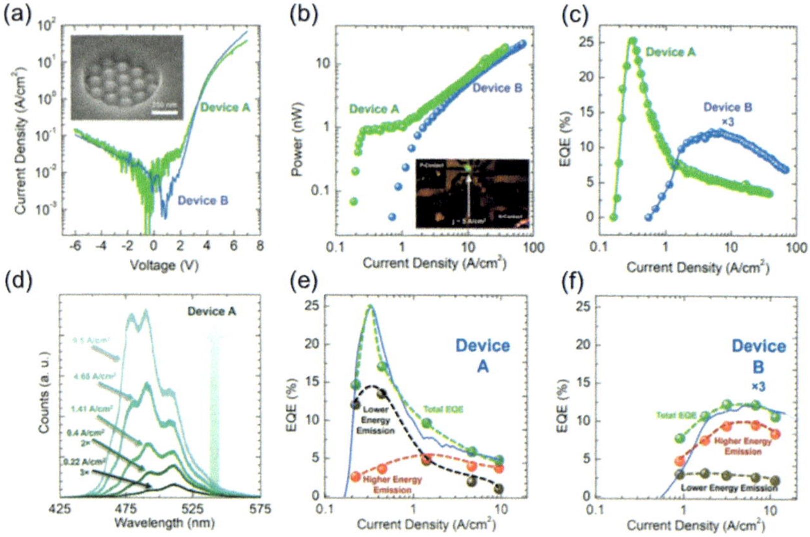

Fig. 7 (a) *J–V* characteristics of Devices A and B. The inset shows an SEM image of a representative injection opening within the nanowire array. (b) L_{op}–J curve for Devices A and B. The inset shows strong green emission from a submicron device. (c) EQE measured at different current densities for Devices A and B. (d) EL spectra measured at different injection currents for Device A. Plot of the relative EQE (dashed green curve) as well as individual contributions of the lower energy emission (black dashed curve) and higher energy emission (red dashed curve) measured from the EL spectra for (e) Device A and (f) Device B, respectively. The solid blue curves in both figures are the measured EQE for each device, overlaid on top (Pandey et al., 2023). *Figures reproduced with permission from Pandey, A., et al., 2023. An ultrahigh efficiency excitonic micro-LED. Nano Lett. 23 (5), 1680–1687.*

no transition in the EL emission peak energy is observed beyond the maximum EQE of 4.1%. This observation aligns well with the prevalence of free electron–hole recombination in Device B. A careful comparison of the EQE curves for Device A and Device B reveals that they become almost identical for injection currents above a few A/cm^2 (as shown in Fig. 7c). In this operational regime, band-to-band recombination dominates for both devices, resembling the behavior of conventional quantum well LED devices.

A comprehensive analysis of the contribution of excitonic emission and band-to-band recombination in relation to the overall EQE has been conducted. Fig. 7e presents the relative EQE for the lower-energy emission peak at approximately 515 nm in Device A (excitonic emission) and the higher-energy emission peak at around 490 nm. Fig. 7f shows the same plots for Device B. The relative EQE for the combined emission from these peaks, represented by the dashed green curves for each device, closely corresponds to the actual measured EQE indicated by the solid blue curves. Notably, the higher-energy emissions (dashed red curves in Fig. 7e and f) exhibit remarkably similar trends for both Devices A and B, peaking at a few A/cm^2. This EQE behavior aligns with conventional quantum well LEDs and can be attributed to the prevalence of band-to-band recombination, in addition to Shockley–Read–Hall recombination and electron overflow. However, the contribution of lower-energy emissions (dashed black curves in Fig. 7e and f) differs fundamentally between Devices A and B.

In the case of Device A, characterized by a smaller diameter, efficient strain relaxation, enhanced faceting, and reduced polarization fields, the low-energy (excitonic) emission significantly outweighs the higher-energy (band-to-band) emission in the low current regime. This substantial disparity contributes to an unprecedented level of efficiency that was previously unattainable for micro-LEDs (Pasayat et al., 2020; Chen et al., 2021; Pieniak et al., 2021; Li et al., 2022; Smith et al., 2020; Wu et al., 2018b; Jiang and Lin, 2013). Conversely, for Device B, the low-energy emission remains negligible across the entire range of measured currents due to the absence of a significant excitonic effect. These investigations and analyses provide unequivocal evidence that the remarkable enhancement in efficiency observed in Device A is not a result of variations in light extraction efficiency. Instead, it stems from the excitonic emission originating from the strain-relaxed semi-polar quantum disks formed in dislocation-free nanowire arrays.

4.3 Low threshold excitonic lasers

The presence of optical gain and the stimulated emission related to excitonic complexes in traditional semiconductors have been investigated and holds significant importance in both fundamental condensed matter physics and potential device applications (Mak et al., 2013; Ross et al., 2013). The existence of optical gain at extremely low densities bears immense significance in the development of nanolasers with exceptionally low thresholds and power consumption (Wang et al., 2020). The presence of optical gain is not reliant on the total quantity of injected carriers but rather on the disparity in population between the trion band and the electron band. Consequently, trion gain can potentially exist even at minuscule carrier density levels. This prospect opens up vast possibilities for manufacturing lasers with extraordinarily low thresholds, reaching down to the single charge level (Wang et al., 2020).

The interaction between excitons and photons in quantum electrodynamics poses a recognized challenge. In the context of an optical microcavity aligned with the exciton's transition energy, the exciton can originate from photon absorption or undergo recombination, emitting a photon within the corresponding mode (Kurizki et al., 2015; Aspelmeyer et al., 2014; Rogers et al., 2014; Kippenberg et al., 2005). This transfer of energy, when performed at a rate surpassing the decay or dephasing of both the exciton and photon, exhibits reversibility and coherence, potentially resulting in the creation of polaritons. Notably, there have been documented instances of polariton laser diodes with minimal threshold current (Ohtani et al., 2019; Schneider et al., 2013; Wei and Zhao, 2014). In addition to engineered photon–exciton coupling in semiconductors, laser devices have also utilized light confinement through Anderson localization in random arrays of nanowires (Zhao et al., 2016b; Li et al., 2015; Zhao et al., 2015). Effective light localization can be achieved with a high probability when the nanocrystals' average size and density fall within appropriate ranges. Standard techniques such as photolithography, surface passivation, planarization, and contact metallization were employed to fabricate nanowire light-emitting devices. Fig. 8a displays the current–voltage characteristics, while the inset reveals minimal leakage current under reverse bias.

To investigate the lasing characteristics of AlGaN nanowires, electrical injection was applied for continuous-wave operation at various temperatures. Fig. 8b depicts the EL spectra of the AlGaN nanowire devices measured at 6 K. Notably, as current densities varied, distinct peaks emerged atop the broad spontaneous emission background of the AlGaN active region. Two specific peaks, centered at 332.7 nm and 334.1 nm, were measured, with the latter peak exhibiting clear dominance. The threshold and linewidth properties of this dominant peak were analyzed. Fig. 8c and d illustrate the plots of integrated intensity and linewidth, respectively, as a function of current density. The nonlinear increase in emission intensity and spectral narrowing near the threshold serve as unequivocal evidence of lasing achievement. At the lasing threshold, the spectral linewidth narrows to a mere 0.2 nm, limited by the setup's spectral resolution. The estimated lasing threshold is approximately 12 A/cm^2, considering factors such as nanowire structure, filling factor, and active lasing area (approximately 10 μm × 10 μm). This threshold is nearly three orders of magnitude lower than that of previously reported GaN-based ultraviolet lasers. The output power is estimated to be around 2 μW with

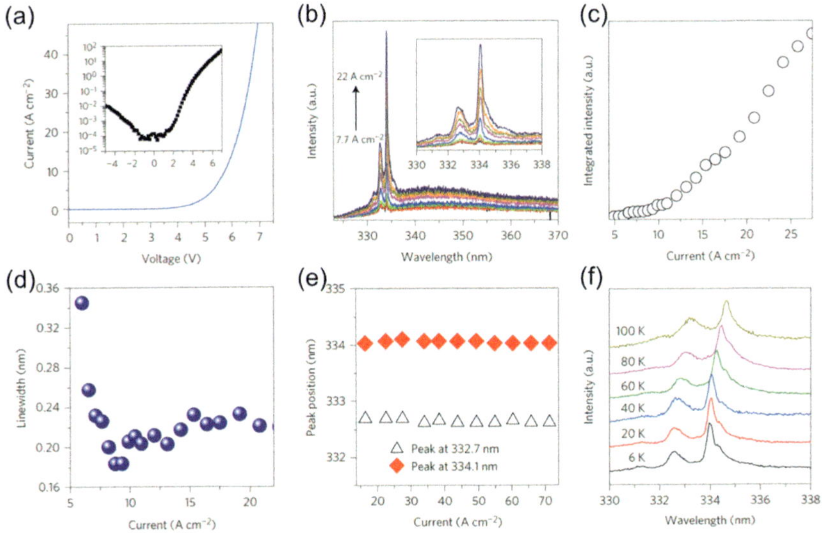

Fig. 8 (a) *I–V* characteristics of the AlGaN nanowire laser; the inset shows the *I–V* curve on a semi-log scale. (b) Emission spectra measured at 6 K under different current densities. The black arrow denotes that the current density increases from 7.7 A/cm^2 to 22 A/cm^2 and the inset shows an enlarged view of the lasing spectra. Integrated intensity (c) and linewidth (d) of the lasing peak at 334.1 nm as a function of injection-current density. (e) Plot of peak position in (b) versus current density. (f) EL spectra measured at different operation temperatures. *Figures reproduced with permission from Li, K.H., Liu, X., Wang, Q., Zhao, S., Mi, Z., 2015. Ultralow-threshold electrically injected AlGaN nanowire ultraviolet lasers on Si operating at low temperature. Nat. Nanotechnol. 10, 140.*

an injection current of 30 A/cm^2. Fig. 8e presents the plot of peak positions as a function of current densities. Remarkably, the lasing wavelengths remain virtually unchanged as the injection current increases from the threshold to approximately six times above the threshold. This suggests that the lasing emission is predominantly governed by the highly stable excitonic transition rather than the emission associated with the electron–hole plasma. The ultralow threshold is attributed to the high-Q optical resonance, nearly defect-free AlGaN nanowires, significantly reduced nonradiative surface recombination facilitated by the AlGaN core-shell structures, and possibly the strong quantum-confinement effect resulting from nanometric compositional fluctuations within the active region. Temperature-dependent measurements were also conducted at a constant current density of approximately 27 A/cm^2. As the temperature increased from 6 to 100 K, both distinct sharp peaks demonstrated redshifts of approximately 0.7 nm, as illustrated in Fig. 8f.

4.4 Summary and prospects

While nano/quantum-structured III-nitrides are under rapid developments, we note that the excitonic properties of III-nitrides are sensitive to the density of threading dislocations in the epitaxial substrates, the quality of interfaces during heteroepitaxy and the n-/p-type doping, etc. With a better understanding and control of the initial nucleation and subsequent epitaxy, it is expected that III-nitrides with larger exciton binding energies can be synthesized at a large scale, which is essential for practical optoexcitonic devices, including quantum light sources, photodetectors, and power devices. Meanwhile, novel structural designs play a key role in the performance of optoexcitonic devices, wherein tunnel junctions, extremely confined quantum structures, and hetero-integration can improve the efficiency, power, and lifetime of optoexcitonic devices. Finally, III-nitride electronic devices that operate based on the exciton transport have also drawn considerable attention, which can potentially lead to the development of next-generation memristor, transistors, and power devices, etc. (Wang et al., 2023a, 2021a, 2021b; Mondal et al., 2022; Wang et al., 2023b, 2023c).

Acknowledgment

This book chapter writing was supported by the National Science Foundation (NSF) (DMR-2118809) and the W. M. Keck Foundation.

Conflict of interest

Some IP related to this work has been licensed to NS Nanotech, Inc., which was co-founded by Z. Mi. The University of Michigan and Mi have a financial interest in the company.

References

Aiello, A., et al., 2019. Deep ultraviolet luminescence due to extreme confinement in monolayer GaN/Al(Ga)N nanowire and planar heterostructures. Nano Lett. 19 (11), 7852–7858.

Aiello, A., Wu, Y., Mi, Z., Bhattacharya, P., 2020. Deep ultraviolet monolayer GaN/AlN disk-in-nanowire array photodiode on silicon. Appl. Phys. Lett. 116 (6), 061104.

Alexeev, E.M., et al., 2019. Resonantly hybridized excitons in moire superlattices in van der Waals heterostructures. Nature 567 (7746), 81–86.

Aspelmeyer, M., Kippenberg, T.J., Marquardt, F., 2014. Cavity optomechanics. Rev. Mod. Phys. 86 (4), 1391–1452.

Baek, H., et al., 2020. Highly energy-tunable quantum light from moire-trapped excitons. Sci. Adv. 6 (37), eaba8526.

Bai, J., et al., 2020. Ultrasmall, ultracompact and ultrahigh efficient InGaN micro light emitting diodes (muLEDs) with narrow spectral line width. ACS Nano 14 (6), 6906–6911.

Bastard, G., Mendez, E.E., Chang, L.L., Esaki, L., 1982. Exciton binding energy in quantum wells. Phys. Rev. B 26 (4), 1974–1979.

Bayerl, D., Kioupakis, E., 2019. Room-temperature stability of excitons and transverse-electric polarized deep-ultraviolet luminescence in atomically thin GaN quantum wells. Appl. Phys. Lett. 115 (13), 131101.

Bayerl, D., et al., 2016. Deep ultraviolet emission from ultra-thin GaN/AlN heterostructures. Appl. Phys. Lett. 109 (24), 241102.

Ben, J., et al., 2021. 2D III-nitride materials: properties, growth, and applications. Adv. Mater. 33 (27), e2006761.

Biswas, M., et al., 2020. Effects of rapid thermal annealing in InGaN/GaN quantum disk-in-GaN nanowire arrays. J. Lumin. 222, 117123.

Bogusławski, P., Rapcewicz, K., Bernholc, J.J., 2000. Surface segregation and interface stability of AlN/GaN, GaN/InN, and AlN/InN {0001} epitaxial systems. Phys. Rev. B 61 (16), 10820–10826.

Calman, E.V., et al., 2018. Indirect excitons in van der Waals heterostructures at room temperature. Nat. Commun. 9 (1), 1895.

Calman, E.V., et al., 2020. Indirect excitons and trions in MoSe2/WSe2 van der Waals heterostructures. Nano Lett. 20 (3), 1869–1875.

Carnevale, S.D., et al., 2012. Polarization-induced pn diodes in wide-band-gap nanowires with ultraviolet electroluminescence. Nano Lett. 12 (2), 915–920.

Cassabois, G., Valvin, P., Gil, B., 2016. Hexagonal boron nitride is an indirect bandgap semiconductor. Nat. Photonics 10 (4), 262–266.

Chen, T.A., et al., 2020. Wafer-scale single-crystal hexagonal boron nitride monolayers on Cu (111). Nature 579 (7798), 219–223.

Chen, Y., et al., 2018. Growth of 2D GaN single crystals on liquid metals. J. Am. Chem. Soc. 140 (48), 16392–16395.

Chen, Z., Yan, S., Danesh, C., 2021. MicroLED technologies and applications: characteristics, fabrication, progress, and challenges. J. Phys. D: Appl. Phys. 54 (12), 123001.

Chiu, M.H., et al., 2015. Determination of band alignment in the single-layer MoS2/WSe2 heterojunction. Nat. Commun. 6, 7666.

Cho, Y.J., et al., 2016. Hexagonal boron nitride tunnel barriers grown on graphite by high temperature molecular beam epitaxy. Sci. Rep. 6, 34474.

Dean, C.R., et al., 2010. Boron nitride substrates for high-quality graphene electronics. Nat. Nanotechnol. 5 (10), 722–726.

Gačević, Ž., Eljarrat, A., Peiró, F., Calleja, E., 2013. Insight into high-reflectivity AlN/GaN Bragg reflectors with spontaneously formed (Al,Ga)N transient layers at the interfaces. J. Appl. Phys. 113 (18), 183106.

Gao, N., et al., 2019. Integral monolayer-scale featured digital-alloyed AlN/GaN superlattices using hierarchical growth units. Cryst. Growth Des. 19 (3), 1720–1727.

Gogneau, N., et al., 2004. Influence of AlN overgrowth on structural properties of GaN quantum wells and quantum dots grown by plasma-assisted molecular beam epitaxy. J. Appl. Phys. 96 (2), 1104–1110.

Grandjean, N., Massies, J., Leroux, M., 1999. Self-limitation of AlGaN/GaN quantum well energy by built-in polarization field. Appl. Phys. Lett. 74 (16), 2361–2363.

Hirayama, H., et al., 2009. 222–282 nm AlGaN and InAlGaN-based deep-UV LEDs fabricated on high-quality AlN on sapphire. Phys. Status Solidi (a) 206 (6), 1176–1182.

Holder, C.O., et al., 2014. Nonpolar III-nitride vertical-cavity surface emitting lasers with a polarization ratio of 100% fabricated using photoelectrochemical etching. Appl. Phys. Lett. 105 (3), 031111.

Hu, Z., et al., 2018. Two-dimensional transition metal dichalcogenides: interface and defect engineering. Chem. Soc. Rev. 47 (9), 3100–3128.

Hwang, D., Mughal, A., Pynn, C.D., Nakamura, S., DenBaars, S.P., 2017. Sustained high external quantum efficiency in ultrasmall blue III–nitride micro-LEDs. Appl. Phys. Express 10 (3), 032101.

Islam, S.M., et al., 2017a. MBE-grown 232–270 nm deep-UV LEDs using monolayer thin binary GaN/AlN quantum heterostructures. Appl. Phys. Lett. 110 (4), 041108.
Islam, S.M., et al., 2017b. Deep-UV emission at 219 nm from ultrathin MBE GaN/AlN quantum heterostructures. Appl. Phys. Lett. 111 (9), 091104.
Jauregui, L.A., et al., 2019. Electrical control of interlayer exciton dynamics in atomically thin heterostructures. Science 366 (6467), 870–875.
Jiang, H.X., Lin, J.Y., 2013. Nitride micro-LEDs and beyond–a decade progress review. Opt. Express 21 (Suppl 3), A475–A484.
Khan, A., Balakrishnan, K., Katona, T., 2008. Ultraviolet light-emitting diodes based on group three nitrides. Nat. Photonics 2 (2), 77–84.
Kim, M.-H., et al., 2007. Origin of efficiency droop in GaN-based light-emitting diodes. Appl. Phys. Lett. 91 (18), 183507.
Kioupakis, E., Rinke, P., Delaney, K.T., Van de Walle, C.G., 2011. Indirect Auger recombination as a cause of efficiency droop in nitride light-emitting diodes. Appl. Phys. Lett. 98 (16), 161107.
Kioupakis, E., Yan, Q., Van de Walle, C.G., 2012. Interplay of polarization fields and Auger recombination in the efficiency droop of nitride light-emitting diodes. Appl. Phys. Lett. 101 (23), 231107.
Kippenberg, T.J., Rokhsari, H., Carmon, T., Scherer, A., Vahala, K.J., 2005. Analysis of radiation-pressure induced mechanical oscillation of an optical microcavity. Phys. Rev. Lett. 95 (3), 033901.
Kurizki, G., et al., 2015. Quantum technologies with hybrid systems. Proc. Natl Acad. Sci. U. S. A. 112 (13), 3866–3873.
Laleyan, D.A., et al., 2018. Effect of growth temperature on the structural and optical properties of few-layer hexagonal boron nitride by molecular beam epitaxy. Opt. Express 26 (18), 23031–23039.
Laleyan, D.A., et al., 2023. Epitaxial hexagonal boron nitride with high quantum efficiency. APL Mater. 11 (5), 051103.
Lee, K.B., et al., 2010a. The origin of the high ideality factor in AlGaN-based quantum well ultraviolet light emitting diodes. Phys. Status Solidi (b) 247 (7), 1761–1763.
Lee, S.H., et al., 2019. Improvements in structural and optical properties of wafer-scale hexagonal boron nitride film by post-growth annealing. Sci. Rep. 9 (1), 10590.
Lee, Y.-J., Chen, C.-H., Lee, C.-J., 2010b. Reduction in the efficiency-droop effect of InGaN green light-emitting diodes using gradual quantum wells. IEEE Photonics Technol. Lett. 22 (20), 1506–1508.
Leroux, M., et al., 1998. Quantum confined Stark effect due to built-in internal polarization fields in (Al,Ga)N/GaN quantum wells. Phys. Rev. B 58 (20), R13371–R13374.
Li, D., Jiang, K., Sun, X., Guo, C., 2018. AlGaN photonics: recent advances in materials and ultraviolet devices. Adv. Opt. Photonics 10 (1), 43.
Li, H., et al., 2019. Study of efficient semipolar (11-22) InGaN green micro-light-emitting diodes on high-quality (11-22) GaN/sapphire template. Opt. Express 27 (17), 24154–24160.
Li, K.H., Liu, X., Wang, Q., Zhao, S., Mi, Z., 2015. Ultralow-threshold electrically injected AlGaN nanowire ultraviolet lasers on Si operating at low temperature. Nat. Nanotechnol. 10, 140.
Li, P., et al., 2022. Red InGaN micro-light-emitting diodes (>620 nm) with a peak external quantum efficiency of 4.5% using an epitaxial tunnel junction contact. Appl. Phys. Lett. 120 (12), 121102.
Liang, Y.H., Towe, E., 2018. Progress in efficient doping of high aluminum-containing group III-nitrides. Appl. Phys. Rev. 5 (1), 011107.
Liu, B., et al., 2011. Composition pulling effect and strain relief mechanism in AlGaN/AlN distributed Bragg reflectors. Appl. Phys. Lett. 98 (26), 261916.

Liu, F., et al., 2020b. Hexagonal BN-assisted epitaxy of strain released GaN films for true green light-emitting diodes. Adv. Sci. (Weinh.) 7 (21), 2000917.
Liu, X., Sun, Y., Malhotra, Y., Wu, Y., Mi, Z., 2021. Monolithic integration of multicolor InGaN LEDs with uniform luminescence emission. Opt. Express 29 (21), 32826–32832.
Liu, X., Wu, Y., Malhotra, Y., Sun, Y., Mi, Z., 2020a. Micrometer scale InGaN green light emitting diodes with ultra-stable operation. Appl. Phys. Lett. 117 (1), 011104.
Liu, Z., Chong, W.C., Wong, K.M., Lau, K.M., 2015. GaN-based LED micro-displays for wearable applications. Microelectronic Eng. 148, 98–103.
Lu, G., et al., 2017. Synthesis of high-quality graphene and hexagonal boron nitride monolayer in-plane heterostructure on Cu-Ni alloy. Adv. Sci. (Weinh.) 4 (9), 1700076.
Maity, A., Grenadier, S.J., Li, J., Lin, J.Y., Jiang, H.X., 2021. Hexagonal boron nitride: epitaxial growth and device applications. Prog. Quantum Electron. 76, 100302.
Mak, K.F., et al., 2013. Tightly bound trions in monolayer MoS2. Nat. Mater. 12 (3), 207–211.
Mazumder, B., et al., 2012. Asymmetric interfacial abruptness in N-polar and Ga-polar GaN/AlN/GaN heterostructures. Appl. Phys. Lett. 101 (9), 091601.
Mengle, K.A., Kioupakis, E., 2019. Impact of the stacking sequence on the bandgap and luminescence properties of bulk, bilayer, and monolayer hexagonal boron nitride. APL Mater. 7 (2), 021106.
Meyaard, D.S., et al., 2013. Identifying the cause of the efficiency droop in GaInN light-emitting diodes by correlating the onset of high injection with the onset of the efficiency droop. Appl. Phys. Lett. 102 (25), 251114.
Mondal, S., et al., 2022. Reconfigurable self-powered deep UV photodetectors based on ultrawide bandgap ferroelectric ScAlN. APL Mater. 10 (12), 121101.
Muller-Caspary, K., et al., 2019. Electrical polarization in AlN/GaN nanodisks measured by momentum-resolved 4D scanning transmission electron microscopy. Phys. Rev. Lett. 122 (10), 106102.
Muziol, G., et al., 2019. Beyond quantum efficiency limitations originating from the piezoelectric polarization in light-emitting devices. ACS Photonics 6 (8), 1963–1971.
Nakarmi, M.L., et al., 2006. Correlation between optical and electrical properties of Mg-doped AlN epilayers. Appl. Phys. Lett. 89 (15), 152120.
Nam, K.B., Nakarmi, M.L., Li, J., Lin, J.Y., Jiang, H.X., 2003. Mg acceptor level in AlN probed by deep ultraviolet photoluminescence. Appl. Phys. Lett. 83 (5), 878–880.
Ohtani, K., et al., 2019. An electrically pumped phonon-polariton laser. Sci. Adv. 5 (7), eaau1632.
Olivier, F., et al., 2017. Influence of size-reduction on the performances of GaN-based micro-LEDs for display application. J. Lumin. 191, 112–116.
Oto, T., Banal, R.G., Kataoka, K., Funato, M., Kawakami, Y., 2010. 100 mW deep-ultraviolet emission from aluminium-nitride-based quantum wells pumped by an electron beam. Nat. Photonics 4 (11), 767–770.
Pandey, A., et al., 2023. An ultrahigh efficiency excitonic micro-LED. Nano Lett. 23 (5), 1680–1687.
Pandey, A., Gim, J., Hovden, R., Mi, Z., 2020b. An AlGaN tunnel junction light emitting diode operating at 255 nm. Appl. Phys. Lett. 117 (24), 241101.
Pandey, A., Shin, W.J., Gim, J., Hovden, R., Mi, Z., 2020a. High-efficiency AlGaN/GaN/AlGaN tunnel junction ultraviolet light-emitting diodes. Photonics Res. 8 (3), 331–337.
Park, A.H., et al., 2015. Efficient stress-relaxation in InGaN/GaN light-emitting diodes using carbon nanotubes. Nanoscale 7 (37), 15099–15105.
Park, S.-H., Chuang, S.-L., 2000. Spontaneous polarization effects in wurtzite GaN/AlGaN quantum wells and comparison with experiment. Appl. Phys. Lett. 76 (15), 1981–1983.

Pasayat, S.S., et al., 2020. Demonstration of ultra-small (<10 μm) 632 nm red InGaN micro-LEDs with useful on-wafer external quantum efficiency (>0.2%) for mini-displays. Appl. Phys. Express 14 (1), 011004.
Pieniak, K., et al., 2021. Quantum-confined Stark effect and mechanisms of its screening in InGaN/GaN light-emitting diodes with a tunnel junction. Opt. Express 29 (2), 1824–1837.
Pynn, C.D., et al., 2016. Green semipolar III-nitride light-emitting diodes grown by limited area epitaxy. Appl. Phys. Lett. 109 (4), 041107.
Ra, Y.-H., et al., 2020. An electrically pumped surface-emitting semiconductor green laser. Sci. Adv. 6 (1), eaav7523.
Rogers, B., Lo Gullo, N., De Chiara, G., Palma, G.M., Paternostro, M., 2014. Hybrid optomechanics for quantum technologies. Quantum Meas. Quantum Metrol. 2 (1).
Ross, J.S., et al., 2013. Electrical control of neutral and charged excitons in a monolayer semiconductor. Nat. Commun. 4, 1474.
Sadaf, S.M., Ra, Y.H., Szkopek, T., Mi, Z., 2016. Monolithically integrated metal/semiconductor tunnel junction nanowire light-emitting diodes. Nano Lett. 16 (2), 1076–1080.
Schneider, C., et al., 2013. An electrically pumped polariton laser. Nature 497 (7449), 348–352.
Schué, L., et al., 2019. Bright luminescence from indirect and strongly bound excitons in h-BN. Phys. Rev. Lett. 122 (6), 067401.
Simon, J., Protasenko, V., Lian, C., Xing, H., Jena, D., 2010. Polarization-induced hole doping in wide-band-gap uniaxial semiconductor heterostructures. Science 327 (5961), 60–64.
Sinito, C., et al., 2019. Absence of quantum-confined Stark effect in GaN quantum disks embedded in (Al,Ga)N nanowires grown by molecular beam epitaxy. Nano Lett. 19 (9), 5938–5948.
Smith, J.M., et al., 2020. Comparison of size-dependent characteristics of blue and green InGaN microLEDs down to 1 μm in diameter. Appl. Phys. Lett. 116 (7), 071102.
Song, S.B., et al., 2021. Deep-ultraviolet electroluminescence and photocurrent generation in graphene/hBN/graphene heterostructures. Nat. Commun. 12 (1), 7134.
Speck, J.S., Chichibu, S.F., 2011. Nonpolar and semipolar group III nitride-based materials. MRS Bull. 34 (5), 304–312.
Stachurski, J., Tamariz, S., Callsen, G., Butte, R., Grandjean, N., 2022. Single photon emission and recombination dynamics in self-assembled GaN/AlN quantum dots. Light Sci. Appl. 11 (1), 114.
Sun, Y., et al., 2019. Ultrahigh Q microring resonators using a single-crystal aluminum-nitride-on-sapphire platform. Opt. Lett. 44 (23), 5679–5682.
Taniyasu, Y., Kasu, M., Makimoto, T., 2006. An aluminium nitride light-emitting diode with a wavelength of 210 nanometres. Nature 441 (7091), 325–328.
Toropov, A.A., et al., 2020. Strongly confined excitons in GaN/AlN nanostructures with atomically thin GaN layers for efficient light emission in deep-ultraviolet. Nano Lett. 20 (1), 158–165.
Tran, N.H., Le, B.H., Zhao, S., Mi, Z., 2017. On the mechanism of highly efficient p-type conduction of Mg-doped ultra-wide-bandgap AlN nanostructures. Appl. Phys. Lett. 110 (3), 032102.
Wang, D., et al., 2023a. Fully epitaxial, monolithic ScAlN/AlGaN/GaN ferroelectric HEMT. Appl. Phys. Lett. 122 (9), 090601.
Wang, D., et al., 2023b. Ultrathin nitride ferroic memory with large ON/OFF ratios for analog in-memory computing. Adv. Mater. 35 (20), e2210628.
Wang, D., et al., 2023c. Thickness scaling down to 5 nm of ferroelectric ScAlN on CMOS compatible molybdenum grown by molecular beam epitaxy. Appl. Phys. Lett. 122 (5), 052101.

Wang, P., et al., 2021a. Oxygen defect dominated photoluminescence emission of ScxAl1−xN grown by molecular beam epitaxy. Appl. Phys. Lett. 118 (3), 032102.

Wang, P., et al., 2021b. N-polar ScAlN and HEMTs grown by molecular beam epitaxy. Appl. Phys. Lett. 119 (8), 082101.

Wang, P., et al., 2022a. Scalable synthesis of monolayer hexagonal boron nitride on graphene with giant bandgap renormalization. Adv. Mater. 34 (21), e2201387.

Wang, P., et al., 2022b. Interfacial modulated lattice-polarity-controlled epitaxy of III-nitride heterostructures on Si(111). ACS Appl. Mater. Interfaces 14 (13), 15747–15755.

Wang, S., Robertson, A., Warner, J.H., 2018a. Atomic structure of defects and dopants in 2D layered transition metal dichalcogenides. Chem. Soc. Rev. 47 (17), 6764–6794.

Wang, Y., et al., 2018b. Wafer-scale synthesis of monolayer WSe2: a multi-functional photocatalyst for efficient overall pure water splitting. Nano Energy 51, 54–60.

Wang, Y., et al., 2019a. A single-junction cathodic approach for stable unassisted solar water splitting. Joule 3 (10), 2444–2456.

Wang, Y., et al., 2019b. An In0.42Ga0.58N tunnel junction nanowire photocathode monolithically integrated on a nonplanar Si wafer. Nano Energy 57, 405–413.

Wang, Y., Wu, Y., Sun, K., Mi, Z., 2019a. A quadruple-band metal–nitride nanowire artificial photosynthesis system for high efficiency photocatalytic overall solar water splitting. Mater. Horiz. 6 (7), 1454–1462.

Wang, Z., et al., 2020. Excitonic complexes and optical gain in two-dimensional molybdenum ditelluride well below the Mott transition. Light Sci. Appl. 9, 39.

Watanabe, K., Taniguchi, T., Kanda, H., 2004. Direct-bandgap properties and evidence for ultraviolet lasing of hexagonal boron nitride single crystal. Nat. Mater. 3 (6), 404–409.

Wei, C., Zhao, Y.S., 2014. Electrically pumped polariton lasers. J. Mater. Chem. C. 2 (13), 2295–2297.

Wen, Q., et al., 2020. Hyperspectral absorption of semiconductor monolayer crystals. Appl. Phys. Lett. 116 (18), 181103.

Wong, M.S., Nakamura, S., DenBaars, S.P., 2020. Review—Progress in high performance III-nitride micro-light-emitting diodes. ECS J. Solid State Sci. Technol. 9 (1), 015012.

Wu, T., et al., 2018b. Mini-LED and Micro-LED: promising candidates for the next generation display technology. Appl. Sci. 8 (9), 1557.

Wu, Y., et al., 2015. High efficiency single Ag nanowire/p-GaN substrate Schottky junction-based ultraviolet light emitting diodes. Appl. Phys. Lett. 106 (5), 051108.

Wu, Y., et al., 2018a. Molecular beam epitaxy and characterization of Mg-doped GaN epilayers grown on Si (0 0 1) substrate through controlled nanowire coalescence. J. Cryst. Growth 498, 109–114.

Wu, Y., et al., 2020a. Monolayer GaN excitonic deep ultraviolet light emitting diodes. Appl. Phys. Lett. 116 (1), 013101.

Wu, Y., et al., 2020b. Controlling defect formation of nanoscale AlN: toward efficient current conduction of ultrawide-bandgap semiconductors. Adv. Electron. Mater. 6 (9), 2000337.

Wu, Y., et al., 2022a. III-nitride nanostructures: emerging applications for Micro-LEDs, ultraviolet photonics, quantum optoelectronics, and artificial photosynthesis. Prog. Quantum Electron. 85, 100401.

Wu, Y., et al., 2022b. InGaN micro-light-emitting diodes monolithically grown on Si: achieving ultra-stable operation through polarization and strain engineering. Light Sci. Appl. 11 (1), 294.

Wu, Y., et al., 2023. Perspectives and recent advances of two-dimensional III-nitrides: material synthesis and emerging device applications. Appl. Phys. Lett. 122 (16), 160501.

Wu, Y., Wang, P., Kioupakis, E., Mi, Z., 2021. Nanoscale AlGaN and BN: molecular beam epitaxy, properties, and device applications. Semiconduct. Semimet. 107, 153–189.

Wu, Y., Wang, Y., Sun, K., Mi, Z., 2019. Molecular beam epitaxy and characterization of AlGaN nanowire ultraviolet light emitting diodes on Al coated Si (0 0 1) substrate. J. Cryst. Growth 507, 65–69.
Yang, Q., Wu, Y., Liu, Y., Pan, C., Wang, Z.L., 2014. Features of the piezo-phototronic effect on optoelectronic devices based on wurtzite semiconductor nanowires. Phys. Chem. Chem. Phys. 16 (7), 2790–2800.
Yang, X.L., Guo, S.H., Chan, F.T., Wong, K.W., Ching, W.Y., 1991. Analytic solution of a two-dimensional hydrogen atom. I. Nonrelativistic theory. Phys. Rev. A 43 (3), 1186–1196.
Zhang, Y., Jamal-Eddine, Z., Rajan, S., 2019a. Recent progress of tunnel junction-based ultra-violet light emitting diodes. Jpn. J. Appl. Phys. 58 (SC), SC0805.
Zhang, Z., et al., 2019b. A 271.8 nm deep-ultraviolet laser diode for room temperature operation. Appl. Phys. Express 12 (12), 124003.
Zhao, S., et al., 2015. An electrically injected AlGaN nanowire laser operating in the ultraviolet-C band. Appl. Phys. Lett. 107 (4), 043101.
Zhao, S., et al., 2016a. Molecular beam epitaxy growth of Al-rich AlGaN nanowires for deep ultraviolet optoelectronics. APL Mater. 4 (8), 086115.
Zhao, S., Liu, X., Wu, Y., Mi, Z., 2016b. An electrically pumped 239 nm AlGaN nanowire laser operating at room temperature. Appl. Phys. Lett. 109 (19), 191106.
Zheng, Z., et al., 2021. Gallium nitride-based complementary logic integrated circuits. Nat. Electron. 4 (8), 595–603.

CHAPTER SEVEN

Mid-Infrared light emitters based on black phosphorus and its alloys

Po-Liang Chen, Shyam Narayan Singh Yadav, and Chang-Hua Liu*
Institute of Photonics Technologies, National Tsing Hua University, Hsinchu, Taiwan
*Corresponding author. e-mail address: changhua@gapp.nthu.edu.tw

Contents

Abstract

The mid-infrared (MIR) spectral region is crucial in scientific and technological fields due to its atmospheric transmission windows. These windows enable unhindered infrared radiation transmission, facilitating applications like remote sensing and thermal imaging. Additionally, the MIR region acts as a fingerprint region for various molecules, enabling precise identification and analysis, particularly for greenhouse gases like methane, carbon dioxide, and nitrous oxide. As a result, MIR spectroscopy has become a powerful tool for environmental monitoring and atmospheric studies. Advancing MIR sensing and imaging technologies has recently focused on developing high-performance emitters operating in the MIR band. However, traditional MIR optoelectronics based on III-V or II-VI superlattice materials present challenges in

Semiconductors and Semimetals, Volume 112
ISSN 0080-8784, https://doi.org/10.1016/bs.semsem.2023.09.006

growth complexity, silicon integration, and cryogenic cooling requirements. The emergence of van der Waals (vdW) materials, including black phosphorus (BP), graphene, and semimetals, has shown immense promise as favorable alternatives to conventional materials. Their layered crystalline structure allows for atomic-level manipulation and seamless integration onto photonic structures, eliminating the need for stringent lattice matching. This chapter highlights recent advancements in utilizing BP and its alloys for MIR emission applications. We explore various approaches, including mechanical strain, electric field, and substitutional doping, to engineer their MIR emission wavelengths and the possibility of integrating BP emitters with photonic structures and flexible substrates, facilitating on-chip sensing, communications, and spectroscopy. Additionally, we discuss the challenges of BP-based emitters and potential future breakthroughs aimed at advancing the development of practical and accessible MIR optoelectronic technologies, harnessing novel innovations to revolutionize the field and expand MIR possibilities.

1. Introduction

The mid-infrared (MIR) spectral region, spanning from approximately 2 to 25 μm, holds immense importance in various scientific and technological domains. One of the key factors contributing to its significance is the presence of atmospheric transmission windows within this band. Specifically, the MIR region encompasses two significant windows of atmospheric transparency: one in the 3–5 μm range and another in the 8–14 μm range (Fig. 1). These windows allow for relatively unobstructed passage of infrared radiation through the Earth's atmosphere, enabling

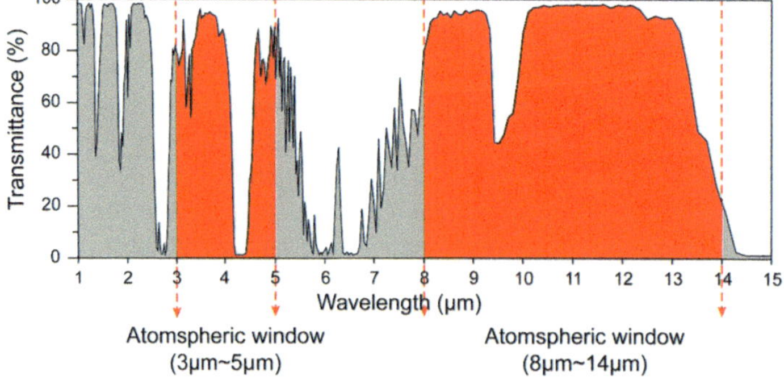

Fig. 1 Atmospheric transmission changes with different wavelengths and fingerprint region.

applications such as remote sensing and thermal imaging (Krier, 2006; Godard, 2007). Additionally, the MIR region serves as a fingerprint region for numerous molecules (Fig. 1). Many substances exhibit distinctive spectral fingerprints within this range, allowing for precise identification and analysis. For example, common greenhouse gases like methane, carbon dioxide, and nitrous oxide exhibit significant absorption features in the MIR region, typically in the 3–8 μm range. This characteristic makes MIR spectroscopy a powerful tool for environmental monitoring, gas sensing, and atmospheric studies (Baker et al., 2014; Stanley, 2012; Jung et al., 2017).

To advance these MIR sensing and imaging technologies, tremendous efforts have been made on developing high-performing detectors and emitters, operated in the MIR band. However, traditional MIR optoelectronics heavily depend on III–V or II–VI superlattice materials, which give rise to various challenges (Krier, 2006; Jung et al., 2017; Soref, 2010; Spott et al., 2017; Lin et al., 2018). Firstly, the growth of these materials is notoriously difficult, requiring atomic-level precision in layer thickness and the addition of multiple buffer layers to mitigate lattice and thermal mismatches at heterointerfaces. As a consequence, complex fabrication processes and increased costs are incurred. Moreover, integrating these superlattice materials with silicon, a widely used substrate material in electronics, poses significant technical hurdles. The incompatibility between III–V or II–VI superlattices and silicon limits the scalability and widespread adoption of MIR optoelectronics. Furthermore, III–V or II–VI-based optoelectronics often require cryogenic cooling to achieve optimal performance, adding further complexity and cost to MIR devices. Overcoming these limitations and pursuing alternative materials and integration techniques are paramount to realizing practical and accessible MIR optoelectronic technologies.

In recent years, a significant advancement has been observed in the field of MIR optoelectronics with the emergence of van der Waals (vdW) materials (Fang et al., 2020; Xia et al., 2014b; Bhimanapati et al., 2015; Elbanna et al., 2023). Notably, materials like graphene, black phosphorus (BP), and semimetals such as $PtTe_2$ and ternary compounds $TaIrTe_4$ have shown immense promise as highly favorable alternatives to conventional III–V or II–VI materials. For instance, these materials exhibit a layered crystalline structure, where each layer consists of atoms held together by strong covalent bonds, while the interactions between layers are mediated by weaker vdW forces. Such characteristic allows for the exfoliation of vdW materials into individual atomic layers. The resulting two-dimensional (2D) materials can be manipulated and utilized at the atomic scale, offering unprecedented control over their electronic and optical properties. Additionally, the weak vdW bonding enables the vertical stacking of different vdW layers, leading to the formation of heterostructures with customizable properties (Liu et al., 2016; Geim and Grigorieva, 2013). Notably, one significant advantage lies in the seamless integration of these materials and their heterostructures onto integrated photonic structures, such as cavities or waveguides (Liu et al., 2019; Youngblood and Li, 2017; Jariwala et al., 2017; Chen et al., 2022). This integration is achieved through a simple transfer process, eliminating the stringent requirements of lattice matching. The flexibility afforded by vdW materials in terms of integration opens up exciting possibilities for the design and fabrication of optoelectronic devices with enhanced performance and expanded functionality, propelling the progress of MIR optoelectronics to new heights.

In this chapter, we will highlight the recent advancements and emerging trends in utilizing BP and its alloys for MIR emission applications. Our focus is on exploring various approaches and design strategies to engineer the MIR emission wavelength. Moreover, we'll emphasize the exciting potential of integrating BP emitters with diverse photonic structures and flexible substrates, enabling the applications for on-chip sensing, communications, and spectroscopy. As exploring the current landscape of BP-based MIR emitters, we'll also discuss the potential future breakthroughs and the challenges that researchers are currently facing in pushing these technologies to new frontiers.

2. Crystal structure and electronic band structure of black phosphorus

Phosphorus exhibits three allotropic forms: white, red, and black phosphorus, with black phosphorus (BP) emerging as the most thermodynamically stable form at room temperature. BP, similar to graphene, possesses a layered structure; however, it sets itself apart from other layered materials containing group IV elements in several ways. In a monolayer of BP, phosphorus atoms exhibit a $3s^2 3p^3$ valence shell configuration with five valence electrons. To facilitate bonding with other atoms, each phosphorus atom undergoes hybridization, leading to the formation of sp^3 hybridized orbitals. This results in covalent bonds between each phosphorus atom and three adjacent phosphorus atoms. The arrangement of these bonds gives rise to a unique puckered structure with an irregular hexagonal structure. Within this structure, two out of the three adjacent phosphorus atoms lie in the plane of the layer, forming a bond length of 0.2224 nm and bond angle of 96.3°, while the third phosphorus atom sits at an angle of 102.1° between the layers, with a bond length of 0.2244 nm, as schematically shown in Fig. 2a, b. Moreover, in a monolayer of BP, two distinct crystal directions, referred to as zigzag and armchair, can be identified. In the bulk form of BP, the layers are stacked in an ABA sequence, with an interlayer distance of approximately 5.3 Å (Xia et al., 2014a, Ling et al., 2015).

This unique crystal structure gives rise to anisotropic optoelectronic properties (Yuan et al., 2015). Angle-resolved photoemission spectroscopy (ARPES) and theoretical band calculations were utilized to investigate the band dispersion anisotropy of bulk BP. The results clearly demonstrate that the valence band maximum and conduction band minimum of bulk BP are situated at the same K point, indicating a direct and narrow bandgap of 0.36 eV, which closely aligns with the experimental characterizations (0.33 eV).

Furthermore, in the k_x direction (corresponding to the armchair direction in real space), holes and electrons exhibit smaller effective masses of approximately $0.12m_0$ and $0.11m_0$, respectively, while in contrast, the valence and conduction bands along the k_y direction (corresponding to the zigzag direction in real space) appear relatively flat, leading to significantly larger effective masses for electrons and holes. As a result, BP displays strong anisotropic electrical transport properties (Ling et al., 2015; Qiao et al., 2014). Additionally, the strong anisotropic optical properties of BP can be found (Fig. 3). This is attributed to the symmetry-forbidden selection rule, as the crystal structure of BP possesses mirror reflection symmetry and inversion symmetry along the zigzag direction (Xia et al., 2014a, Yuan et al., 2015; Qiao et al., 2014).

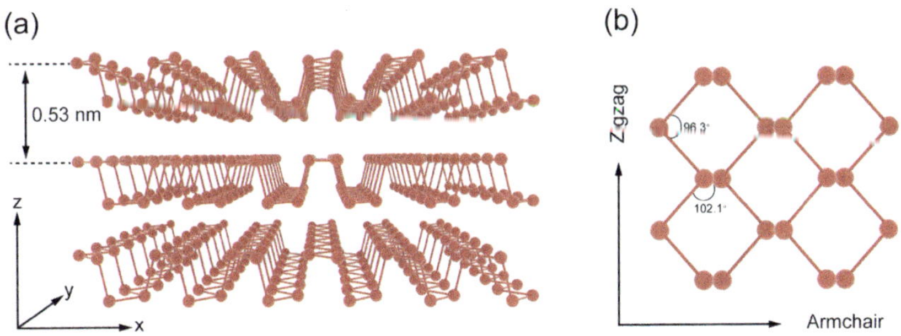

Fig. 2 (a) Side view and (b) top view of crystal structure of black phosphorus.

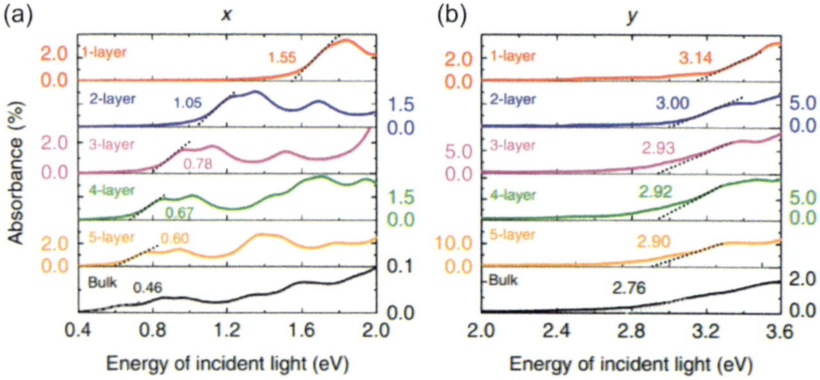

Fig. 3 Optical absorption spectra of few-layer BP for light incident polarized along the (a) armchair and (b) zigzag crystal orientation of black phosphorus, respectively. *With permission from Qiao, J., Kong, X., Hu, Z.-X., Yang, F., Ji, W., 2014. High-mobility transport anisotropy and linear dichroism in few-layer black phosphorus. Nat. Commun. 5, 4475. © copyright 2014 by The Author(s).*

When reducing the thickness of BP, it is notable that these anisotropic optoelectronic properties are still observed; however, the bandgap of BP undergoes significant variation due to the quantum confinement effect of charge carriers in the out-of-plane direction (Zhang et al., 2017a, Wang et al., 2015; Tran et al., 2014; Castellanos-Gomez, 2015). The confinement effect gives rise to the emergence of quantum-well-like subbands within both the conduction and valence band complexes in ultrathin flakes of BP. Furthermore, as the thickness of BP decreases, the bandgap experiences an increase, reaching approximately 2 eV for monolayer BP. Notably, despite this variation, the bandgap remains direct and is located at the same momentum point within the Brillouin zone. The unique characteristic of BP makes it an excellent candidate for studying excitonic physics and harnessing its polarization-sensitive absorption, emission, and photodetection capabilities (Fang et al., 2020; Elbanna et al., 2023; Chang et al., 2020a). Further elaboration on these capabilities will be provided in subsequent sections of the chapter.

3. Tuning electronic band structure of black phosphorus

The operational spectral range of an optoelectronic device is primarily determined by the bandgap of its photoactive material. In the case of bulk BP, it exhibits a direct and narrow bandgap of approximately 0.33 eV, rendering it well-suited for MIR optoelectronics. However, in order to extend its functionality to a broader range of wavelengths, it becomes imperative to adjust the bandgap of BP. This can be achieved through some common techniques, including strain engineering, applying electric fields, and introducing substitutional doping. In this section, we will delve into these methods and their impact on modifying the bandgap of BP.

3.1 Substitutional doping

Extrinsic-atom-based substitution doping involves the intentional introduction of non-phosphorus atoms into the crystal lattice of BP as replacements for phosphorus atoms. Various studies have successfully achieved substitutional doping by directly adding low melting atoms (Fig. 4), such as arsenic (As), selenium (Se), antimony (Sb), and tellurium (Te) during the growth process of BP crystals (Hu et al., 2020; Lei et al., 2017; Xu et al., 2019). These alloying techniques enable the modification of BP's crystal structure, resulting in the emergence of new properties and improved performance of BP-based devices. Throughout this discussion, our primary focus will be on exploring the potential of arsenic-doped BP (b-AsP) due to its characteristic of possessing a significantly reduced bandgap, making it a highly promising candidate for various MIR applications.

B-AsP, a compound composed of phosphorus and arsenic atoms, maintains the same orthorhombic lattice and puckered honeycomb structure as pristine BP, thereby preserving its inherent anisotropic optoelectronic properties (Fig. 5a) (Amani et al., 2017; Yuan et al., 2018). Notably, the bandgap of bulk b-AsP can be effectively tuned and is influenced by the

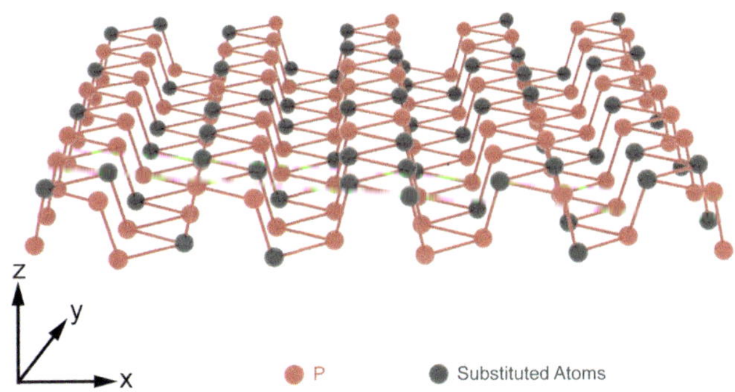

Fig. 4 Schematic of the crystal structures of black phosphorus-based alloy. Brown represents phosphorus atoms and black represents substituted atoms.

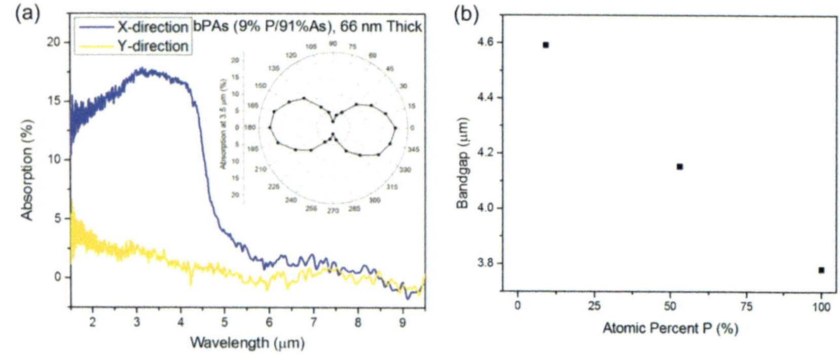

Fig. 5 (a) Characterization the absorption spectra of b-AsP along its armchair (*x*-direction) and zigzag (*y*-direction) crystal orientation. (b) The bandgap of b-AsP alloys with different concentrations of phosphorus. *With permission from Amani, M., Regan, E., Bullock, J., Ahn, G.H., Javey, A., 2017. Mid-wave infrared photoconductors based on black phosphorus-arsenic alloys. ACS Nano 11, 11724–11731. © copyright 2017 by the American Chemical Society.*

concentration of arsenic. Researchers (Liu et al., 2015a) have utilized infrared absorption spectroscopy to investigate the effects of varying arsenic concentrations in b-As_xP_{1-x} alloys, revealing a significant reduction in the bandgap. Specifically, when the arsenic concentration ranges from 0% to 40%, the bandgap undergoes a remarkable reduction from 0.33 eV to approximately 0.17 eV. As the arsenic concentration increases further, up to 83%, the reduction in the bandgap becomes less pronounced, reaching approximately 0.15 eV (corresponding to an absorption wavelength of approximately 8.2 μm). Other studies (Amani et al., 2017) have also employed Fourier-transform infrared spectroscopy (FTIR) techniques to characterize b-As_xP_{1-x} alloys and discovered a reduction in the bandgap to 0.269 eV (corresponding to 4.6 μm) for b-$A_{0.91}P_{0.9}$, as illustrated in Fig. 5b. These findings highlight the potential of b-AsP alloys as promising alternatives to existing III–V or II–VI technologies, offering exciting prospects for a wide range of optoelectronic applications.

3.2 Electric field

When 2D material is subjected to an electric field perpendicular to its layers, two significant phenomena come into play, namely the Franz–Keldysh effect and the quantum-confined Stark effect. These phenomena have a notable impact on the bandgap of ultrathin 2D flake and can induce changes in interband transition processes (Chaves et al., 2020; Scharf et al., 2016). To generate the desired out-of-plane electric field, an effective approach is to utilize the surface doping technique. For instance, potassium metal atoms have been deposited onto few-layer BP, resulting in the transfer of charges from the potassium atoms to the underlying BP (Kim et al., 2015). This process leads to the formation of a 2D electron gas near the surface of BP and the creation of a vertical electric field. By employing Density Functional Theory (DFT) calculations and Angular Resolved Photoelectron Spectroscopy (ARPES) measurements, it was observed that the bandgap increased to approximately 0.6 eV as the valence band maximum of BP shifted downwards, while the conduction band minimum remained above the Fermi energy. As the doping concentration was further increased, the conduction band minimum gradually descended below the Fermi energy, causing the bandgap to decrease gradually. Eventually, BP transforms into a band-inverted semimetal with anisotropic dispersion. The related phenomena are attributed to the giant Stark effect.

An alternative method involves utilizing electrostatic gating to generate an out-of-plane electric field (Liu et al., 2017b). In the study, they observed a significant Stark effect in few-layer BP (consisting of 11 layers), enabling the electrical tuning of the BP bandgap from 310 ± 20–200 ± 20 meV (corresponding to a ~35.5% modulation), as confirmed through low-temperature scanning tunneling microscopy and theoretical calculations (Fig. 6a, b). Additionally, different research group demonstrated field effect modulation of inter-subband transitions by investigating the MIR transmission of few-layer BP (Whitney et al., 2017). This gate-tunable phenomenon was attributed to Pauli-blocking and quantum-confined Franz–Keldysh effects.

These surface doping and electrostatic gating techniques offer wide and continuous tunability of the bandgap of BP. Consequently, they can be utilized for the development of tunable MIR optoelectronics. It is worth noting that the BP flakes employed in these techniques should be few-layer thick to minimize the screening of the induced vertical field.

3.3 Strain engineering

Strain is a powerful tool for manipulating the optoelectronic properties of semiconductors and offers a means to precisely control and reversibly tune the bandgap. This technique has been extensively employed in the Si, Ge, and III–V semiconductor industry in the last few decades (Jacobsen et al., 2006; Lee et al., 2004; Chu et al., 2009). In the realm of 2D materials, the deformation capacity is even more impressive, enabling them to withstand higher levels of elastic strain without fracturing. This exceptional deformability is particularly notable in black BP due to its unique puckered structure (Yang et al., 2021; Sun and Liu, 2018).

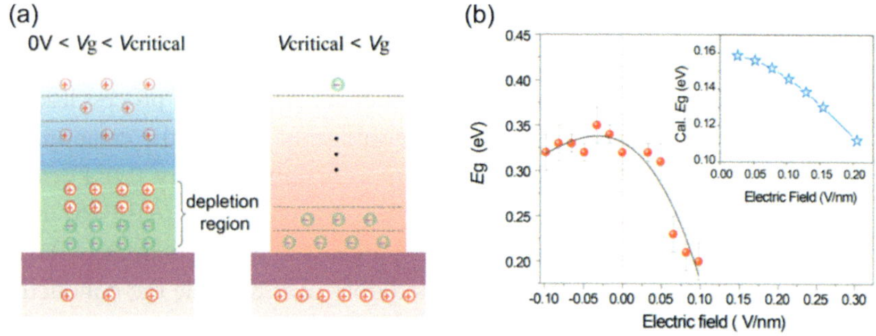

Fig. 6 (a) Schematic illustration of the gate-induced vertical electric field in 11-layer BP. (b) The bandgap of an 11-layer BP sample is measured under the influence of an applied electrical field. The inset illustrates the calculated bandgap of a BP sample with the same thickness, considering different strengths of electrical fields. *With permission from Liu, Y., Qiu, Z., Carvalho, A., Bao, Y., Xu, H., Tan, S.J.R., et al., 2017b. Gate-tunable giant stark effect in few-layer black phosphorus. Nano Lett. 17, 1970–1977. © copyright 2017 by the American Chemical Society.*

A study (Huang et al., 2019a) utilizing infrared spectroscopy to investigate the influence of in-plane biaxial strain on the energy band structure and vdW interaction in few-layer black phosphorus (2–10 layers). The results reveal that applying biaxial tensile (compressive) strain on BP can lead to the increase (decrease) of bandgap, and the shift rate of bandgap of bulk BP is larger than 200 meV per percentage of applied strain (Fig. 7a, b). Remarkably, the rate of bandgap shift in bulk BP exceeds 200 meV per percentage of applied strain, qualitatively aligning with other reports (Zhang et al., 2017b, Quereda et al., 2016). Theoretical calculations further suggest that the puckered lattice structure of BP confers mechanical resilience, enabling the material to withstand significant strain modulation. The strain-induced changes in interatomic distances, bond angles, and interlayer coupling contribute to the tunability of the bandgap (Quereda et al., 2016). These results shed light on the potential of strain engineering in BP for tailored MIR optoelectronic applications.

4. Photoluminescence and optically pumped lasers based on black phosphorus

In recent years, there has been significant interest in harnessing the potential of 2D materials and vdW materials for the development of atomically-thin emitters. Early research efforts primarily focused on utilizing

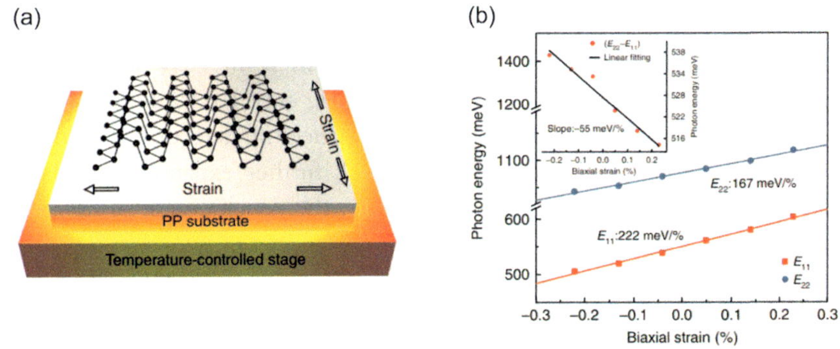

Fig. 7 (a) A schematic illustration of the experimental setup utilized for applying in-plane biaxial strain on BP. (b) The change in measured absorption peak energies (i.e., E_{11} and E_{22} peak energies) of BP are plotted as a function of biaxial strain. *With permission from Huang, S., Zhang, G., Fan, F., Song, C., Wang, F., Xing, Q., et al., 2019a. Strain-tunable van der waals interactions in few-layer black phosphorus. Nat. Commun. 10, 2447. © copyright 2019 by The Author(s).*

direct bandgap monolayer transition-metal dichalcogenides (MX_2, where M = Mo, W and X = S, Se) (Wang et al., 2018a, Wang et al., 2018b) and few-layer BP (Wang et al., 2015) as emission materials. These materials have emission wavelengths that fall within the visible to near-infrared (NIR) regions, making them easily detectable using commercially available spectrometers equipped with charge-coupled-device (CCD) array silicon or InGaAs detectors.

In contrast, the investigation of MIR emission from 2D or vdW materials has lagged behind. This is primarily attributed to the unique challenges associated with measuring MIR emission. This is largely due to the challenges associated with measuring MIR emission, which requires specialized techniques and equipment to ensure accurate detection and analysis. Specifically, MIR measurements demand the use of specialized detectors such as cooled mercury–cadmium–telluride (MCT) or indium antimonide (InSb) detectors that are sensitive to the MIR wavelength range. Additionally, dedicated spectrometers with appropriate optics and filters are necessary to selectively isolate the desired MIR signals (Razeghi and Nguyen, 2014; Rogalski, 2017). These measurements demand careful attention to instrument calibration and optimization to ensure reliable and accurate results.

Despite these challenges, there is growing recognition of the importance of MIR emission in various applications such as sensing and spectroscopy

(Jung et al., 2017; Schliesser et al., 2012). Therefore, efforts to advance the understanding and utilization of MIR emissions from vdW materials are gaining momentum. Researchers are exploring novel techniques, such as FTIR and advanced imaging methods, to enhance the detection and analysis of MIR emissions. These advancements will further expand the scope of atomically-thin emitters and pave the way for the development of advanced MIR devices and systems.

4.1 Photoluminescence of black phosphorus

An extensive examination of the thickness and temperature dependence of MIR photoluminescence (PL) in thin-film BP was first presented in 2019 (Chen et al., 2019). The investigation encompassed the PL spectra of BP flakes at 80 K, revealing a notable increase in the bandgap from 0.308 eV for 46 nm BP to 0.441 eV for 4.5 nm BP (Fig. 8a, b). The PL emission from the 46-nm-thick BP exhibited significant intensity, approximately seven times lower than that of an indium arsenide (InAs) multiple quantum well

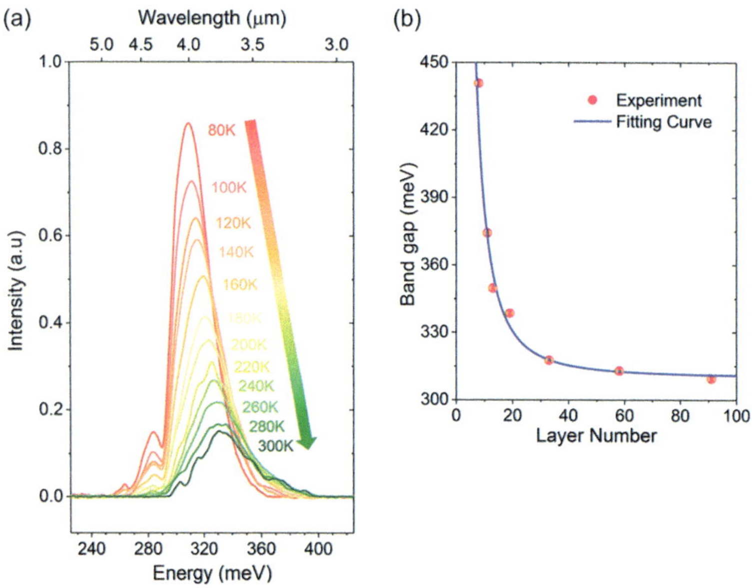

Fig. 8 (a) The PL spectrum of the 46-nm-thick BP shows a clear blueshift at higher temperatures, contrary to traditional semiconductors. (b) The bandgap of BP shows dependence on its thickness. *With permission from Chen, C., Chen, F., Chen, X., Deng, B., Eng, B., Jung, D., et al., 2019. Bright mid-infrared photoluminescence from thin-film black phosphorus. Nano Lett. 19, 1488–1493. © copyright 2019 by the American Chemical Society.*

structure. Furthermore, as the temperature was raised to 300 K, the PL spectrum underwent a blueshift, reaching 2686 cm^{-1} with a bandgap of 0.334 eV, thus displaying anomalous temperature dependence that aligned with theoretical predictions (Villegas et al., 2016; Huang et al., 2020).

Building upon this foundation, another study conducted a comprehensive investigation into the quantum yield of PL in BP with varying thicknesses (Higashitarumizu et al., 2023). Their findings shed light on the nature of radiative emission in bulk BP, which primarily arises from free carrier recombination. As the thickness of BP decreases from bulk to approximately 4 nm, an observed decrease in the PL quantum yield is attributed to heightened surface carrier recombination. Notably, the surge in the PL quantum yield is observed as the thickness scales further, reaching an average value of approximately 30% for monolayers. This phenomenon is attributed to the prevalence of radiative excitonic recombination.

4.2 Tuning photoluminescence emission wavelength of black phosphorus

As outlined in Section 3, effective manipulation of the electronic band structure of black phosphorus (BP) can be achieved through various techniques such as mechanical strain, the Franz–Keldysh effect, Stark effect, and alloying. These approaches have gained significant attention as researchers explore the potential of extending the MIR emission wavelength from BP into the longer MIR region. For instance, a dual-gate hBN/BP/hBN structure was used to induce a vertical displacement field, resulting in the Franz–Keldysh effect or Stark effect (Chen et al., 2020). Through the application of a moderate displacement field of up to 0.48 V/nm, researchers were able to achieve continuous tuning of the PL peak in a ∼20-layer BP flake from 3.7 to 7.7 μm (Fig. 9a, b). However, it is important to note that the increased displacement field leads to a decrease in dipole oscillator strength, resulting in reduced PL intensity.

In order to achieve strain tuning of BP's PL, an alternative approach involves applying mechanical strain through stretching or compressing flexible substrates such as polyethylene terephthalate (PET), polyethylene terephthalate glycol (PETG), or polydimethylsiloxane (PDMS) with BP positioned on top (Kim et al., 2021). This strain engineering technique enables precise control over the emission wavelength of BP, ranging from 2.3 μm to 5.5 μm, with a remarkable modulation rate of 1.70 μm per percentage of applied strain. Notably, this represents the highest modulation rate reported for direct bandgap semiconductors, and the wide-range

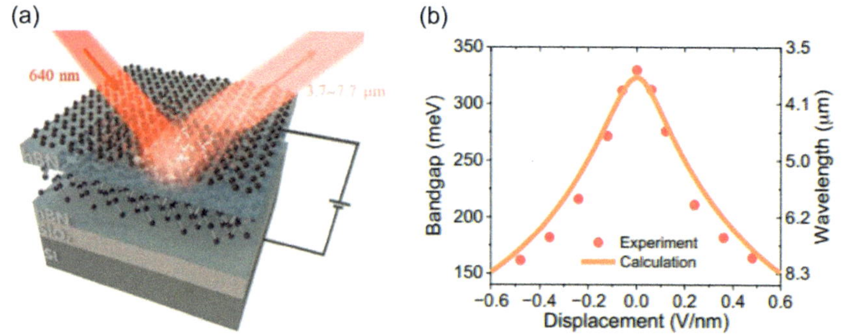

Fig. 9 (a) A schematic of the dual-gate device for tunning emission wavelength of BP. (b) Electric field-induced bandgap/wavelength tuning of the 20-layer BP. *With permission from Chen, C., Lu, X., Deng, B., Chen, X., Guo, Q., Li, C., Ma, C., et al., 2020. Widely tunable mid-infrared light emission in thin-film black phosphorus. Sci. Adv. 6, eaay6134. © Copyright 2020 by The Authors.*

and reversible tunability of MIR emissions achieved through strain engineering opens up possibilities for multiplexed nondispersive gas sensing.

The extensive research on the MIR emission properties of BP has yielded valuable insights, and the exploration of MIR emission from BP alloys emerged as a new area of investigation in 2023 (Chen et al., 2023), revealing that the incorporation of arsenic into BP induces a red-shift in the PL emissions. Specifically, alloys such as bulk $As_{0.25}P_{0.75}$ and b-$As_{0.46}P_{0.54}$ exhibited PL peaks at approximately 4 μm and 4.6 μm, respectively (Fig. 10a, b). This red-shift is attributed to a reduction in the bandgap resulting from an increased As/P ratio. Furthermore, owing to the puckered crystal structure shared by $As_{0.25}P_{0.75}$ and b-$As_{0.46}P_{0.54}$, their emissions display a feature of linear dichroism, characterized by a degree of PL linear polarization exceeding 80%. This finding highlights the potential of utilizing b-AsP alloys to develop linearly polarized MIR sources.

While significant progress has been made in understanding BP alloys, the investigation of MIR emission from other BP-based materials with smaller band gaps, such as b-$As_{0.83}P_{0.17}$ and black phosphorus carbide, remains relatively unexplored. These materials likely exhibit MIR emission wavelengths falling within the second MIR atmospheric transmission window (Liu et al., 2015a, Yang et al., 2022), ranging from 8 to 14 μm. Exploring the MIR emission properties of these materials represents an exciting frontier for future research, offering opportunities to advance MIR optoelectronics.

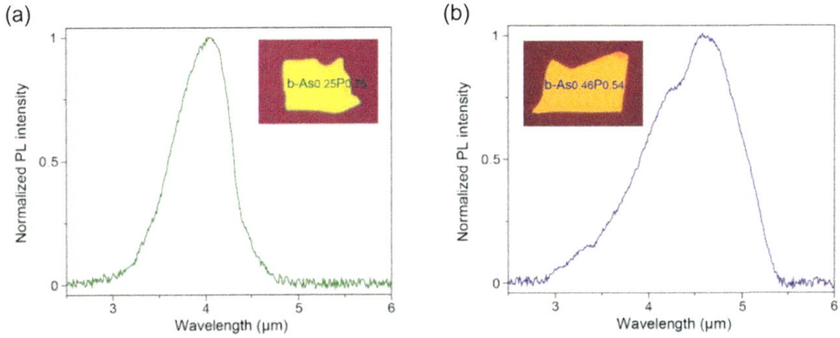

Fig. 10 (a) The PL emission spectra were measured from two flakes: (a) a 50 nm-thick b-$As_{0.25}P_{0.75}$ flake, and (b) a 35 nm-thick b-$As_{0.46}P_{0.54}$ flake. *With permission from Chen, P.-L., Chang, T.-Y., Chen, P.-S., Chan, A. H.-Y., Rosyadi, A. S., Lin, Y.-J., et al., 2023. Van der waals heterostructure mid-infrared emitters with electrically controllable polarization states and spectral characteristics. ACS Nano 17, 10181–10190. © copyright 2023 by the American Chemical Society.*

4.3 Optically pumped MIR laser based on black phosphorus

Extensive research has been conducted on optically-pumped lasing from 2D materials, with a particular emphasis on transition-metal dichalcogenides, like MoS_2, WSe_2 and $MoTe_2$. Previous studies have integrated these materials with different types of optical resonators, such as nanobeams, microdisk resonators, and photonic crystals, resulting in lasing behavior predominantly observed in the visible to NIR range (Wu et al., 2015; Salehzadeh et al., 2015; Li et al., 2017; Ye et al., 2015). Confirmation of lasing has been achieved through observations of superlinear intensity increase, spectral linewidth narrowing, and coherence times. Efforts have also been made to enhance the electroluminescence (EL) of TMDCs by incorporating them with photonic crystal cavities (Liu et al., 2017a) and dielectric distributed Bragg reflectors (Del Pozo-Zamudio et al., 2020), leading to improved emission characteristics. However, the development of electrically-pumped nanolasers using TMDCs remains a challenge. On the other hand, BP, known for its remarkable light–matter interaction properties (Elbanna et al., 2023; Wang and Lan, 2016), holds great potential for efficient light amplification and stimulated emission, making it an attractive candidate for laser applications in the MIR region. BP-based MIR optically-pumped lasers have been successfully realized using various optical resonators, including Fabry–Perot cavities and distributed feedback structures, offering advantages such as low-threshold operation and tunable emission wavelengths within the MIR range.

For instance, a recent study focused on harnessing the potential of BP nanoflakes as the gain medium in a wavelength-tunable MIR laser (Zhang et al., 2020). They achieved this by integrating 100-nm-thick BP nanoflakes into an open Si/SiO_2 distributed Bragg reflector (DBR) cavity and employing a femtosecond laser with a wavelength of 1520 nm for pumping. The results revealed a distinctive emission line at 3611 nm with nearly perfect linear polarization. The emission peak exhibited a narrow full-width at half-maximum (FWHM) of 7 nm, which is a characteristic feature of cavity mode emission. The lasing threshold was found to be $1.3\ mJ/cm^2$ per pulse, corresponding to an average power of $1.3\ W/cm^2$. Additionally, one notable aspect of the study was the ability to tune the laser emission wavelength by controlling the thickness of the BP layer, thereby adjusting the length of the cavity. This enabled wavelength tuning from 3425 to 4068 nm, corresponding to the fifth- and sixth-order cavity modes. In a different investigation, researchers successfully achieved the development of a surface-emitting laser operating in the MIR range (Huang et al., 2019b). Their approach involved the incorporation of BP as the gain material within an integrated SiO_2/Si_3N_4 open microcavity situated on a silicon substrate (Fig. 11a). Optically pumping the device with a Ti:sapphire laser operating at 740 nm resulted in the observation of lasing behavior. This achievement was confirmed by two key findings: first, the nonlinear response observed in the peak intensity as a function of pumping power, indicating the onset of stimulated emission; and second, the

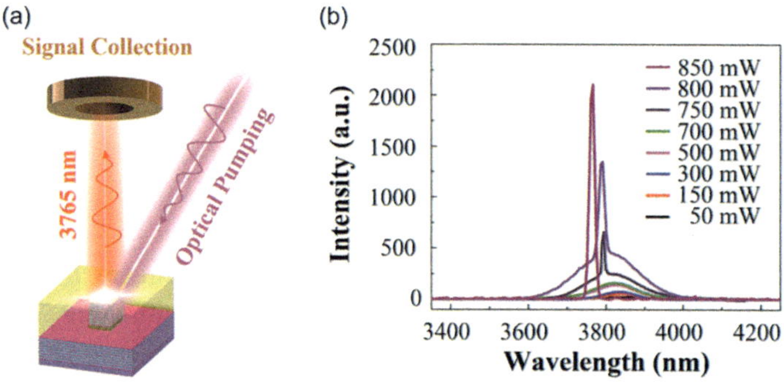

Fig. 11 (a) A schematic of an optically pumped MIR laser featuring BP integrated with SiO_2/Si_3N_4 open microcavity. (b) Emission spectra of the BP laser device obtained at room temperature under different pumping powers. *With permission from Huang, Y., Ning, J., Chen, H., Xu, Y., Wang, X., Ge, X., et al., 2019b. Mid-infrared black phosphorus surface-emitting laser with an open microcavity. ACS Photonics 6, 1581–1586. © copyright 2019 by the American Chemical Society.*

significant reduction in emission linewidth to ~9 nm, indicating the establishment of laser action. It is noteworthy that the lasing behavior was observed at a relatively high excitation power of 800 mW (Fig. 11b). Expanding upon similar device concepts, researchers further integrated b-As_xP_{1-x} with DBR cavities to successfully demonstrate optically-pumped nanolasers (Zhang et al., 2023). By manipulating the stoichiometry ($x = 0$–0.5), the bandgap of b-AsP compounds could be selectively tuned from 0.35 to 0.25 eV. This allowed for the selective modification of MIR lasing wavelengths from 3.42 to 4.65 μm, effectively covering the range of the first atmospheric window.

5. MIR light emitting diodes based on black phosphorus and their applications

A deep understanding of the PL properties of semiconductors is crucial for the development of efficient electroluminescent (EL) devices. Previous studies provide valuable insights into the fundamental PL and photocarrier characteristics of BP, such as emission wavelength, carrier lifetime and radiative recombination mechanisms (Chen et al., 2019; Miao et al., 2018; Ge et al., 2015; Zereshki et al., 2018). The understanding enables the optimization of material properties, device structures, and fabrication techniques, leading to improved EL performance and the realization of practical applications in various fields such as sensing and optoelectronics. This section will discuss the recent advancements and progress in the field of BP-based light emitting diodes (LEDs), highlighting the significant breakthroughs achieved in understanding and harnessing the EL properties of BP.

5.1 MIR light emitting diodes based on black phosphorus

In 2020, three studies emerged as the first demonstrations of EL utilizing BP as the active material. One research group made a contribution by developing a heterojunction structure (Wang et al., 2020), which involved stacking a 10-nm-thick MoS_2 layer onto a 70-nm-thick BP layer. Through the incorporation of p- and n-doping in BP and MoS_2, respectively, a type-II band alignment can be formed at the interface (Fig. 12a). Operating the heterojunction under forward bias, the induced current was primarily attributed to the thermionic crossing of the barrier by electrons from MoS_2 to BP, resulting in EL generation with a peak wavelength of 3.68 μm at room temperature (Fig. 12b). Additionally, due to the anisotropic crystal

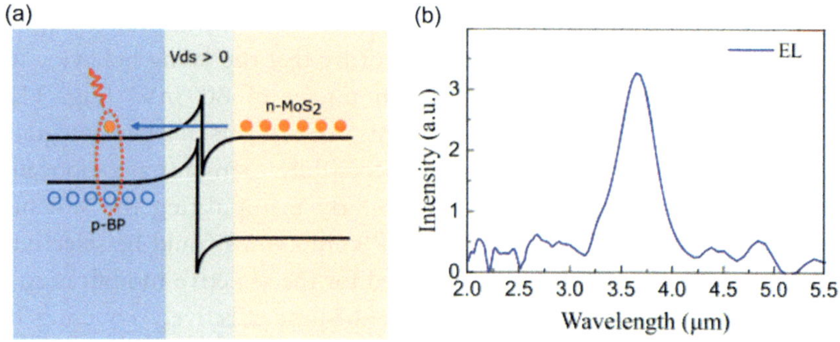

Fig. 12 (a) A schematic of band diagram of BP/MoS_2 heterostructure MIR LED. (b) EL spectrum measured from the BP/MoS_2 heterostructure LED. *With permission from (Wang et al., 2020). © copyright 2020 by the American Chemical Society.*

structure of BP, the MIR emission is linearly polarized with a degree of polarization of approximately 50%. These findings were further supported by an independent study (Zong et al., 2020). But it is worth noting that the band minimum of BP and MoS_2 are at distinct locations in the K space of the Brillouin zone. This poses challenges in achieving efficient transport across the heterojunction, necessitating the involvement of phonons. Furthermore, the injection and transport of electrons could be influenced by the MoS_2/metal Schottky contact and resistive MoS_2 channel. Such factors may have implications for the quantum efficiency of emitters.

In another notable development (Chang et al., 2020b), researchers presented an alternative approach for developing MIR LEDs using BP-based vdW heterostructures. Their design involved using two graphite contacts to sandwich an 87-nm-thick BP flake, as schematically shown in Fig. 13a. The BP layer served as the emitting layer, while the graphite layers acted as highly conducting electrodes. When a voltage was applied across the graphite contacts, EL was generated due to ambipolar charge injection. The resulting EL exhibited a similar emission peak wavelength of approximately 3.64 μm (Fig. 13b) and a linear dichroic feature. However, it should be noted that the observed EL spectral linewidth was broader than that of BP's PL, suggesting that the injected carriers occupied higher energy states in the momentum space of BP (Li et al., 2015). Furthermore, the use of graphite as a carrier injection layer provided several advantages (Chang et al., 2016; Chang et al., 2014), including high conductivity, high transparency in the MIR region, and low parasitic resistances with both BP and metal electrodes. These features allowed the developed MIR LEDs to

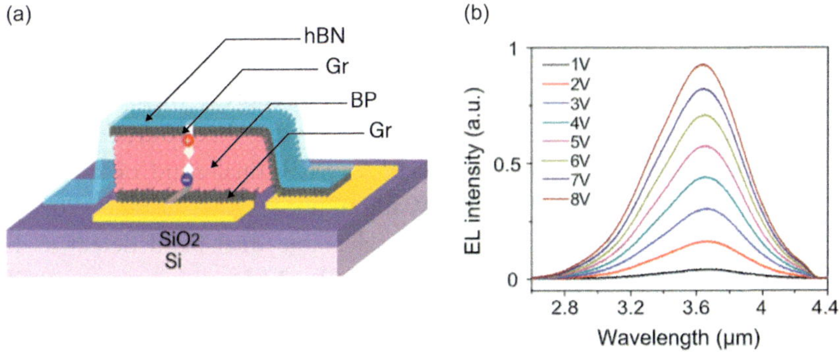

Fig. 13 (a) A schematic of MIR LED, composed of vertically stacked hexagonal boron nitride/graphene/BP/graphene heterostructures. Electrons and holes injected from two graphene contacts recombine in the BP flake, resulting in electroluminescence. (b) Bias dependence of EL spectra. *With permission from Chang, T.-Y., Chen, Y., Luo, D.-I., Li, J.-X., Chen, P.-L., Lee, S., et al., 2020b. Black phosphorus mid-infrared light-emitting diodes integrated with silicon photonic waveguides. Nano Lett. 20, 6824–6830. © copyright 2020 by the American Chemical Society.*

sustain high current density (up to 5.58×10^5 mA/cm^2), offer high MHz range modulation speed, ensure stable operation for at least 100 h, and achieve a notable extrinsic quantum efficiency of approximately 0.8%.

5.2 Bias-selectable MIR light emitting diodes using black phosphorus-based heterostructures

Over the past few decades, considerable efforts have been dedicated to the realization of bias-selectable dual-band photodetectors, sophisticated optoelectronic devices designed to detect light in two distinct wavelength bands (Haddadi et al., 2017; Rogalski et al., 2017; Rogalski, 2002). These photodetectors consist of two back-to-back connected p–n junctions, with each junction sensitive to different wavelength ranges. This configuration enables independent biasing of each junction, allowing for the selective detection of light in specific spectral regions. When operating, the application of a bias voltage to one junction results in forward-biasing, effectively collecting and detecting photons within the desired wavelength range. Simultaneously, the other junction remains reverse-biased, acting as a barrier to prevent unwanted light from contributing to the photocurrent. By adjusting the polarity of the bias voltage, the photodetector's sensitivity to different wavelength bands can be dynamically controlled, offering a versatile and tunable solution for dual-band sensing applications. Traditionally, these bias-selectable detectors have been achieved using III–V or

II–VI materials as photoactive elements. However, such implementations often encounter challenges related to complex device structures and thermal or lattice mismatches at heterointerfaces. To address these limitations, researchers have turned to the advantages of vdW heterostructures, successfully demonstrating bias-selectable NIR/MIR dual-band or dual-color MIR detectors (Chang et al., 2022; Wang et al., 2023; Wu et al., 2022).

Recently, the development of bias-selectable dual-band detectors has inspired the advancement of bias-selectable LEDs, sharing a similar underlying concept. While both devices employ multiple junctions within their structures, their key distinction lies in their operation modes. The dual-band detector utilizes junctions operating in the reverse bias region, while the LED incorporates junctions operating in the forward bias region. The related development is depicted in Fig. 14, where the schematic of a vdW heterostructures bias-selectable LED is illustrated (Chen et al., 2023). This design comprises two back-to-back p–n junctions, formed by stacking anisotropic light-emitting materials, namely BP and b-$As_{0.46}P_{0.54}$, with the MoS_2. A key characteristic of MoS_2 lies in its substantial valence band discontinuity but almost negligible conduction band offset with the adjacent BP and b-$As_{0.46}P_{0.54}$ layer. This feature allows electrons to flow unimpeded throughout the emitter. But the movement of holes is

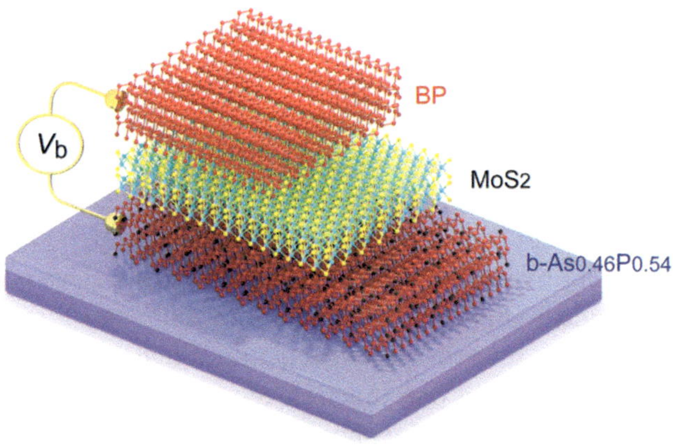

Fig. 14 Device structure of the bias-selectable MIR LED, composed of vertically stacked BP/MoS_2/b-AsP heterostructures. *With permission from Chen, P.-L., Chang, T.-Y., Chen, P.-S., Chan, A. H.-Y., Rosyadi, A. S., Lin, Y.-J., et al., 2023. Van der waals heterostructure mid-infrared emitters with electrically controllable polarization states and spectral characteristics. ACS Nano 17, 10181–10190. © copyright 2023 by the American Chemical Society.*

effectively obstructed by the barrier at the BP/MoS_2 (b-$As_{0.46}P_{0.54}$/MoS_2) interface when a positive (negative) bias voltage is applied across the two light emitting layers, with b-$As_{0.46}P_{0.54}$ connected to the ground (Chang et al., 2022; Bullock et al., 2018; Deng et al., 2014). This strategic arrangement ensures that the location of electron–hole pair recombination is linked to the polarity of the applied bias, thereby offering the possibility of tuning the MIR emission wavelength (Fig. 15a, b). Furthermore, the authors demonstrate that operating the emitter under the polarity-switched pulse mode, with two pulses of opposite polarity in a cycle, results in time-averaged EL exhibiting the characteristics of both BP and b-$As_{0.46}P_{0.54}$. The emission shows the feature of broad spectral coverage, spanning the entire first MIR atmospheric window (λ: 3–5 μm), and is electrically tunable, with versatile spectral shape, offering a versatile and adaptive solution for various MIR applications.

5.3 Leveraging integrability of black phosphorus light emitting diodes for advanced MIR optoelectronics

2D or vdW materials offer a crucial advantage in optoelectronic applications: their integrability onto diverse platforms. Unlike III–V or II–VI devices, the vdW nature of these materials enables seamless integration with different substrates and device architectures, eliminating challenges related to lattice or thermal mismatches (Liu et al., 2019; Youngblood and Li, 2017). In the field of MIR research, significant progress has been achieved

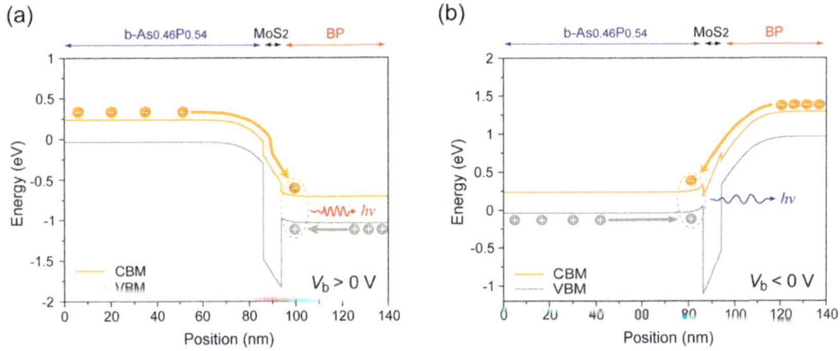

Fig. 15 (a) Energy band diagrams of the bias-selectable MIR LED, operated in (a) forward, and (b) reverse bias region, respectively. *With permission from Chen, P.-L., Chang, T.-Y., Chen, P.-S., Chan, A. H.-Y., Rosyadi, A. S., Lin, Y.-J., et al., 2023. Van der waals heterostructure mid-infrared emitters with electrically controllable polarization states and spectral characteristics. ACS Nano 17, 10181–10190. © copyright 2023 by the American Chemical Society.*

in the integration of MIR LEDs with various photonics structures or flexible device substrates. This integration has resulted in hybrid platforms that not only provide enhanced functionality and improved performance but also offer a wide range of applications in on-chip sensing, communications, and spectroscopy.

As an example, conventional MIR BP-based LEDs are often fabricated on SiO_2/Si substrates, which can lead to significant outcoupling losses. This occurs because the emitted light tends to primarily radiate towards the high refractive index substrate rather than into the air (Wang et al., 2020). However, an effective solution to this challenge involves transferring BP-based LEDs, consisting of ITO/BP/MoS_2, onto an optical cavity comprising gold and Al_2O_3 films (Gupta et al., 2022). By harnessing the interference effect (Fig. 16a), the emitted MIR light is redirected into the upper half space, resulting in an enhancement in the outcoupling power by one order of magnitude (Fig. 16b). The demonstrated cavity-integrated BP-LEDs exhibit external quantum efficiencies and wall-plug efficiencies of up to 4.43% and 1.78%, respectively.

Another excellent example of harnessing the integrability of vdW materials lies in the development of on-chip MIR light sources (Chang et al., 2020b). In this work, a BP LED, consisting of graphene/BP/graphene heterostructures, was directly transferred onto a MIR waveguide

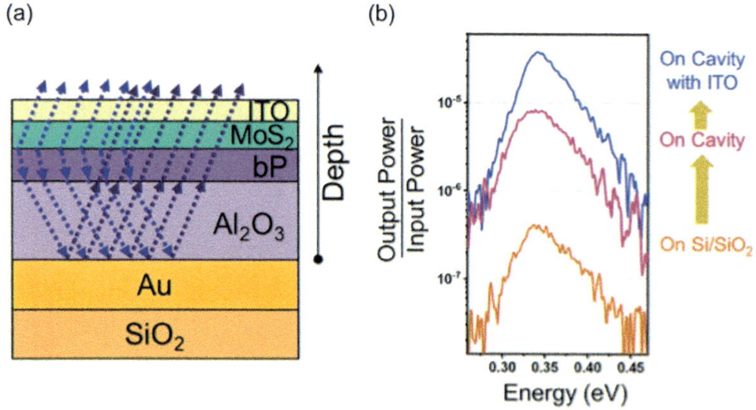

Fig. 16 (a) BP-based MIR LED integrated with an open cavity, leading to the enhanced outcoupling efficiency as depicted in (b). *With permission from Gupta, N., Kim, H., Azar, N. S., Uddin, S. Z., Lien, D.-H., Crozier, K.B., et al. 2022. Bright mid-wave infrared resonant-cavity light-emitting diodes based on black phosphorus. Nano Lett., 22, 1294–1301. © copyright 2022 by the American Chemical Society.*

with two grating couplers at its ends, specifically designed for optimal performance in the 3–4 μm MIR region (Fig. 17a). By applying a bias voltage to the BP LED and conducting scanning EL measurements, researchers observed three bright spots. The central and brighter spot corresponded to the position of the BP LED, while the two additional emission spots were located at the grating couplers. The observed results provide compelling evidence of the successful evanescent coupling of MIR EL from the BP LED to the silicon photonic waveguide (Fig. 17b). This coupled light efficiently propagates through the waveguide, reaching the coupler, where it is redirected into free space. To achieve even higher coupling efficiency, researchers suggest that the advanced photonic structures like slot waveguides or photonic crystal waveguides (Bie et al., 2017) can be used to integrated with BP LEDs.

The vdW nature of BP LEDs also allows for their integration with flexible substrates like PETG or PET, providing enhanced versatility and adaptability in device design. By applying mechanical strain to the flexible substrates, which house the MIR LEDs comprising BP/MoS_2 heterostructures, the BP bandgap can be dynamically altered, leading to a tunable EL emission wavelength ranging from approximately 2.5 to 4 μm (Kim et al., 2021). This exceptional tunability of the emission wavelength

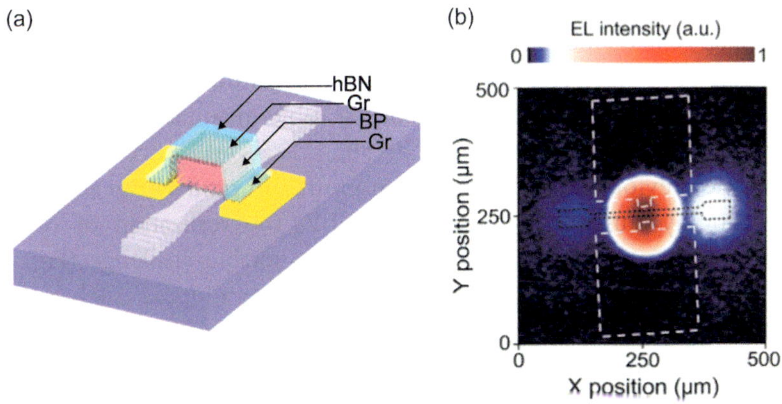

Fig. 17 (a) BP-based MIR LED integrated with a silicon waveguide with grating couplers at two ends. (b) The scanning EL map indicates the emission of BP LED can be evanescently coupled into the waveguide and then propagate on the silicon chip. *With permission from Chang, T.-Y., Chen, Y., Luo, D.-I., Li, J.-X., Chen, P.-L., Lee, S., et al., 2020b. Black phosphorus mid-infrared light-emitting diodes integrated with silicon photonic waveguides. Nano Lett. 20, 6824–6830. © copyright 2020 by the American Chemical Society.*

presents exciting prospects for multiplexed nondispersive infrared gas sensing applications. With just a single light source, this flexible and strain-tunable BP LED can effectively detect and identify various target gases, such as carbon dioxide, methane, and water vapor, by discerning their distinct spectral signatures in the MIR region.

6. Summary and outlook

In this chapter, we have highlighted BP and its alloys as an exciting platform to explore excitonic physics in narrow direct gap materials and advance MIR emitters. The unique vdW nature of these materials has facilitated their seamless integration with diverse device platforms, unlocking enhanced functionality and opening new possibilities for various applications. While BP-based emitters hold great promise for advancing MIR emission applications, several challenges must be addressed to fully unlock their potential (Fig. 18).

Firstly, the air instability of BP and b-AsP necessitates the implementation of advanced packaging techniques to prevent the materials from degradation and maintain their performance over time. Developing reliable encapsulation methods is essential to ensure the long-term stability and reliability of these devices in practical applications (Peng et al., 2020; Wood et al., 2014). Secondly, achieving large-scale and uniform thick BP and b-AsP materials poses a current challenge, calling for the exploration of novel materials growth methods to enable scalable manufacturing processes (Eswaraiah et al., 2016; Liu et al., 2015b).

While previous research has shown promising results in generating EL and optically-pumped lasing based on BP, the realization of on-chip electrically-pumped MIR nanolasers poses a notable challenge (Fig. 18). To overcome this, multiple approaches could be adopted, such as using high-Q resonators to enhance light–matter interactions, improving the quality of BP to reduce defect densities, and engineering heterostructures to enhance carrier injection efficiency. Notably, several studies have successfully demonstrated waveguide-integrated BP modulators and detectors (Chen et al., 2022; Li et al., 2020; Huang et al., 2021). These advancements, combined with the potential development of MIR BP-based on-chip lasers, represent a significant step towards harnessing BP-based vdW heterostructures for MIR silicon photonics.

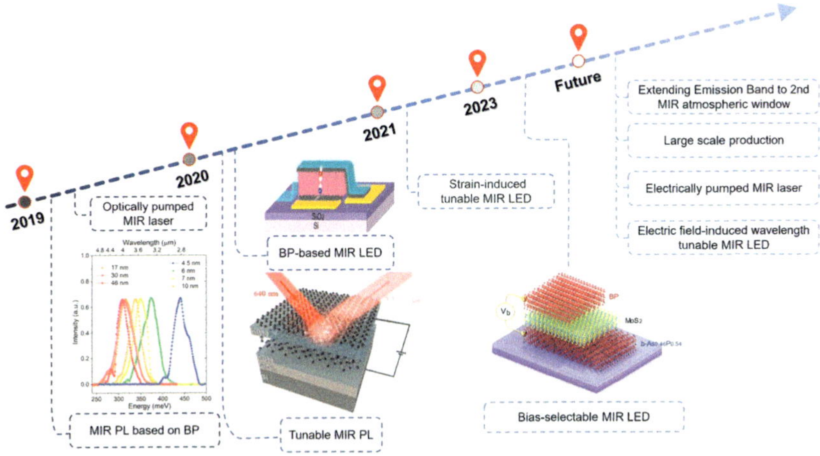

Fig. 18 Roadmap for BP-based MIR emitters. *With permission from Chen, C., Chen, F., Chen, X., Deng, B., Eng, B., Jung, D., et al., 2019. Bright mid-infrared photoluminescence from thin-film black phosphorus. Nano Lett., 19, 1488–1493; Chang, T.-Y., Chen, Y., Luo, D.-I., Li, J.-X., Chen, P.-L., Lee, S., et al., 2020b. Black phosphorus mid-infrared light-emitting diodes integrated with silicon photonic waveguides. Nano Lett. 20, 6824–6830; Chen, C., Lu, X., Deng, B., Chen, X., Guo, Q., Li, C., et al., 2020. Widely tunable mid-infrared light emission in thin-film black phosphorus. Sci. Adv. 6, eaay6134; Chen, P.-L., Chang, T.-Y., Chen, P.-S., Chan, A. H.-Y., Rosyadi, A. S., Lin, Y.-J., et al., 2023. Van der waals heterostructure mid-infrared emitters with electrically controllable polarization states and spectral characteristics. ACS Nano 17, 10181–10190. © copyright 2019 by the American Chemical Society, © copyright 2020 by the American Chemical Society, © Copyright 2020 by The Authors, © copyright 2023 by the American Chemical Society.*

Another notable area of investigation involves a dual-gating configuration with few-layer BP sandwiched between bottom and top hBN flakes. This setup has demonstrated the generation of bright and wide-tunable PL spanning from 3.7 to 7.7 μm as discuss in Section 4.2 (Chen et al., 2020). However, achieving electrically induced widely tunable MIR EL remains a challenging task (Fig. 18). Successfully accomplishing these feats would lead to the realization of miniature MIR tunable LEDs, with broad applications in tunable light sources, spectroscopy, and on-chip sensing.

Furthermore, current demonstrations of PL or EL from BP devices have been confined to the first MIR atmospheric windows (3–5 μm). Extending the emission wavelength to cover the second MIR atmospheric windows (8–14 μm) would significantly broaden the range of applications, particularly in areas such as gas sensing, thermal imaging, and free-space optical communications (Fig. 18). The use of BP alloys, heavily doped with arsenic or carbon, offers a potential avenue to achieve this goal (Liu et al., 2015a; Yang et al., 2022; Wang and Lan, 2016; Tan et al., 2018), and this requires further investigation. As we continue to explore these frontiers, the potential for BP and b-AsP devices to revolutionize MIR emission technologies remains both promising and exciting.

References

Amani, M., Regan, E., Bullock, J., Ahn, G.H., Javey, A., 2017. Mid-wave infrared photoconductors based on black phosphorus-arsenic alloys. ACS Nano 11, 11724–11731.

Baker, M.J., Trevisan, J., Bassan, P., Bhargava, R., Butler, H.J., Dorling, K.M., et al., 2014. Using fourier transform ir spectroscopy to analyze biological materials. Nat. Protoc. 9, 1771–1791.

Bhimanapati, G.R., Lin, Z., Meunier, V., Jung, Y., Cha, J., Das, S., et al., 2015. Recent advances in two-dimensional materials beyond graphene. ACS Nano 9, 11509–11539.

Bie, Y.-Q., Grosso, G., Heuck, M., Furchi, M.M., Cao, Y., Zheng, J., et al., 2017. A $MoTe_2$-based light-emitting diode and photodetector for silicon photonic integrated circuits. Nat. Nanotechnol. 12, 1124–1129.

Bullock, J., Amani, M., Cho, J., Chen, Y.-Z., Ahn, G.H., Adinolfi, V., et al., 2018. Polarization-resolved black phosphorus/molybdenum disulfide mid-wave infrared photodiodes with high detectivity at room temperature. Nat. Photonics 12, 601–607.

Castellanos-Gomez, A., 2015. Black phosphorus: narrow gap, wide applications. J. Phys. Chem. Lett. 6, 4280–4291.

Chang, T.-Y., Chen, P.-L., Chen, P.-S., Li, W.-Q., Li, J.-X., He, M.-Y., et al., 2022. Van der waals heterostructure photodetectors with bias-selectable infrared photoresponses. ACS Appl. Mater. Interfaces 14, 32665–32674.

Chang, T.-Y., Chen, P.-L., Yan, J.-H., Li, W.-Q., Zhang, Y.-Y., Luo, D.-I., et al., 2020a. Ultra-broadband, high speed, and high-quantum-efficiency photodetectors based on black phosphorus. ACS Appl. Mater. Interfaces 12, 1201–1209.

Chang, T.-Y., Chen, Y., Luo, D.-I., Li, J.-X., Chen, P.-L., Lee, S., et al., 2020b. Black phosphorus mid-infrared light-emitting diodes integrated with silicon photonic waveguides. Nano Lett. 20, 6824–6830.

Chang, Y.-C., Liu, C.-H., Liu, C.-H., Zhang, S., Marder, S.R., Narimanov, E.E., et al., 2016. Realization of mid-infrared graphene hyperbolic metamaterials. Nat. Commun. 7, 10568.

Chang, Y.-C., Liu, C.-H., Liu, C.-H., Zhong, Z., Norris, T.B., 2014. Extracting the complex optical conductivity of mono- and bilayer graphene by ellipsometry. Appl. Phys. Lett. 104, 261909.

Chaves, A., Azadani, J.G., Alsalman, H., Da Costa, D.R., Frisenda, R., Chaves, A.J., et al., 2020. Bandgap engineering of two-dimensional semiconductor materials. NPJ 2D Mater. Appl. 4, 29.

Chen, C., Chen, F., Chen, X., Deng, B., Eng, B., Jung, D., et al., 2019. Bright mid-infrared photoluminescence from thin-film black phosphorus. Nano Lett. 19, 1488–1493.

Chen, C., Lu, X., Deng, B., Chen, X., Guo, Q., Li, C., et al., 2020. Widely tunable mid-infrared light emission in thin-film black phosphorus. Sci. Adv. 6 eaay6134.

Chen, P.-L., Chang, T.-Y., Chen, P.-S., Chan, A.H.-Y., Rosyadi, A.S., Lin, Y.-J., et al., 2023. Van der waals heterostructure mid-infrared emitters with electrically controllable polarization states and spectral characteristics. ACS Nano 17, 10181–10190.

Chen, P.-L., Chen, Y., Chang, T.-Y., Li, W.-Q., Li, J.-X., Lee, S., et al., 2022. Waveguide-integrated van der waals heterostructure mid-infrared photodetector with high performance. ACS Appl. Mater. Interfaces 14, 24856–24863.

Chu, M., Sun, Y., Aghoram, U., Thompson, S.E., 2009. Strain: a solution for higher carrier mobility in nanoscale MOSFETS. Annu. Rev. Mater. Res. 39, 203–229.

Del Pozo-Zamudio, O., Genco, A., Schwarz, S., Withers, F., Walker, P.M., Godde, T., et al., 2020. Electrically pumped WSe_2-based light-emitting van der waals heterostructures embedded in monolithic dielectric microcavities. 2D Mater. 7, 031006.

Deng, Y., Luo, Z., Conrad, N.J., Liu, H., Gong, Y., Najmaei, S., et al., 2014. Black phosphorus–monolayer MoS_2 van der waals heterojunction p–n diode. ACS Nano 8, 8292–8299.

Elbanna, A., Jiang, H., Fu, Q., Zhu, J.-F., Liu, Y., Zhao, M., et al., 2023. 2D material infrared photonics and plasmonics. ACS Nano 17, 4134–4179.

Eswaraiah, V., Zeng, Q., Long, Y., Liu, Z., 2016. Black phosphorus nanosheets: synthesis, characterization and applications. Small 12, 3480–3502.

Fang, Y., Ge, Y., Wang, C., Zhang, H., 2020. Mid-infrared photonics using 2D materials: status and challenges. Laser Photonics Rev. 14, 1900098.

Ge, S., Li, C., Zhang, Z., Zhang, C., Zhang, Y., Qiu, J., et al., 2015. Dynamical evolution of anisotropic response in black phosphorus under ultrafast photoexcitation. Nano Lett. 15, 4650–4656.

Geim, A.K., Grigorieva, I.V., 2013. Van der waals heterostructures. Nature 499, 419–425.

Godard, A., 2007. Infrared (2–12 μm) solid-state laser sources: a review. C. R. Phys. 8, 1100–1128.

Gupta, N., Kim, H., Azar, N.S., Uddin, S.Z., Lien, D.-H., Crozier, K.B., et al., 2022. Bright mid-wave infrared resonant-cavity light-emitting diodes based on black phosphorus. Nano Lett. 22, 1294–1301.

Haddadi, A., Dehzangi, A., Chevallier, R., Adhikary, S., Razeghi, M., 2017. Bias–selectable nBn dual–band long–/very long–wavelength infrared photodetectors based on $InAs/InAs_{1-x}Sb_x/AlAs_{1-x}Sb_x$ type–II superlattices. Sci. Rep. 7, 3379.

Higashitarumizu, N., Uddin, S.Z., Weinberg, D., Azar, N.S., Reaz Rahman, I.K.M., Wang, V., et al., 2023. Anomalous thickness dependence of photoluminescence quantum yield in black phosphorous. Nat. Nanotechnol. 18, 507–513.

Hu, H., Shi, Z., Khan, K., Cao, R., Liang, W., Tareen, A.K., et al., 2020. Recent advances in doping engineering of black phosphorus. J. Mater. Chem. A 8, 5421–5441.

Huang, L., Dong, B., Yu, Z.G., Zhou, J., Ma, Y., Zhang, Y.W., et al., 2021. Mid-infrared modulators integrating silicon and black phosphorus photonics. Mater. Today Adv. 12, 100170.

Huang, S., Wang, F., Zhang, G., Song, C., Lei, Y., Xing, Q., et al., 2020. From anomalous to normal: temperature dependence of the band gap in two-dimensional black phosphorus. Phys. Rev. Lett. 125, 156802.

Huang, S., Zhang, G., Fan, F., Song, C., Wang, F., Xing, Q., et al., 2019a. Strain-tunable van der waals interactions in few-layer black phosphorus. Nat. Commun. 10, 2447.

Huang, Y., Ning, J., Chen, H., Xu, Y., Wang, X., Ge, X., et al., 2019b. Mid-infrared black phosphorus surface-emitting laser with an open microcavity. ACS Photonics 6, 1581–1586.

Jacobsen, R.S., Andersen, K.N., Borel, P.I., Fage-Pedersen, J., Frandsen, L.H., Hansen, O., et al., 2006. Strained silicon as a new electro-optic material. Nature 441, 199–202.

Jariwala, D., Marks, T.J., Hersam, M.C., 2017. Mixed-dimensional van der waals heterostructures. Nat. Mater. 16, 170–181.

Jung, D., Bank, S., Lee, M.L., Wasserman, D., 2017. Next-generation mid-infrared sources. J. Opt. 19, 123001.

Kim, H., Uddin, S.Z., Lien, D.-H., Yeh, M., Azar, N.S., Balendhran, S., et al., 2021. Actively variable-spectrum optoelectronics with black phosphorus. Nature 596, 232–237.

Kim, J., Baik, S.S., Ryu, S.H., Sohn, Y., Park, S., Park, B.-G., et al., 2015. Observation of tunable band gap and anisotropic dirac semimetal state in black phosphorus. Science 349, 723–726.

Krier, A., 2006. Mid-infrared Semiconductor Optoelectronics. Springer Series in Optical Sciences Springer-Verlag.
Lee, M.L., Fitzgerald, E.A., Bulsara, M.T., Currie, M.T., Lochtefeld, A., 2004. Strained Si, SiGe, and Ge channels for high-mobility metal-oxide-semiconductor field-effect transistors. J. Appl. Phys. 97, 011101.
Lei, W., Liu, G., Zhang, J., Liu, M., 2017. Black phosphorus nanostructures: recent advances in hybridization, doping and functionalization. Chem. Soc. Rev. 46, 3492–3509.
Li, D., Cheng, R., Zhou, H., Wang, C., Yin, A., Chen, Y., et al., 2015. Electric-field-induced strong enhancement of electroluminescence in multilayer molybdenum disulfide. Nat. Commun. 6, 7509.
Li, J., Liu, C., Chen, H., Guo, J., Zhang, M., Dai, D., 2020. Hybrid silicon photonic devices with two-dimensional materials. Nanophotonics 9, 2295–2314.
Li, Y., Zhang, J., Huang, D., Sun, H., Fan, F., Feng, J., et al., 2017. Room-temperature continuous-wave lasing from monolayer molybdenum ditelluride integrated with a silicon nanobeam cavity. Nat. Nanotechnol. 12, 987–992.
Lin, H., Luo, Z., Gu, T., Kimerling, L.C., Wada, K., Agarwal, A., et al., 2018. Mid-infrared integrated photonics on silicon: a perspective. Nanophotonics 7, 393–420.
Ling, X., Wang, H., Huang, S., Xia, F., Dresselhaus, M.S., 2015. The renaissance of black phosphorus. Proc. Natl. Acad. Sci. 112, 4523–4530.
Liu, B., Köpf, M., Abbas, A.N., Wang, X., Guo, Q., Jia, Y., et al., 2015a. Black arsenic–phosphorus: layered anisotropic infrared semiconductors with highly tunable compositions and properties. Adv. Mater. 27, 4423–4429.
Liu, C.-H., Clark, G., Fryett, T., Wu, S., Zheng, J., Hatami, F., et al., 2017a. Nanocavity integrated van der waals heterostructure light-emitting tunneling diode. Nano Lett. 17, 200–205.
Liu, C.-H., Zheng, J., Chen, Y., Fryett, T., Majumdar, A., 2019. Van der waals materials integrated nanophotonic devices [invited]. Opt. Mater. Express 9, 384–399.
Liu, H., Du, Y., Deng, Y., Ye, P.D., 2015b. Semiconducting black phosphorus: synthesis, transport properties and electronic applications. Chem. Soc. Rev. 44, 2732–2743.
Liu, Y., Qiu, Z., Carvalho, A., Bao, Y., Xu, H., Tan, S.J.R., et al., 2017b. Gate-tunable giant stark effect in few-layer black phosphorus. Nano Lett. 17, 1970–1977.
Liu, Y., Weiss, N.O., Duan, X., Cheng, H.-C., Huang, Y., Duan, X., 2016. Van der waals heterostructures and devices. Nat. Rev. Mater. 1, 16042.
Miao, X., Zhang, G., Wang, F., Yan, H., Ji, M., 2018. Layer-dependent ultrafast carrier and coherent phonon dynamics in black phosphorus. Nano Lett. 18, 3053–3059.
Peng, L., Abbasi, N., Xiao, Y., Xie, Z., 2020. Black phosphorus: degradation mechanism, passivation method, and application for in situ tissue regeneration. Adv. Mater. Interfaces 7, 2001538.
Qiao, J., Kong, X., Hu, Z.-X., Yang, F., Ji, W., 2014. High-mobility transport anisotropy and linear dichroism in few-layer black phosphorus. Nat. Commun. 5, 4475.
Quereda, J., San-Jose, P., Parente, V., Vaquero-Garzon, L., Molina-Mendoza, A.J., Agraït, N., et al., 2016. Strong modulation of optical properties in black phosphorus through strain-engineered rippling. Nano Lett. 16, 2931–2937.
Razeghi, M., Nguyen, B.-M., 2014. Advances in mid-infrared detection and imaging: a key issues review. Rep. Prog. Phys. 77, 082401.
Rogalski, A., 2002. Infrared detectors: an overview. Infrared Phys. Technol. 43, 187–210.
Rogalski, A. Next decade in infrared detectors. Proc.SPIE, 2017. 104330L.
Rogalski, A., Martyniuk, P., Kopytko, M., 2017. InAs/GaSb type-II superlattice infrared detectors: future prospect. Appl. Phys. Rev. 4, 031304.
Salehzadeh, O., Djavid, M., Tran, N.H., Shih, I., Mi, Z., 2015. Optically pumped two-dimensional MoS_2 lasers operating at room-temperature. Nano Lett. 15, 5302–5306.

Scharf, B., Frank, T., Gmitra, M., Fabian, J., Žutić, I., Perebeinos, V., 2016. Excitonic stark effect in MoS_2 monolayers. Phys. Rev. B 94, 245434.
Schliesser, A., Picqué, N., Hänsch, T.W., 2012. Mid-infrared frequency combs. Nat. Photonics 6, 440–449.
Soref, R., 2010. Mid-infrared photonics in silicon and germanium. Nat. Photonics 4, 495–497.
Spott, A., Stanton, E.J., Volet, N., Peters, J.D., Meyer, J.R., Bowers, J.E., 2017. Heterogeneous integration for mid-infrared silicon photonics. IEEE J. Sel. Top. Quantum Electron. 23, 1–10.
Stanley, R., 2012. Plasmonics in the mid-infrared. Nat. Photonics 6, 409–411.
Sun, Y., Liu, K., 2018. Strain engineering in functional 2-dimensional materials. J. Appl. Phys. 125, 082402.
Tan, W.C., Huang, L., Ng, R.J., Wang, L., Hasan, D.M.N., Duffin, T.J., et al., 2018. A black phosphorus carbide infrared phototransistor. Adv. Mater. 30, 1705039.
Tran, V., Soklaski, R., Liang, Y., Yang, L., 2014. Layer-controlled band gap and anisotropic excitons in few-layer black phosphorus. Phys. Rev. B 89, 235319.
Villegas, C.E.P., Rocha, A.R., Marini, A., 2016. Anomalous temperature dependence of the band gap in black phosphorus. Nano Lett. 16, 5095–5101.
Wang, J., Rousseau, A., Yang, M., Low, T., Francoeur, S., Kéna-Cohen, S., 2020. Mid-infrared polarized emission from black phosphorus light-emitting diodes. Nano Lett. 20, 3651–3655.
Wang, J., Verzhbitskiy, I., Eda, G., 2018a. Electroluminescent devices based on 2D semiconducting transition metal dichalcogenides. Adv. Mater. 30, 1802687.
Wang, S., Ashokan, A., Balendhran, S., Yan, W., Johnson, B.C., Peruzzo, A., et al., 2023. Room temperature bias-selectable, dual-band infrared detectors based on lead sulfide colloidal quantum dots and black phosphorus. ACS Nano 17, 11771–11782.
Wang, X., Jones, A.M., Seyler, K.L., Tran, V., Jia, Y., Zhao, H., et al., 2015. Highly anisotropic and robust excitons in monolayer black phosphorus. Nat. Nanotechnol. 10, 517–521.
Wang, X., Lan, S., 2016. Optical properties of black phosphorus. Adv. Opt. Photonics 8, 618–655.
Wang, Z., Jingjing, Q., Wang, X., Zhang, Z., Chen, Y., Huang, X., et al., 2018b. Two-dimensional light-emitting materials: Preparation, properties and applications. Chem. Soc. Rev. 47, 6128–6174.
Whitney, W.S., Sherrott, M.C., Jariwala, D., Lin, W.-H., Bechtel, H.A., Rossman, G.R., et al., 2017. Field effect optoelectronic modulation of quantum-confined carriers in black phosphorus. Nano Lett. 17, 78–84.
Wood, J.D., Wells, S.A., Jariwala, D., Chen, K.-S., Cho, E., Sangwan, V.K., et al., 2014. Effective passivation of exfoliated black phosphorus transistors against ambient degradation. Nano Lett. 14, 6964–6970.
Wu, P., Ye, L., Tong, L., Wang, P., Wang, Y., Wang, H., et al., 2022. Van der waals two-color infrared photodetector. Light Sci. Appl. 11, 6.
Wu, S., Buckley, S., Schaibley, J.R., Feng, L., Yan, J., Mandrus, D.G., et al., 2015. Monolayer semiconductor nanocavity lasers with ultralow thresholds. Nature 520, 69–72.
Xia, F., Wang, H., Jia, Y., 2014a. Rediscovering black phosphorus as an anisotropic layered material for optoelectronics and electronics. Nat. Commun. 5, 4458.
Xia, F., Wang, H., Xiao, D., Dubey, M., Ramasubramaniam, A., 2014b. Two-dimensional material nanophotonics. Nat. Photonics 8, 899–907.
Xu, Y., Shi, Z., Shi, X., Zhang, K., Zhang, H., 2019. Recent progress in black phosphorus and black-phosphorus-analogue materials: properties, synthesis and applications. Nanoscale 11, 14491–14527.

Yang, S., Chen, Y., Jiang, C., 2021. Strain engineering of two-dimensional materials: Methods, properties, and applications. InfoMat 3, 397–420.

Yang, S., Peng, J., Huang, H., Li, Z., Dong, H., Wu, F., 2022. The tunable bandgap of phosphorus-arsenic alloys for mid-and long-infrared regime photodetectors. Mater. Sci. Semiconduct. 144, 106552.

Ye, Y., Wong, Z.J., Lu, X., Ni, X., Zhu, H., Chen, X., et al., 2015. Monolayer excitonic laser. Nat. Photonics 9, 733–737.

Youngblood, N., Li, M., 2017. Integration of 2D materials on a silicon photonics platform for optoelectronics applications. Nanophotonics 6, 1205–1218.

Yuan, H., Liu, X., Afshinmanesh, F., Li, W., Xu, G., Sun, J., et al., 2015. Polarization-sensitive broadband photodetector using a black phosphorus vertical p–n junction. Nat. Nanotechnol. 10, 707–713.

Yuan, S., Shen, C., Deng, B., Chen, X., Guo, Q., Ma, Y., et al., 2018. Air-stable room-temperature mid-infrared photodetectors based on hBN/black arsenic phosphorus/hBN heterostructures. Nano Lett. 18, 3172–3179.

Zereshki, P., Wei, Y., Ceballos, F., Bellus, M.Z., Lane, S.D., Pan, S., et al., 2018. Photocarrier dynamics in monolayer phosphorene and bulk black phosphorus. Nanoscale 10, 11307–11313.

Zhang, G., Huang, S., Chaves, A., Song, C., Özçelik, V.O., Low, T., et al., 2017a. Infrared fingerprints of few-layer black phosphorus. Nat. Commun. 8, 14071.

Zhang, J., Xie, M., Zhang, Y., Wang, J., Zhao, X., Chen, C., et al., 2023. Wide-wavelength tunable mid-infrared lasing based on black arsenic phosphorus. Adv. Opt. Mater. 11, 2300278.

Zhang, Y., Wang, S., Chen, S., Zhang, Q., Wang, X., Zhu, X., et al., 2020. Wavelength-tunable mid-infrared lasing from black phosphorus nanosheets. Adv. Mater. 32, 1808319.

Zhang, Z., Li, L., Horng, J., Wang, N.Z., Yang, F., Yu, Y., et al., 2017b. Strain-modulated bandgap and piezo-resistive effect in black phosphorus field-effect transistors. Nano Lett. 17, 6097–6103.

Zong, X., Hu, H., Ouyang, G., Wang, J., Shi, R., Zhang, L., et al., 2020. Black phosphorus-based van der waals heterostructures for mid-infrared light-emission applications. Light Sci. Appl. 9, 114.

Index

Note: Page numbers followed by "*f*" indicate figures and "*t*" indicate tables.

N

O

P

Q

R

S

U

V

W

X

Z

9780443193927